Introduction to Emergency Management

Second Edition

Introduction to Emergency Management

Second Edition

George D. Haddow

Jane A. Bullock

With Contributions by
Damon P. Coppola

ELSEVIER
BUTTERWORTH
HEINEMANN

Amsterdam • Boston • Heidelberg • London • New York • Oxford • Paris
San Diego • San Francisco • Singapore • Sydney • Tokyo

Elsevier Butterworth–Heinemann
30 Corporate Drive, Suite 400, Burlington, MA 01803, USA
Linacre House, Jordan Hill, Oxford OX2 8DP, UK

∞ Recognizing the importance of preserving what has been written,
Elsevier prints its books on acid-free paper whenever possible. ∞

Library of Congress Cataloging-in-Publication Data
Haddow, George D.
   Introduction to emergency management / George D. Haddow,
Jane A. Bullock. — 2nd ed.
      p. cm.
   Includes bibliographical references and index.
   ISBN 0-7506-7961-1 (hardcover : alk. paper) 1. Emergency
management.   2. Emergency management—United States.
I. Bullock, Jane A.   II. Title.
   HV551.2.H3 2006
   363.34′8′0973—dc22

        2005014579

British Library Cataloguing-in-Publication Data
A catalogue record for this book is available from the British
Library.

ISBN 13: 978-0-7506-7961-9
ISBN 10: 0-7506-7961-1

For information on all Elsevier Butterworth–Heinemann publications
visit our Web site at www.books.elsevier.com

Printed in the United States of America
06 07 08 09 10   10 9 8 7 6 5 4 3 2 1

Table of Contents

Foreword

In 1993, when I took over leadership of the Federal Emergency Management Agency (FEMA), emergency management was not a very well known or respected discipline. Many in the profession were hold-over from the days of civil defense and most elected officials did not see the value of emergency management until they had a major disaster in their community; and even then the value was transitory. Throughout the 1990s, as the United States and the world experienced an unprecedented number of severe disasters, the critical role emergency management plays in protecting the social and economic stability of our communities was evidenced. Emergency management began to grow beyond the response environment and focus on risk analysis, communications, risk prevention/mitigation and social and economic recovery. This required a new skill base for emergency managers and colleges and universities added courses and degrees in emergency management to their offerings. This resulted in a better educated, multidisciplinary, proactive approach to emergency management. Emergency managers were valued members of a community's leadership. Emergency management became an important profession. It allowed me as Director of FEMA, to work with our State, local and private partners to build one of the most respected emergency management systems in the world.

While the events of September 11th have altered our lives, and the profession has evolved since I left FEMA, one thing is clear. There is no time in our recent history when the need for and understanding of the discipline of emergency management has been more important. The current risk environment we live in, from potential bioterrorist threats, increasingly severe hurricanes and more frequent damaging earthquakes, has dramatically increased the skills and knowledge required to be an effective emergency manager in today's world.

Introduction to Emergency Management is the authoritative guide on today's discipline of emergency management. It takes the reader through the historical context of emergency management to the present day evolution into the world of homeland security. The book focuses on the elements of an emergency management process while providing the policy underpinnings that support that process. It examines the new environment of terrorism and its implications for emergency management. While focusing on the current changes happening to United States system for emergency management, it provides readers with a solid background in international practices and policies for disaster management/homeland security. The book gives the reader practical, real world experiences through documented case studies and provides extensive references and internet sites for follow up research.

My philosophy about emergency management has always been that we need to take a common-sense, practical approach to reducing the risks we face and protecting our citizens and our communities. We need to identify our risks, educate and communicate to our people about those risks, prepare as best we can for the risks, and then, together, form partnerships to take action to reduce those risks. This approach applies whether we are dealing with a flood, a tornado, a hazardous materials spill, a wildfire, a potential suicide bomb explosion, or a pandemic flu outbreak. The authors of this book were my Deputy Chief of Staff and my Chief of Staff, respectively, when I was Director of FEMA. Together we worked to apply this approach to making our citizens and communities more disaster resistant and safer throughout the world. As you read and learn from this book, I hope you will keep those ideals in mind.

James Lee Witt
CEO of International Code Congress
James Lee Witt Associates

Introduction

There is no country, no community, and no person immune to the impacts of disasters. Disasters, however, can be and have been prepared for, responded to, recovered from, and have had their consequences mitigated to an ever-increasing degree. The profession and the academic discipline that addresses this "management" of disasters is called Emergency Management. This book, *Introduction to Emergency Management*, is designed to provide the reader with a comprehensive foundation on the background, components, and systems involved in the management of disasters and other emergencies. Herein are detailed current practices, strategies, and the key players involved in emergency management both within the United States and around the world. The intent is to provide the reader with a working knowledge of how the functions of emergency management operate and the influence they can have on everyday life.

The capacity and capabilities for emergency management vary significantly between the countries of the world. The U.S. system, led by the Federal Emergency Management Agency (FEMA), has evolved into one of the most effective and emulated nationally-based systems, and has been chosen to be the principal focus of this book. Emergency management in the United States has experienced every form of disaster: natural, man-made, and political. The lessons learned from these experiences, the changes made in response to these events, and how the system continues to evolve because of new threats, provide a solid landscape to examine what emergency management is. The book maintains a U.S. focus in order to most effectively capitalize upon the practical experience and knowledge of the book's authors. Ultimately, such a focus becomes the logical choice because of FEMA's well-deserved reputation as today's most recognized, equipped, and funded emergency management system. By all measures, it is the leading emergency management organization in the world.

However, this book is not exclusively focused on FEMA. State and local emergency management organizations are the subjects of many of the included case studies, and their collaborative affiliations with FEMA are discussed at length throughout the text. In fact, it is the states that are given responsibility for public health and safety under the U.S. Constitution. The federal government becomes involved only after the state government has requested assistance or when it is apparent that the state agencies are or will be unable to fulfill their basic functions. The federal government is the primary source of the funding for public health and safety programs, with the states and communities as the primary recipients—resulting in a strong federal presence in emergency management. The competition for oftentimes scarce resources, coupled with the immediate priorities of state and local governments, has ensured a strong federal influence in emergency management—a trend that may be changing, as we will discuss in later chapters.

A comprehensive chapter is included that describes emergency management activities in the international sector. When the ability of an individual nation or a region as a whole to respond to a disaster is exceeded, the world's nations must join together to intervene and assist to manage the event. With greater frequency, events such as the 2004 Asian Earthquake and Tsunami have highlighted the need for a more robust international emergency management system, and governments across the globe have focused more attention on the issue. A detailed case study of the response to the 2001 earthquake in Gujarat, India, is provided to illustrate these systems.

No discussion of emergency management in the United States would be complete without addressing the significant changes that have occurred as a result of the September 11 terrorist attacks. Consideration of this issue is applied throughout all the chapters of this book, and a full chapter is dedicated entirely to discussing the emerging terrorist threat and the resulting implications to the U.S. emergency management system.

A brief summary of the contents of the book's ten chapters follows:

- Chapter 1, The Historical Context of Emergency Management, includes a brief discussion of the historical, organizational, and legislative evolution of emergency management in the United States by tracing the major changes triggered by disasters or other human or political events, leading up to and beyond the creation of the Department of Homeland Security.
- Chapter 2, Natural and Technological Hazards and Risk Assessment, identifies and defines the hazards confronting emergency management.
- Chapter 3, The Disciplines of Emergency Management: Mitigation, discusses what the function of mitigation is and what the strategies and programs applied by emergency management or other disciplines to reduce the impacts of disaster events are.
- Chapter 4, The Disciplines of Emergency Management: Response, focuses on the essential functions and processes of responding to a disaster event.
- Chapter 5, The Disciplines of Emergency Management; Recovery, describes the broad range of government and voluntary programs available to assist individuals and communities in rebuilding in the aftermath of a disaster.
- Chapter 6, The Disciplines of Emergency Management: Preparedness, catalogues the broad range of programs and processes that comprise the preparedness function of modern emergency management.
- Chapter 7, The Disciplines of Emergency Management: Communications, breaks from the more traditional approach to emergency management and focuses on why communications with the public, with the media, and with partners is critical to emergency management of the twenty-first century.
- Chapter 8, International Disaster Management, provides an overview of current activity in international emergency management through an examination of selected international organizations.

- Chapter 9, Emergency Management and the New Terrorist Threat, describes how the events of September 11 have altered the traditional perceptions of emergency management.
- Chapter 10, The Future of Emergency Management, provides insights, speculations, conclusions, and recommendations from the authors on where emergency management is or should be headed in the future.

Our goal in writing this book was to provide readers with an understanding of emergency management, insight into how events have shaped the discipline, and thoughts about the future direction of emergency management. Additionally, with the recent creation of the Department of Homeland Security and its subsequent absorption of FEMA and 21 other federal agencies, the need for a collective resource of this kind has never been so vital. In the end, we hope it will educate, inform, and possibly encourage individuals to actively participate in the practice of emergency management in their professions and communities.

Acknowledgments

This book could not have been completed without the assistance of two valuable groups of partners. First, the authors are grateful to the Institute for Crisis, Disaster, and Risk Management at The George Washington University and its codirectors, Dr. Jack Harrald and Dr. Joseph Barbera, for their support. The Institute's Greg Shaw was helpful in many ways. Second, we want to acknowledge the contributions of a group of graduate students at the Institute for their hard work in conducting the preliminary research and draft for this book. This group made significant contributions to the book and includes Kathryn A. Allen, James A. Cooke, Damon P. Coppola, Elizabeth N. Halford, Robert D. Hulshouser, Nadeem U. Khan, Mariette M. Larrick, Stephen Marquette, Michele R. Novack, George Nunez, John C. Peyrebrune, Jeannette C. Rood, Christian M. Salmon, Monica M. Severson, and Carmen A. Whitson. Two students deserve special acknowledgment for their interest, commitment, and quality of their efforts: Damon P. Coppola and Alfredo Lagos.

Finally, the authors wish to thank their respective spouses, Dick Bullock and Kim Haddow, for their enduring good humor and patience.

1. The Historical Context of Emergency Management

INTRODUCTION

Emergency management has ancient roots. Early hieroglyphics depict cavemen trying to deal with disasters. The Bible speaks of the many disasters that befell civilizations. In fact, the account of Moses parting the Red Sea could be interpreted as the first attempt at flood control. As long as there have been disasters, individuals and communities have tried to do something about them; however, organized attempts at dealing with disasters did not occur until much later in modern history. The purpose of this chapter is to discuss the historical, organizational, and legislative history of modern emergency management in the United States. Some of the significant events and people that have shaped the emergency management discipline over the years will be reviewed. Understanding the history and evolution of emergency management is important because at different times, the concepts of emergency management have been applied differently. The definition of emergency management can be extremely broad and all-encompassing. Unlike other more structured disciplines, it has expanded and contracted in response to events, Congressional desires, and leadership styles.

A simple definition is that *emergency management* is the discipline dealing with risk and risk avoidance. Risk represents a broad range of issues and includes an equally diverse set of players. The range of situations that could possibly involve emergency management or the emergency management system is extensive. This supports the premise that emergency management is integral to the security of everyone's daily lives and should be integrated into daily decisions and not just called on during times of disasters.

Emergency management is an essential role of government. The Constitution tasks the states with responsibility for public health and safety—hence the responsibility for public risks—with the federal government in a secondary role. The federal role is to help when the state, local, or individual entity is overwhelmed. This fundamental philosophy continues to guide the government function of emergency management.

Based on this strong foundation, the validity of emergency management as a government function has never been in question. Entities and organizations fulfilling the emergency management function existed at the state and local level long before the federal government became involved. But as events occurred, as political philosophies changed, and as the nation developed, the federal role in emergency management steadily increased.

EARLY HISTORY: 1800–1950

In 1803, a Congressional Act was passed to provide financial assistance to a New Hampshire town that had been devastated by fire. This is the first example of the federal government becoming involved in a local disaster. It was not until the administration of Franklin Roosevelt began to use government as a tool to stimulate the economy that a significant investment in emergency management functions was made by the federal government.

During the 1930s, the Reconstruction Finance Corporation and the Bureau of Public Roads were both given authority to make disaster loans available for repair and reconstruction of certain public facilities after disasters. The Tennessee Valley Authority was created during this time to produce hydroelectric power and, as a secondary purpose, to reduce flooding in the region.

A significant piece of emergency management legislation was passed during this time. The Flood Control Act of 1934 gave the U.S. Army Corps of Engineers increased authority to design and build flood control projects. This act has had a significant and long-lasting impact on emergency management in this country. This act reflected a philosophy that man could control nature, thereby eliminating the risk of floods. Although this program would promote economic and population growth patterns along the nation's rivers, history has proven that this attempt at emergency management was shortsighted and costly.

THE COLD WAR AND THE RISE OF CIVIL DEFENSE: 1950s

The next notable timeframe for the evolution of emergency management occurs during the 1950s. The era of the Cold War presented the principal disaster risk as the potential for nuclear war and nuclear fallout. Civil Defense programs proliferated across communities during this time. Individuals and communities were encouraged to build bomb shelters to protect themselves and their families from nuclear attack from the Soviet Union.

Almost every community had a civil defense director, and most states had someone who represented civil defense in their state government hierarchy. By profession, these individuals were usually retired military personnel, and their operations received little political or financial support from their state or local governments. Equally often, the civil defense responsibility was in addition to other duties.

Federal support for these activities was vested in the Federal Civil Defense Administration (FCDA), an organization with little staff or financial resources whose main role was to provide technical assistance. In reality, the local and state civil defense directors were the first recognized face of emergency management in the United States.

A companion office to the FCDA, the Office of Defense Mobilization was established in the Department of Defense (DoD). The primary functions of this office were to allow for quick mobilization of materials and production and stockpiling of critical materials in the event of a war. It included a function called *emergency*

Figure 1-1 Midwest Floods, June 1994. Homes, businesses, and personal property were all destroyed by the high flood levels. A total of 534 counties in nine states were declared for federal disaster aid. As a result of the floods, 168,340 people registered for federal assistance. FEMA News Photo.

preparedness. In 1958, these two offices were merged into the Office of Civil and Defense Mobilization.

The 1950s were a quiet time for large-scale natural disasters. Hurricane Hazel, a Category 4 hurricane, inflicted significant damage in Virginia and North Carolina in 1954; Hurricane Diane hit several mid-Atlantic and northeastern states in 1955; and Hurricane Audrey, the most damaging of the three storms, struck Louisiana and North Texas in 1957. Congressional response to these disasters followed a familiar pattern of ad hoc legislation to provide increased disaster assistance funds to the affected areas.

As the 1960s started, three major natural disaster events occurred. In a sparsely populated area of Montana, the Hebgen Lake earthquake, measuring 7.3 on the Richter scale, brought attention to the fact that the nation's earthquake risk went beyond the California borders. Also in 1960, Hurricane Donna hit the west coast of Florida, and Hurricane Carla blew into Texas in 1961. The incoming Kennedy administration decided to make a change to the federal approach. In 1961 it created the Office of Emergency Preparedness inside the White House to deal with natural disasters. Civil Defense responsibilities remained in the Office of Civil Defense within the DoD.

NATURAL DISASTERS BRING CHANGES TO EMERGENCY MANAGEMENT: 1960s

As the 1960s progressed, the United States would be struck by a series of major natural disasters. The Ash Wednesday storm in 1962 devastated more than 620 miles of shoreline on the East Coast, producing more than $300 million in damages. In 1964, an earthquake measuring 9.2 on the Richter scale in the Prince William Sound, Alaska, became front-page news throughout America and the world. This quake generated a tsunami that affected beaches as far down the Pacific Coast as California and killed 123 people. Hurricane Betsey struck in 1965, and Hurricane Camille in 1969, killing and injuring hundreds of people and causing hundreds of millions of dollars in damage along the Gulf Coast.

As with previous disasters, the response was passage of ad hoc legislation for funds; however, the financial losses resulting from Hurricane Betsey's path across Florida and Louisiana started a discussion of insurance as a protection against future floods and a potential method to reduce continued government assistance after disasters. Congressional interest was prompted by the unavailability of flood protection insurance on the standard homeowner policy. Where this type of insurance was available, it was cost prohibitive. These discussions eventually led to passage of the National Flood Insurance Act of 1968, which created the National Flood Insurance Program (NFIP).

Congressman Hale Boggs of Louisiana is appropriately credited with steering this unique legislation through Congress. Unlike previous emergency management/disaster legislation, this bill sought to do something about the risk before the disaster struck. It brought the concept of *community-based mitigation* into the practice of emergency management. In simple terms, when a community joined the NFIP, in exchange for making federally subsidized, low-cost flood insurance available to its citizens, the community had to pass an ordinance restricting future development in its floodplains. The federal government also agreed to help local communities by producing maps of their community's floodplains.

The NFIP began as a voluntary program as part of a political compromise that Boggs reached with then-Senator Tom Eagleton of Missouri. As a voluntary program, few communities joined. After Hurricane Camille struck the Louisiana, Alabama, and Mississippi coasts in 1969, the goals of the NFIP to protect people's financial investments and to reduce government disaster expenditures were not being met. It took Hurricane Agnes devastating Florida for a change to occur.

George Bernstein, brought down from New York by President Nixon to run the Federal Insurance Administration (FIA) within the Department of Housing and Urban Development (HUD), proposed linking the mandatory purchase of flood insurance to all homeowner loans backed by federal mortgages. This change created an incentive for communities to join the NFIP because a significant portion of the home mortgage market was federally backed. This change became the Flood Insurance Act of 1972.

It is important to note how local and state governments choose to administer this flood risk program. Civil defense departments usually had responsibility to deal with risks and disasters. Although the NFIP dealt with risk and risk avoidance, responsibilities for the NFIP were sent to local planning departments and State Departments of Natural Resources. This reaction is one illustration of the fragmented and piecemeal approach to emergency management that evolved during the 1960s and 1970s.

THE CALL FOR A NATIONAL FOCUS ON EMERGENCY MANAGEMENT: 1970s

In the 1970s, responsibility for emergency management functions was evident in more than five federal departments and agencies, including the Department of Commerce (weather, warning, and fire protection), the General Services Administration (continuity of government, stockpiling, federal preparedness), the Treasury Department (import investigation), the Nuclear Regulatory Commission (power plants), and HUD (flood insurance and disaster relief).

With passage of the Disaster Relief Act of 1974, prompted by the previously mentioned hurricanes and the San Fernando earthquake of 1971, HUD possessed the most significant authority for natural disaster response and recovery through the NFIP under the FIA and the Federal Disaster Assistance Administration (disaster response, temporary housing, and assistance). On the military side, there existed the Defense Civil Preparedness Agency (nuclear attack) and the U.S. Army Corps of Engineers (flood control); however, taking into account the broad range of risks and potential disasters, more than 100 federal agencies were involved in some aspect of risk and disasters.

This pattern continued down to the state and, to a lesser extent, local levels. Parallel organizations and programs added to confusion and turf wars, especially during disaster response efforts. The states and the governors grew increasingly frustrated over this fragmentation. In the absence of one clear federal lead agency in emergency management, a group of state Civil Defense directors led by Lacy Suiter of Tennessee and Erie Jones of Illinois launched an effort through the National Governor's Association to consolidate federal emergency management activities in one agency.

With the election of a fellow state governor, President Jimmy Carter of Georgia, the effort gained steam. President Carter came to Washington committed to streamlining all government agencies and seeking more control over key administrative processes. The state directors lobbied the National Governor's Association (NGA) and Congress for a consolidation of federal emergency management functions. When the Carter administration proposed such an action, it met with a receptive audience in the Senate. Congress already had expressed concerns about the lack of a coherent federal policy and the inability of states to know whom to turn to in the event of an emergency.

The federal agencies involved were not as excited about the prospect. A fundamental law of bureaucracy is a continued desire to expand control and authority, not to lose control. In a consolidation of this sort, there would be losers and winners. There was a question of which federal department/agency should house the new consolidated structure. As the debate continued, the newly organized National Association of State Directors of Emergency Preparedness championed the creation of a new independent organization, an idea that was quickly supported by the Senate.

In the midst of these discussions, an accident occurred at the Three Mile Island Nuclear Power Plant in Pennsylvania, which added impetus to the consolidation effort. This accident brought national media attention to the lack of adequate off-site preparedness around commercial nuclear power plants and the role of the federal government in responding to such an event.

On June 19, 1978, President Carter transmitted to the Congress the Reorganization Plan Number 3 (3 CFR 1978, 5 U.S. Code 903). The intent of this plan was to consolidate emergency preparedness, mitigation, and response activities into one federal emergency management organization. The President stated that the plan would establish the Federal Emergency Management Agency (FEMA) and that the FEMA Director would report directly to the President.

Reorganization Plan Number 3 transferred the following agencies or functions to FEMA: National Fire Prevention Control Administration (Department of Commerce), Federal Insurance Administration (HUD), Federal Broadcast System (Executive Office of the President), Defense Civil Preparedness Agency (DoD), Federal Disaster Assistance Administration (HUD), and the Federal Preparedness Agency (GSA).

Additional transfers of emergency preparedness and mitigation functions to FEMA were:

- Oversight of the Earthquake Hazards Reduction Program (Office of Science and Technology Policy)
- Coordination of dam safety (Office of Science and Technology Policy)
- Assistance to communities in the development of readiness plans for severe weather-related emergencies
- Coordination of natural and nuclear disaster warning systems
- Coordination of preparedness and planning to reduce the consequences of major terrorist incidents

Reorganization Plan Number 3 articulated several fundamental organizational principles:

> First, Federal authorities to anticipate, prepare for, and respond to major civil emergencies should be supervised by one official responsible to the President and given attention by other officials at the highest levels. Second, an effective civil defense system requires the most efficient use of all available resources. Third, whenever possible, emergency responsibilities should be extensions of federal agencies. Fourth, federal hazard mitigation activities should be closely linked with emergency preparedness and response functions.

Subsequent to Congressional review and concurrence, the Federal Emergency Management Agency was officially established by Executive Order 12127 of March 31, 1979 (44 FR 19367, 3 CFR, Comp., p. 376). A second Executive Order, 12148, mandated reassignment of agencies, programs, and personnel into the new entity FEMA.

Creating the new organization made sense, but integrating the diverse programs, operations, policies, and people into a cohesive operation was a much bigger task than realized when the consolidation began. It would take extraordinary leadership and a common vision. The consolidation also created immediate political problems. By consolidating these programs and the legislation that created them, FEMA would have to answer to 23 committees and subcommittees in Congress with oversight of its programs. Unlike most other federal agencies, it would have no organic legislation to support its operations and no clear champions to look to during the Congressional appropriations process.

In addition, President Carter had problems finding a director for this new organization. No large constituent group was identified with emergency management. Furthermore, the administration was facing major problems with Congress and the public because of the Iranian hostage crisis. President Carter finally reached into his own cabinet and asked John Macy, then head of the Office of Personnel Management (OPM), to become director of FEMA.

John Macy's task was to unify an organization that was not only physically separated—parts of the agency were located in five different buildings around Washington—but also philosophically separate. Programs focused on nuclear war preparations were combined with programs focused on a new consciousness of the environment and floodplain management. Macy focused his efforts by emphasizing the similarities between natural hazards preparedness and civil defense by developing a new concept called the Integrated Emergency Management System (IEMS). This system was an all-hazards approach that included direction, control, and warning as functions common to all emergencies from small, isolated events, to the ultimate emergency of nuclear attack.

For all his good efforts, FEMA continued to operate as individual entities pursuing their own interests and answering to their different Congressional bosses. It was a period of few major disasters, so virtually nobody noticed this problem of disjointedness.

CIVIL DEFENSE REAPPEARS AS NUCLEAR ATTACK PLANNING: 1980s

The early and mid-1980s saw FEMA facing many challenges but no significant natural disasters. The absence of the need for a coherent federal response to disasters, as was called for by Congress when it approved the establishment of FEMA, allowed FEMA to continue to exist as an organization of many parts.

In 1982, President Reagan appointed Louis O. Guiffrida as director of FEMA. Mr. Guiffrida, a California friend of Ed Meese, one of the President's closest advisors, had a background in training and terrorism preparedness at the state government level. General Guiffrida proceeded to reorganize FEMA consistent with administration policies and his background. Top priority was placed on government preparedness for a nuclear attack. Resources within the agency were realigned, and additional budget authority was sought to enhance and elevate the national security responsibilities of the agency. With no real role for the states in these national security activities, the state directors who had lobbied for the creation of FEMA saw their authority and federal funding declining.

Guiffrida also angered one of the only other visible constituents of the agency—the fire services community. Guiffrida diminished the authority of the U.S. Fire Administration by making it part of FEMA's Directorate of Training and Education. The newly acquired campus at Emmitsburg, Maryland, was intended to become the preeminent National Emergency Training Center (NETC).

During Guiffrida's tenure, FEMA faced several unusual challenges that stretched its authority, including asserting FEMA into the lead role for continuity of civilian government in the aftermath of a nuclear attack, managing the federal response to the contamination at Love Canal and Times Beach, Missouri, and the Cuban refugee crisis. Although Guiffrida managed to bring the agency physically together in a new headquarters building in Southwest Washington, severe morale problems persisted.

Dislike of Guiffrida's style and questions about FEMA's operations came to the attention of U.S. Representative Al Gore of Tennessee, who then served on the House Science and Technology Committee. As the Congressional hearings proceeded, the Department of Justice and a grand jury began investigations of senior political officials at FEMA. These inquiries led to the resignation of Guiffrida and top aides in response to a variety of charges, including misuse of government funds, but the shake-up marked a milestone of sorts: FEMA and emergency management had made it into the comic strip "Doonesbury."

President Reagan then selected General Julius Becton to be director of FEMA. General Becton was a retired military general and had been the director of the Office of Foreign Disaster Assistance in the State Department. General Becton is credited uniformly with restoring integrity to the operations and appropriations of the agency. From a policy standpoint, he continued to emphasize the programs of his predecessor but in a less visible manner. Becton expanded the duties of FEMA when he was asked by the DoD to take over the program dealing with the off-site cleanup of chemical stockpiles on DoD bases. This program was fraught with problems, and bad feelings existed between the communities and the bases over the funds available to the communities for the cleanup. FEMA had minimal technical expertise to administer this program and was dependent on the DoD and the Army for the funding. This situation led to political problems for the agency and did not lead to significant advancements in local emergency management operations, as promised by the DoD.

At one point in his tenure, General Becton ranked the programs in FEMA by level of importance. Of the more than 20 major programs, the earthquake, hurricane, and flood programs ranked near the bottom. This priority seems logical based on the absence of any significant natural hazards, but this situation is noteworthy in the context that it continued the pattern of isolating resources for national security priorities without recognizing the potential of a major natural disaster.

This issue was raised by then Senator Al Gore in hearings on FEMA's responsibilities as lead agency for the National Earthquake Hazards Reduction Program (NEHRP). Senator Gore, reacting to a scientific report that said there could be up to 200,000 casualties from an earthquake occurring on the New Madrid fault, believed that FEMA's priorities were misplaced. The legislation that created the NEHRP called on FEMA to develop a plan for how the federal government would respond to a catastrophic earthquake. This Federal Response Plan would later become the operating Bible for all the federal agencies response operations. Senator Gore concluded that FEMA needed to spend more time working with its federal, state, and local partners on natural hazards planning.

AN AGENCY IN TROUBLE: 1989–1992

As Congress debated, and finally passed, major reform of federal disaster policy as part of the Stewart McKinney–Robert Stafford Act, the promise of FEMA and its ability to support a national emergency management system remained in doubt.

As the 1980s closed, FEMA was an agency in trouble. It suffered from severe morale problems, disparate leadership, and conflicts with its partners at the state and local level over agency spending and priorities. In 1989, two devastating natural disasters called the continued existence of FEMA into question. In September, Hurricane Hugo slammed into North Carolina and South Carolina after first hitting Puerto Rico and the Virgin Islands. It was the worst hurricane in a decade, with more than $15 billion in damages and 85 deaths. FEMA was slow to respond, waiting for the process to work and for the governors to decide what to do. Sen. Ernest Hollings (D-SC) personally called the FEMA director and asked for help, but the Agency moved slowly. Hollings went on national television to berate FEMA in some of the most colorful language ever, calling the agency the "sorriest bunch of bureaucratic jackasses."

Less than a month later, the Bay Area of California was rocked by the Loma Prieta earthquake as the 1989 World Series got under way in Oakland Stadium. FEMA was not prepared to respond, but it was lucky. Although FEMA had spent the last decade focused on nuclear attack planning, FEMA's state partners in emergency management, especially in California, had been preparing for a more realistic risk—an earthquake. Damages were high, but few lives were lost. This outcome was a testament to good mitigation practices in building codes and construction that were adopted in California, and some good luck relative to the time when the earthquake hit.

A few years later, FEMA was not so lucky. In August 1992, Hurricane Andrew struck Florida and Louisiana, and Hurricane Iniki struck Hawaii within months of each other. FEMA wasn't ready, and neither were FEMA's partners at the state level. The agency's failure to respond was witnessed by Americans all across the country as major news organizations followed the crisis. The efficacy of FEMA as the national emergency response agency was in doubt. President Bush dispatched then Secretary of Transportation Andrew Card to take over the response operation and sent in the military.

It was not just FEMA that failed in Hurricane Andrew; it was the process and the system. In Hurricane Andrew, FEMA recognized the need to apply all its resources to the response and began to use its national security assets for the first time in a natural disaster response—but it was too late. Starting with Hurricane Hugo, public concern over natural disasters was high. People wanted, and expected, government to be there to help in their time of need. FEMA seemed incapable of carrying out the essential government function of emergency management.

In the aftermath of Hurricanes Andrew and Iniki, there were calls for abolishing FEMA. Investigations by the General Accounting Office (GAO) and other government and nongovernmental watchdog groups called for major reforms. None of this was lost on the incoming Clinton administration. As Governor of Arkansas, President Clinton had experience responding to several major flooding disasters, and realized how important an effective response and quick recovery were to communities and to voters. At his side throughout these disasters was James Lee Witt, former county judge and administrator of Yell County, and later, the state director for Emergency Management in Arkansas.

THE WITT REVOLUTION: 1993–2001

When President Clinton nominated James Lee Witt to be director of FEMA, he breathed life back into FEMA and brought a new style of leadership to the troubled agency. Witt was the first director of FEMA with emergency management experience. He was from the constituency who had played a major role in creating FEMA but had been forgotten—the state directors. With Witt, President Clinton had credibility and, more important, a skilled politician who knew the importance of building partnerships and serving customers.

Witt came in with a mandate to restore the trust of the American people that their government would be there for them during times of crisis. He initiated sweeping reforms inside and outside the agency. Inside FEMA, he reached out to all employees, implemented customer service training, and reorganized the agency to break down bottlenecks. He supported application of new technologies to the delivery of disaster services and focused on mitigation and risk avoidance. Outside the agency, he strengthened the relationships with state and local emergency managers and built new ones with Congress, within the administration, and with the media. Open communications internally and externally was one of the hallmarks of the Witt years at FEMA.

Figure 1-2   Northridge Earthquake, CA, January 17, 1994. Many roads, including bridges and elevated highways, were damaged by the 6.7 magnitude earthquake. Approximately 114,000 residential and commercial structures were damaged and 72 deaths were attributed to the earthquake. Damage costs were estimated at $25 billion. FEMA News Photo.

Witt's leadership and the changes he made were quickly tested as the nation experienced an unprecedented series of natural disasters. The Midwest floods in 1993 resulted in major disaster declarations in nine states. The Midwest floods called into question the value of some of the flood control measures initiated long ago as part of the 1930s Army Corps of Engineers' legislation. FEMA's successful response to these floods brought the opportunity to change the focus of postdisaster recovery by initiating the largest voluntary buyout and relocation program to date in an effort to move people out of the floodplain and out of harm's way.

The Northridge, California earthquake quickly followed the Midwest floods in 1994. Northridge tested all the new streamlined approaches and technology advancements for delivery of services and created some more. Throughout the next several years, FEMA and its state and local partners would face every possible natural hazard, including killer tornadoes, ice storms, hurricanes, floods, wildfires, and drought.

When President Clinton elevated Witt as director of FEMA to be a member of his Cabinet, the value and importance of emergency management was recognized. Witt used this promotion as an opportunity to lobby the nation's governors to include their state emergency management directors in their Cabinets.

The Oklahoma City bombing in April 1995 represented a new phase in the evolution of emergency management. This event, following the first bombing of the World Trade Center in New York City in 1992, raised the issue of America's preparedness for terrorism events. Because emergency management responsibilities are defined by risks and the consequences of those risks, responding to terrorist threats was included. The Oklahoma City bombing tested this thesis and set the stage for interagency disagreements over which agency would be in charge of terrorism.

Figure 1-3 Franklin, VA, September 21, 1999. Hurricane Floyd left the downtown section of Franklin, VA, under six feet of water. The water has begun to recede, as shown by the high-water marks, but hazards still include propane tanks, gas tanks, chemical barrels, and pesticides. Photo by Liz Roll/FEMA News Photo.

The Nunn-Lugar legislation of 1995 left the question open as to who would be the lead agency in terrorism. Many people fault FEMA leadership for not quickly claiming that role, and the late 1990s were marked by several different agencies and departments having a role in terrorism planning. The question of who is the first responder to a terrorism incident—fire, police, emergency management, or emergency medical services—was closely examined, without any clear answers. The state directors were looking for FEMA to claim the leadership role. In an uncharacteristic way, the leadership of FEMA vacillated on this issue. Terrorism was certainly part of the all-hazards approach to emergency management championed by FEMA, but the resources and technologies needed to address specific issues such as biochemical warfare and weapons of mass destruction events seemed well beyond the reach of the current emergency management structure.

While this debate continued, FEMA took an important step in its commitment to disaster mitigation by launching a national initiative to promote a new community-based approach called Project Impact: Building Disaster-Resistant Communities. This project was designed to mainstream emergency management and mitigation practices into every community in America. It went back to the roots of emergency management. It asked a community to identify risks and establish a plan to reduce those risks. It asked communities to establish partnerships that included all the stakeholders in the community, including, for the first time, the business sector.

The goal of Project Impact was to incorporate decisions about risk and risk avoidance into the community's everyday decision-making processes. By building a disaster-resistant community, the community would promote sustainable economic development, protect and enhance its natural resources, and ensure a better quality of life for its citizens. Project Impact had ambitious goals and was well received by the communities and Congress. It was designed to create a broader constituency, a grassroots campaign, for emergency management issues.

As the decade ended without any major technological glitches from Y2K, FEMA was recognized as the preeminent emergency management system in the world. It was emulated in other countries, and Witt became an ambassador for emergency management overseas. Hurricane Mitch saw a change in American foreign policy toward promoting and supporting community-based mitigation projects. State and local emergency management programs had grown and their value recognized and supported by society. Private-sector and business continuity programs were flourishing.

The role and responsibility and the partnerships supporting emergency management had significantly increased, and its budget and stature had grown. Good emergency management became a way to get economic and environmental issues on the table; it became a staple of discussion relative to a community's quality of life.

The profession of emergency management was attracting a different type of individual. Political and management skills were critical, and candidates for state, local, and private emergency management positions were now being judged on their training and experience rather than their relationship to the community's political leadership. Undergraduate and advanced degree programs in emergency management were flourishing at more than 65 national colleges and universities. It was now a respected, challenging, and sought-after profession.

TERRORISM BECOMES MAJOR FOCUS: 2001

With the election of George W. Bush, a new FEMA director, Joe Allbaugh, was named to head the agency. As a former Chief of Staff to Governor Bush in Texas and President Bush's campaign manager in the 2000 presidential race, Allbaugh had a close personal relationship with the President. As demonstrated by Witt and Clinton, this was viewed as a positive for the agency. His lack of emergency management background was not an issue during his confirmation hearings.

Allbaugh got off to a rocky start when the administration decided to eliminate funding for the popular Project Impact. Immediately after this decision was announced, the 6.8 magnitude Nisqually earthquake shook Seattle, Washington. Seattle happened to be one of the most successful Project Impact communities. The mayor of Seattle appeared on national television and credited Project Impact as responsible for why there was almost no damage from the quake. Later that evening, Vice President Dick Cheney was asked why the program was being eliminated, and he replied by saying there were questions about its effectiveness. As FEMA's budget proceeded through the appropriations process, Congress put funding back into the Project Impact.

As part of major reorganization of the agency, Allbaugh recreated the Office of National Preparedness (ONP). This office was first established in the 1980s during the Guiffrida reign for planning for World War III, and eliminated by Witt in 1992. This action raised some concerns among FEMA's constituents and FEMA staff. However, this time the mission of the office was focused upon terrorism.

In a September 10, 2001 speech, Director Allbaugh spoke about his priorities as being firefighters, disaster mitigation, and catastrophic preparedness. These words seem prophetic in light of the events of September 11. As the events of that tragic day unfolded, FEMA activated the Federal Response Plan and response operations proceeded as expected in New York and in Virginia. Most of the agency's senior leaders, including the director, were in Montana attending the Annual Meeting of the National Emergency Management Association (NEMA), an organization that represents state emergency management directors. The strength of the U.S. emergency management system was proven, however, as hundreds of response personnel initiated their operations within just minutes of the onset of events.

The Creation of the Department of Homeland Security: 2001–2004

Almost immediately following the terrorist attacks, the President created by executive order the Office of Homeland Security within the White House. The same day that announcement was made, Pennsylvania Governor Tom Ridge was sworn in to lead the office with the rank of Assistant to the President. The office had only 120 employees, and what was derided as a prohibitively small budget in light of the gravity of the events the nation had just witnessed, and began to be seen as just another government bureaucracy.

In March of 2002, President Bush signed Homeland Security Presidential Directive-3 (HSPD-3), which stated that:

> The Nation requires a Homeland Security Advisory System to provide a comprehensive and effective means to disseminate information regarding the risk of terrorist acts to Federal, State, and local authorities and to the American people. Such a system would provide warnings in the form of a set of graduated "Threat Conditions" that would increase as the risk of the threat increases. At each Threat Condition, Federal departments and agencies would implement a corresponding set of "Protective Measures" to further reduce vulnerability or increase response capability during a period of heightened alert.

> This system is intended to create a common vocabulary, context, and structure for an ongoing national discussion about the nature of the threats that confront the homeland and the appropriate measures that should be taken in response. It seeks to inform and facilitate decisions appropriate to different levels of government and to private citizens at home and at work.

What resulted was the widely recognizable five-color coded Homeland Security Advisory System. The Homeland Security Advisory System repeatedly has raised and lowered the nation's alert levels between Elevated (yellow) and High (orange) several times since the system's inception, but has done so with less frequency as standards for such movements have been established.

On November 25, 2002, President Bush signed into law the Homeland Security Act of 2002 (HS Act) (Public Law 107-296), and announced that former Pennsylvania Governor Tom Ridge would become secretary of a new Department of Homeland Security (DHS) to be created through this legislation. This act, which authorized the greatest federal government reorganization since President Harry Truman joined the various branches of the armed forces under the Department of Defense, was charged with a three-fold mission of protecting the United States from further terrorist attacks, reducing the nation's vulnerability to terrorism, and minimizing the damage from potential terrorist attacks and natural disasters.

The sweeping reorganization into the new department, which officially opened its doors on January 24, 2003, joined together over 179,000 federal employees from 22 existing federal agencies under a single, cabinet-level organization. The legislation also included several changes within other federal agencies that were only remotely affiliated with DHS.

The creation of DHS was the culmination of an evolutionary legislative process that began largely in response to criticism that increased federal intelligence interagency cooperation could have prevented the September 11 terrorist attacks. The White House and Congress both had recognized that a Homeland Security czar would require both a staff and a large budget in order to succeed, and thus began deliberations to create a new cabinet-level department that would fuse many of the security-related agencies dispersed throughout the federal government.

For several months during the second half of 2002, Congress jockeyed between different versions of the Homeland Security bill in an effort to establish legislation that was passable yet effective. Lawmakers were particularly mired on the issue of the rights of employees—an issue that prolonged the legal process considerably. Furthermore, efforts to incorporate many of the intelligence-gathering and investigative law enforcement agencies, namely the National Security Agency (NSA), the Federal

Bureau of Investigation (FBI), and the Central Intelligence Agency (CIA), into the legislation failed.

Despite these delays and setbacks, after the 2002 midterm elections, the Republican seats gained in both the House and Senate gave the President the leverage he needed to pass the bill without further deliberation (H.R., 299-121 on November 13, 2002; Senate, 90-9 on November 19, 2002). Although the passage of this act represented a significant milestone, the implementation phase presented a tremendous challenge; a concern expressed by several leaders from the agencies that were to be absorbed. On November 25, 2002, President Bush submitted his Reorganization Plan (as required by the legislation), which mapped out the schedule, methodology, and budget for the monumental task.

Beginning March 1, 2003, almost all the federal agencies named in the Act began their move, whether literally or symbolically, into the new department. Those remaining followed on June 1, 2003, with all incidental transfers completed by September 1, 2003. Although a handful of these agencies remained intact after the move, most were fully incorporated into one of four new directorates; Border and Transportation Security (BTS), Information Analysis and Infrastructure Protection (IAIP), Emergency Preparedness and Response (EP&R), and Science and Technology (S&T). A fifth directorate, Management, incorporated parts of the existing administrative and support offices within the merged agencies.

Secretary Ridge was given exactly one year to develop a comprehensive structural framework for DHS, and to name new leadership for all five directorates and other offices created under the legislation.

In addition to the creation of the Department of Homeland Security, the HS Act made several changes to other federal agencies and their programs, and created several new programs. A list of the most significant is presented here:

- Established a National Homeland Security Council within the Executive Office of the President, which assesses U.S. objectives, commitments, and risks in the interest of Homeland Security; oversees and reviews federal homeland security policies; and makes recommendations to the President.
- Transferred the Bureau of Alcohol, Tobacco, and Firearms (ATF) from the Department of the Treasury to the Department of Justice (DOJ).
- Explicitly prohibits both the creation of a national ID card and the proposed Citizen Corps Terrorism Information and Prevention System (Operation TIPS, which encouraged transportation workers, postal workers, and public utility employees to identify and report suspicious activities linked to terrorism and crime). The Act also reaffirmed the Posse Comitatus Act, which prohibits the use of the Armed Forces in law enforcement activities except under Constitutional or Congressional authority (the Coast Guard is exempt from this act).
- The Arming Pilots Against Terrorism Act, incorporated into the HS Act, allowed pilots to defend aircraft cockpits with firearms or other "less-than-lethal weapons" against acts of criminal violence or air piracy, and provides anti-terrorism training to flight crews.
- The Critical Infrastructure Information Act (2002), incorporated in the HS Act, exempts certain components of critical infrastructure from Freedom of Information Act (FOIA) regulations.

- The Johnny Michael Spann Patriot Trusts, created to provide support for surviving spouses, children, or dependent parents, grandparents, or siblings of various federal employees who die in the line of duty as result of terrorist attacks, military operations, intelligence operations, or law enforcements operations.

On November 30, 2004, following the Presidential elections, DHS Secretary Ridge announced his resignation. After an initial nomination of NYPD commissioner Bernard Kerik for the position, which was withdrawn due to questions about an undocumented immigrant he employed at his home, Federal Judge Michael Chertoff was named to lead the agency. On February 16, 2005, Michael Chertoff was unanimously confirmed by the Senate to lead the Department of Homeland Security.

On July 13th, 2005, DHS Secretary Michael Chertoff released a six-point agenda that would be used to guide a reorganization of the Department aimed at streamlining its efforts. The agenda followed an initial review that Chertoff initiated immediately upon assuming the leadership position. The review was designed to closely examine the Department in order to discover ways in which leadership could better manage risk in terms of threat, vulnerability and consequence; prioritize policies and operational missions according to this risk-based approach; and establish a series of preventive and protective steps that would increase security at multiple levels. According to the six-point agenda, changes that will occur at DHS will focus on:

- Increasing overall preparedness, particularly for catastrophic events;
- Creating better transportation security systems to move people and cargo more securely and efficiently;
- Strengthening border security and interior enforcement and reforming immigration processes;
- Enhancing information sharing (with partners);
- Improving financial management, human resource development, procurement and information technology within the department; and
- Realigning the department's organization to maximize mission performance.

As part of the proposed reorganization, virtually all of the remaining preparedness capabilities in FEMA, including the U.S. Fire Administration, will be moved to the new Office of Preparedness. The exception is the Emergency Management Institute (EMI). Although EMI training function was always considered part of preparedness, senior level FEMA officials argued that its courses supported the response and recovery functions of FEMA. The new FEMA Office will focus exclusively on response and recovery.

Under the initial DHS organization, the Emergency Preparedness and Response Directorate (EP&R) contained most of the pre-DHS FEMA functions and staff. Under the Chertoff reorganization, EP&R is eliminated and the Director of FEMA, formerly the Under Secretary for EP&R, now becomes an Office Director. The reorganization is somewhat unclear regarding who will actually be in charge in a disaster since responsibility for the new National Incident Management System (NIMS) is actually vested in the Director of Operations Coordination.

The reorganization raises several policy issues including whether the "all hazards" approach has been abandoned in exchange for a focus on catastrophic events, such as a nuclear war, as evidenced through the creation of a Domestic Nuclear Detection Office. Mitigation, the cornerstone of emergency management, is not even recognized although the National Flood Insurance Program and the other natural hazards mitigation efforts will be part of the FEMA Office.

This latest reorganization has returned the structure of Federal emergency management and disaster assistance functions back to pre-FEMA status. The responsibilities and capabilities for mitigation, preparedness, response and recovery will now be spread out among several entities within the Department of Homeland Security. Policy decisions have been exercised to focus most of the human and financial resources on catastrophic threats of nuclear attack, bioterrorism, and terrorism.

This situation is very similar to the one that existed prior to the creation of FEMA in 1979: Federal emergency management and disaster assistance capabilities were located in numerous agencies and programs scattered across the Federal government and the White House. The policy focus of these agencies at this time was on nuclear attack planning.

The question remains. Will history repeat itself?

THE FUTURE: 2005 AND BEYOND

In the aftermath of the terrorist attacks on September 11, FEMA and the newly formed Department of Homeland Security, together with partners in emergency management, fire, police, and public health at the state and local government levels, have been charged with expanding and enhancing our nation's emergency management system. In the years following the creation of the Department of Homeland Security, billions of dollars has been allocated from the federal government to state and local governments in order to expand existing programs and establish new ones designed to meet the new terrorism threat.

Most notably within the United States, but also in many other countries around the world, a budgetary focus upon the preparedness for and prevention of terrorist attacks has emerged and even increased. In the years that followed the attacks in Washington and Virginia, there have been advancements in transportation security and commerce security, large increases in budgetary allowances for first responder terrorism training and related equipment acquisitions, the emergence of homeland security management structures at the state and local levels, a widespread public cognition of and preparedness for the terrorism threat, and many other positive changes. Whether as a result of these changes or in absence of any significant attempts, there have been no major terrorist attacks within the borders of the United States since those initial attacks in 2001.

The focus on terrorism has, expectedly, altered much of the focus that once existed on the mitigation of and preparedness for natural and technological hazards, which by their very nature, are much more likely to occur. In fact, during this same time period that followed the events of September 11, the nation experienced severe flooding, extensive wildfires, record-breaking hurricanes, tornadoes, earthquakes, volcanic activity, drought, avalanches, ice storms, severe winter storms, and many more major and minor disaster events.

The environment of emergency management has continued to grow, and the quality, skill base, technical demands, and caliber of its practitioners have only increased. The hyper-attention that is given to the terrorist threat has provided an unexpected opportunity to expand that base. The goal of this textbook is to provide the background and working knowledge of the emergency management disciplines, and how they are applied to any profession and in everyday life.

As has often occurred following previous defining events, the environment for emergency management will absorb the event and evolve to reflect its impacts. History has begun to repeat itself, and a focal shift to a more national approach to the problem has occurred, with an emphasis upon preparedness through training and equipment. The resilience of the system allows for these midstream corrections. The long-term viability and measure of the influence of emergency management will continue to depend on its value to all citizens in all communities, every day, not just during times of crisis.

2. Natural and Technological Hazards and Risk Assessment

INTRODUCTION

A *hazard* is defined as a "source of danger that may or may not lead to an emergency or disaster and is named after the emergency/disaster that could be so precipitated." *Risk* is defined as "susceptibility to death, injury, damage, destruction, disruption, stoppage and so forth." *Disaster* is defined as an "event that demands substantial crisis response requiring the use of governmental powers and resources beyond the scope of one line agency or service" (National Governors Association, 1982).

Hazard identification is the foundation of all emergency management activities. When hazards react with the human or built environments, the risks associated with that hazard can be assessed. Understanding the risk posed by identified hazards is the basis for preparedness planning and mitigation actions. Risk, when realized, such as in the event of an earthquake, tornado, flood, and so on, becomes a disaster that prompts emergency response and recovery activities. All emergency management activities are predicated on the identification and assessment of hazards and risks.

This chapter discusses the full range of existing hazards, both natural and technological. For each hazard, a brief description of the hazard and its effects is provided. Also included in this chapter is a discussion of risk assessment.

Much of the information for this chapter was acquired from the U.S. Federal Emergency Management Agency Web site, www.fema.gov, and also from FEMA's book *Multi-Hazard Identification and Risk Assessment: A Cornerstone of the National Mitigation Strategy*. Included in Appendix B are organizations' Web site addresses to reference for more in-depth information on a particular hazard.

NATURAL HAZARDS

Natural hazards are those hazards that exist in the natural environment and pose a threat to human populations and communities. Human development has often exacerbated natural hazards. Building communities in the floodplain or on barrier islands increases the potential damage caused by flooding and storm surge. Building a school on a known earthquake fault increases the potential that the school will be destroyed by an earthquake. How humans can better live with hazards is the principal topic of Chapter 3, "The Disciplines of Emergency Management: Mitigation."

Floods

Floods can be slow- or fast-rising, but generally develop over a period of days. Floods usually occur from large-scale weather systems generating prolonged rainfall or onshore winds. Other causes of flooding include locally intense thunderstorms, snowmelt, ice jams, and dam failures. Floods are capable of undermining buildings and bridges, eroding shorelines and riverbanks, tearing out trees, washing out access routes, and causing loss of life and injuries. Flash floods usually result from intense storms dropping large amounts of rain within a brief period. Flash floods occur with little or no warning and can reach full peak in only a few minutes (see Figure 2-1).

Floods are the most frequent and widespread disaster in many countries around the world. Historically, human development has congregated around rivers and ports, and transportation of goods has most commonly been conducted by water. This relationship has resulted in greater exposure to floods. For example, FEMA estimates that more than nine million households and $390 billion in property are at risk from flooding in the United States alone. Flood losses paid by FEMA's National Flood Insurance Program in the 1990s totaled in the billions of dollars (see Table 2-1).

Governments in many countries maintain river and stream gauges to measure floodwater elevations and to provide information on rising water for use in sandbagging and dyke construction, and to warn populations of an impending flood.

Figure 2-1   Midwest Floods, June 1994. Homes, businesses, and personal property were all destroyed by the high flood levels. A total of 534 counties in nine states were declared for federal disaster aid. As a result of the floods, 168,340 people registered for federal assistance. FEMA News Photo.

Table 2-1 Top Ten U.S. Flood Disasters, 1900–2004

Event	Date	# Paid Losses	Amount of Paid Losses
Tropical storm Allison	June 2001	30,270	$1,094,828,120
Hurricane Ivan	Sept. 2004	17,797	$647,727,548
Louisiana flood	May 1995	31,264	$584,140,014
Hurricane Isabel	Sept. 2003	19,465	$449,348,685
Hurricane Floyd	Sept. 1999	18,601	$438,896,736
Nor'easter	Dec. 1992	24,677	$341,866,823
Hurricane Opal	Oct. 1995	9,913	$399,674,203
Midwest floods	June 1993	10,257	$271,325,626
Texas flood	Oct. 1994	6,152	$216,632,721
Hurricane Fran	Sept. 1996	9,883	$213,646,336

Source: www.fema.gov

THE GREAT MIDWEST FLOODS OF 1993: RECOVERY COSTS

- A total of 534 counties in nine states were declared for federal disaster aid for the 1993 Midwest floods. As a result of the floods, 168,340 people registered for federal assistance.
- According to the Galloway Report in June 1994, estimated federal response and recovery costs included more than $4.2 billion in direct federal assistance, $1.3 billion in federal flood insurance payments, and more than $621 million in federal loans to individuals, businesses, and communities.

Of those totals, an estimated $1.69 billion was provided by the USDA for food stamps/commodities, crop loss payments, and other emergency farm grant and loan programs; $597 million by SBA for loans to homeowners, renters, and businesses; $500 million by HUD for housing and community grants; $200 million by DOC for economic development programs; $253 million by USACE for flood control and other emergency operations; $75 million by HHS for various public health services; $100 million by DOEd for schools and student aid; $64.6 million by DOL for employment training and temporary job assistance; $146.7 million by DOT for federal highway repairs, rail freight assistance, and other transportation and emergency services; $34 million by EPA for environmental abatement, control, and cleanup projects; and $41.2 million by DOI for various construction, survey, and cultural restoration programs.

FEMA's costs currently total $1.17 billion, including $371 million in grants to individuals and families for temporary housing, home repairs, unemployment payments, and other disaster-related expenses; $539.5 million to states and local governments for public property restoration and cleanup work; $167.6 million for property acquisitions and other hazard mitigation projects; and $29.2 million to other federal agencies for delivery of emergency supplies and other mission-assigned work.

Note: All funding amounts are in current CY2000 dollars, unadjusted for inflation.
Source: FEMA, www.fema.gov

Earthquakes

An earthquake is a sudden, rapid shaking of the earth caused by the breaking and shifting of rock beneath the earth's surface. This shaking can cause buildings and bridges to collapse; disrupt gas, electric, and phone service; and sometimes trigger landslides, avalanches, flash floods, fires, and huge, destructive ocean waves (tsunamis). Buildings with foundations resting on unconsolidated landfill, old waterways, or other unstable soil are most at risk. Earthquakes can occur at any time of the year (see Figure 2-2).

Specific active seismic zones have been identified around the globe. Millions of people live in these seismic zones and are exposed to the threat of an earthquake daily. The damage caused by an earthquake can be extensive, especially to incompatible building types and construction techniques. Also, earthquakes usually ignite fires, which can spread rapidly among damaged buildings if the water system has been disabled and fire services cannot access the site of the fire. Thousands of residents of Kobe, Japan, perished in the fires caused by the 1995 earthquake in that city because fire trucks and personnel were unable to get to the fires because of debris from fallen and damaged buildings blocking the streets (see Table 2-2).

Earthquakes are sudden events despite scientists' and soothsayers' best efforts to predict when they will occur. Seismic sensing technology can track seismic activity but has yet to accurately predict when a major seismic shift will occur that causes an earthquake. The effects of earthquakes are commonly described by the Richter scale.

Figure 2-2   Northridge Earthquake, California, January 17, 1994. Buildings, cars, and personal property all were destroyed when the earthquake struck. Approximately 114,000 residential and commercial structures were damaged and 72 deaths were attributed to the earthquake. Damage costs were estimated at $25 billion. FEMA News Photo.

Table 2-2 Estimated Earthquake Losses, 1987–1997

Date	Location	Amount
November 24, 1987	Southern California	$4 million
October 18, 1989	Northern California	$5.6 million
February 28, 1990	Southern California	$12.7 million
April 25, 1991	Northern California	$66 million
June 28, 1992	Southern California	$92 million
January 17, 1994	Southern California	$13–20 billion

Source: United States Geological Survey (USGS).

THE RICHTER SCALE

The Modified Mercalli Intensity scale also measures the effects of earthquakes. The intensity of a quake is evaluated according to the observed severity of the quake at specific locations. The Mercalli scale rates the intensity on a Roman numeral scale that ranges from I to XII.

Modified Mercalli	Damage Sustained	Richter Scale
I–IV Instrumental to Moderate	No damage.	≤4.3
V Rather Strong	Damage negligible. Small, unstable objects displaced or upset; some dishes and glass broken.	4.4–4.8
VI Strong	Damage slight. Windows, dishes, glassware broken. Furniture moved or overturned. Weak plaster and masonry cracked.	4.9–5.4
VII Very Strong	Damage slight to moderate in well-built structures; considerable in poorly built structures. Furniture and weak chimneys broken. Masonry damaged. Loose bricks, tiles, plaster, and stones will fall.	5.5–6.1
VIII Destructive	Structural damage considerable, particularly to poorly built structures. Chimneys, monuments, towers, elevated tanks may fail. Frame houses moved. Trees damaged. Cracks in wet ground and steep slopes.	6.2–6.5

continues

Modified Mercalli	Damage Sustained	Richter Scale
IX Ruinous	Structural damage severe; some buildings will collapse. General damage to foundations. Serious damage to reservoirs. Underground pipes broken. Conspicuous cracks in ground; liquefaction.	6.6–6.9
X Disastrous	Most masonry and frame structures/ foundations destroyed. Some well-built wooden structures and bridges destroyed. Serious damage to dams, dikes, embankments. Sand and mud shifting on beaches and flat land.	7.0–7.3
XI Very Disastrous	Few or no masonry structures remain standing. Bridges destroyed. Broad fissures in ground. Underground pipelines completely out of service. Widespread earth slumps and landslides.	7.4–8.1
XII Catastrophic	Damage nearly total. Large rock masses displaced. Lines of sight and level distorted.	>8.1

Source: FEMA, www.fema.gov

Hurricanes

All hurricanes start as tropical waves that grow in intensity and size to tropical depressions, which in turn grow to be tropical storms. A tropical storm is a warm-core tropical cyclone in which the maximum sustained surface wind speed ranges from 39 miles per hour (mph) to less than 74 mph. Tropical cyclones are defined as a low-pressure area of closed-circulation winds that originates over tropical waters. Winds rotate counterclockwise in the northern hemisphere and clockwise in the southern hemisphere.

A hurricane is a tropical storm with winds that have reached a constant speed of 74 mph or more. Hurricane winds blow in a large spiral around a relatively calm center known as the "eye." The eye is generally 20 to 30 miles wide, and the storm may extend outward for 400 miles. As a hurricane approaches, the skies will begin to darken and winds will strengthen. As a hurricane nears land, it can bring torrential rains, high winds, and storm surges. A single hurricane can last for more than two weeks over open waters and can run a path across the entire length of the eastern seaboard.

Figure 2-3 Hurricane Andrew, Florida, August 24, 1992. An aerial view showing damage from one of the most destructive hurricanes in America's history. One million people were evacuated and 54 died in this hurricane. FEMA News Photo.

Hurricane season runs annually from June 1 through November 30. August and September are peak months during the hurricane season. Hurricanes are commonly described using the Saffir-Simpson scale (see Figure 2-3).

THE SAFFIR-SIMPSON SCALE

1 Wind Speed: 74–95 mph
   Storm Surge: 4–5 feet above normal
   Primary damage to unanchored mobile homes, shrubbery, and trees. Some coastal flooding and minor pier damage. Little damage to building structures.

2 Wind Speed: 96–110 mph
   Storm Surge: 6–8 feet above normal
   Considerable damage to mobile homes, piers, and vegetation. Coastal and low-lying areas escape routes flood 2–4 hours before arrival of hurricane center. Buildings sustain roofing material, door, and window damage. Small craft in unprotected mooring break moorings.

3 Wind Speed: 111–130 mph
   Storm Surge: 9–12 feet above normal
   Mobile homes destroyed. Some structural damage to small homes and utility buildings. Flooding near coast destroys smaller structures; larger structures damaged by floating debris. Terrain continuously lower than 5 feet above sea level (ASL) may be flooded up to six miles inland.

4 Wind Speed: 131–155 mph
   Storm Surge: 13–18 feet above normal

continues

Extensive curtainwall failures, with some complete roof structure failure on small residences. Major erosion of beaches. Major damage to lower floors of structures near the shore. Terrain continuously lower than 10 feet ASL may flood (and require mass evacuations) up to six miles inland.

5 Wind Speed: Over 155 mph

Storm Surge: Over 18 feet above normal

Complete roof failure on many homes and industrial buildings. Some complete building failures. Major damage to lower floors of all structures located less than 15 feet ASL and within 500 yards of the shoreline. Massive evacuation of low-ground, residential areas may be required.

Source: FEMA

Table 2-3 Top Ten Costliest Hurricanes in the United States, 1900–2003, Ranked by FEMA Relief Costs

Hurricane	Year	Category	Damage
Hurricane Georges AL, FL, LA, MS, PR, VI	1998	4	$2.255B
Hurricane Andrew FL, LA	1992	4	$1.814B
Hurricane Hugo NC, SC, PR, VI	1989	4	$1.307B
Hurricane Floyd CT, DE, FL, ME, MD, NH, NJ, NY, NC, PA, SC, VT, VA	1999	3	$1.054B
Hurricane Fran MD, NC, PA, SC, VA, WV	1996	3	$620.9M
Hurricane Isabel DC, DE, MD, NC, VA, WV	2003	2	$558.4M
Hurricane Marilyn PR, VI	1995	2	$461.8M
Hurricane Iniki HI	1992	4	$257.5
Hurricane Lilly LA	2002	4	$262.6M
Hurricane Frederic AL, FL, MS	1979	3	$225.6M

Source: www.fema.gov

Hurricanes are capable of causing great damage and destruction over vast areas. Hurricane Floyd in 1999 first threatened the states of Florida and Georgia, made landfall in North Carolina, and damaged sections of South Carolina, North Carolina, Virginia, Maryland, Delaware, New Jersey, New York, Connecticut, Massachusetts, and Maine. The damage was so extensive in each of these states that they all qualified for federal disaster assistance. More recently, Hurricane Mitch brought death and destruction to Nicaragua, Guatemala, El Salvador, and Honduras.

In recent years, significant advances have been made in hurricane tracking technology and computer models. The National Hurricane Center in Miami, Florida now tracks tropical waves from the moment they form off the coast of West Africa through their development as a tropical depression. Once the tropical depression grows to the strength of a tropical storm, the Hurricane Center assigns the storm a name. Once sustained wind speed of the tropical storm exceeds 74 mph, it becomes a hurricane. The Hurricane Center uses aircraft to observe and collect meteorological data on the hurricane and to track its movements across the Atlantic Ocean. It also uses several sophisticated computer models to predict the storm's path. These predictions are used by local and state emergency officials to make evacuation decisions and to predeploy response and recovery resources (see Table 2-3).

Historically, storm surge and high winds have been the principal contributors to the loss of life and injuries and the property and infrastructure damage caused by hurricanes. In recent years, inland flooding caused by hurricane rainfall has resulted in loss of life and severe property damage. Hurricanes also cause significant damage to the natural environment. Storm surge from hurricanes can result in severe beach erosion on barrier islands. Inland flooding from Hurricane Floyd inundated waste ponds on hog farms in North Carolina, washing the hog waste into the Cape Fear River, which eventually dumped these materials into the ocean.

Storm Surges

Storm surges are storms that generate the large waves on the coast that cause coastal flooding and erosion. They are most common from late fall to early spring but can develop year-round. They are usually associated with extra-tropical cyclones (nor'easters) in the North Atlantic Ocean and the Gulf of Mexico, and severe winter low-pressure systems in the North Pacific Ocean and the Gulf of Alaska.

Tornadoes

A tornado is a rapidly rotating vortex or funnel of air extending groundward from a cumulonimbus cloud. Approximately 1,000 tornadoes are spawned by thunderstorms each year. Most tornadoes remain aloft, but the danger is when they touch the ground. A tornado can lift and move huge objects, destroy or move whole buildings long distances, and siphon large volumes from bodies of water. Tornadoes follow the path of least resistance. People living in valleys have the greatest exposure to damage. Tornadoes commonly are described using the Fujita-Pearson Tornado scale.

THE FUJITA-PEARSON TORNADO SCALE

F-0: 40–72 mph, chimney damage, tree branches broken
F-1: 73–112 mph, mobile homes pushed off foundation or overturned
F-2: 113–157 mph, considerable damage, mobile homes demolished, trees uprooted
F-3: 158–205 mph, roofs and walls torn down, trains overturned, cars thrown
F-4: 207–260 mph, well-constructed walls leveled
F-5: 261–318 mph, homes lifted off foundation and carried considerable distances, autos thrown as far as 100 meters

---

In the United States, the most susceptible states to tornadoes are Texas, Oklahoma, Arkansas, Missouri, and Kansas. Together these states occupy what is commonly known as "tornado alley." In recent years, however, tornadoes have struck in cities that are not regularly frequented by tornadoes, including Miami, Nashville, and Washington, D.C. Tornado season is generally March through August, although tornadoes can occur at any time of year. They tend to occur in the afternoon and evening: more than 80 percent of all tornadoes strike between noon and midnight.

Tornadoes can have winds of up to 300 mph and possess tremendous destructive force. Damage is incurred only when the tornado touches down, but tornadoes can touch down in more than one place. The tornado that struck the Washington, D.C. metropolitan area in 2001 first touched down in Alexandria, Virginia, just south of the District of Columbia, went airborne over the district, and touched down again in College Park, Maryland, just north of the district (see Figure 2-4).

Building collapse and flying debris are the principal causes of death and injuries by tornadoes. Early warning is the key to surviving in the path of a tornado. Doppler radar and other meteorological tools are improving the amount of advance warning time available before a tornado strikes. Improved communications and new technologies have also been critical to giving people advance warning of a tornado.

Buildings that are directly in the path of a tornado have little chance of surviving; however, new "safe room" technology developed by FEMA and Texas A&M University offers families and communities a method for surviving the tornado even if your home or community facility does not. A safe room can be built into an existing or new home for a small cost (estimated between $3,000 to $5,000) that will survive a tornado's high winds and flying debris. Your home may be destroyed, but anyone in the safe room will survive. Similar technology is being developed for community shelters.

Although reducing the loss of life and injuries is the principal goal of tornado preparedness and mitigation activities, new technologies in building design and con-

Figure 2-4   College Park, Maryland, September 25, 2001. Rescue workers clean up the debris left by the tornado that killed two people and left more than $16.5 million in damages. Photo by Jocelyn Augustino/FEMA News Photo.

Table 2-4   The 25 Deadliest U.S. Tornadoes

Date	Place	Deaths
1. March 18, 1925	Tri-State (MO, IL, IN)	689
2. May 6, 1840	Natchez, MS	317
3. May 27, 1896	St. Louis, MO	255
4. April 5, 1936	Tupelo, MS	216
5. April 6, 1936	Gainesville, GA	203
6. April 9, 1947	Woodward, OK	181
7. April 24, 1980	Amite, LA; Purvis, MS	143
8. June 12, 1899	New Richmond, WI	117
9. June 8, 1953	Flint, MI	115
10. May 11, 1953	Waco, TX	114
11. May 18, 1902	Goliad, TX	114
12. March 23, 1913	Omaha, NE	103
13. May 26, 1917	Mattoon, IL	101
14. June 23, 1944	Shinnston, WV	100
15. April 18, 1880	Marshfield, MO	99
16. June 1, 1903	Gainesville & Holland, GA	98
17. May 9, 1927	Poplar Bluff, MO	98
18. May 10, 1905	Snyder, OK	97
19. April 24, 1908	Natchez, MS	91
20. June 9, 1953	Worcester, MA	90
21. April 20, 1920	Starkville, MI; Waco, AL	88
22. June 28, 1924	Lorain & Sandusky, OH	85
23. May 25, 1955	Udall, KS	80
24. Sept. 29, 1927	St. Louis, MO	79
25. March 27, 1890	Louisville, KY	76

Source: National Storm Prediction Center, NOAA.

struction are being developed by FEMA and others to reduce the damage to buildings and structures not located directly in the path of a tornado. Some of the same wind-resistant construction techniques used effectively in high-risk hurricane areas are being incorporated into building renovation and construction in tornado-prone areas.

Wildfires

Wildland fires are classified into three categories: (1) a *surface fire* is the most common type and burns along the floor of a forest, moving slowly and killing or damaging trees; (2) a *ground fire* usually is started by lightning and burns on or below the forest floor; and (3) a *crown fire* spreads rapidly by wind and moves quickly by jumping along the tops of trees. Wildland fires usually are signaled by dense smoke that fills the area for miles around.

As residential areas expand into relatively untouched wildlands, people living in these communities increasingly are threatened by forest fires. Protecting structures in the wildland from fire poses special problems and can stretch firefighting resources to the limit. If heavy rains follow a fire, other natural disasters can occur, including landslides, mudflows, and floods. Once ground cover has been burned away, little is left to hold soil in place on steep slopes and hillsides. A major wildland fire can leave a large amount of scorched and barren land. These areas may not return to prefire conditions for decades. If the wildland fire destroyed the ground cover, then erosion becomes one of several potential problems.

Types of wildland fires include the following:

- *Wildland fires.* Fueled almost exclusively by natural vegetation, they typically occur in national forests and parks, where federal agencies are responsible for fire management and suppression.
- *Interface or intermix fires.* Urban/wildland fires in which vegetation and the built environment provide fuel.
- *Firestorms.* Events of such extreme intensity that effective suppression is virtually impossible, firestorms occur during extreme weather and generally burn until conditions change or the available fuel is exhausted.
- *Prescribed fires and prescribed natural fires.* Fires that are intentionally set or selected natural fires that are allowed to burn for beneficial purposes.

Severe drought conditions and the buildup of large quantities of dead trees and vegetation on the forest floors recently have led to a significant increase in wildfires in the United States. In the summer of 2002, several major wildfires raged across the country, principally in the western states. These fires consumed approximately six million acres of forestland, and 20 firefighters lost their lives fighting these fires.

Landslides

Landslides occur when masses of rock, earth, or debris move down a slope. Landslides may be very small or very large, and they can move at slow to very high speeds. Many landslides have been occurring over the same terrain since prehistoric times. They are activated by storms and fires and by human modification of the land. New landslides occur as a result of rainstorms, earthquakes, volcanic eruptions, and various human activities.

Mudflows (or debris flows) are rivers of rock, earth, and other debris saturated with water. They develop when water rapidly accumulates in the ground, such as during heavy rainfall or rapid snowmelt, changing the earth into a flowing river of mud or "slurry." A slurry can flow rapidly down slopes or through channels and can strike with little or no warning at avalanche speeds. A slurry can travel several miles from its source, growing in size as it picks up trees, cars, and other materials along the way.

Lateral spreads are large elements of distributed, lateral displacement of materials. They occur in rock, but they can also occur in fine-grained, sensitive soils such as quick clays. Loose granular soils commonly produce lateral spreads through liquefaction. Liquefaction can occur spontaneously, presumably because of changes in pore-water pressures or in response to vibrations such as those produced by strong earthquakes.

Falls occur when masses of rock or other material detach from a steep slope or cliff and descend by freefall, rolling, or bouncing. Topples consist of the forward rotation of rocks or other materials about a pivot point on a hill slope.

Tsunamis

A tsunami is a series of waves generated by an undersea disturbance such as an earthquake. From the area of the disturbance, the waves will travel outward in all directions, much like the ripples caused by throwing a rock into a pond. As the waves approach the shallow coastal waters, they appear normal and the speed decreases. Then, as the tsunami nears the coastline, it may grow to great height and smash into the shore, causing much destruction.

Areas at greatest risk are less than 50 feet above sea level and within one mile of the shoreline. Tsunamis arrive as a series of successive "crests" (high water levels) and "troughs" (low water levels). These successive crests and troughs can occur anywhere from 5 to 90 minutes apart. They usually occur 10 to 45 minutes apart. The wave speed in the open ocean will average 450 miles per hour. Tsunamis reaching heights of more than 100 feet have been recorded. Most deaths during a tsunami are a result of drowning. Associated risks include flooding, polluted water supplies, and damaged gas lines.

2004 INDIAN OCEAN TSUNAMI

On December 26, 2004, following an earthquake off the coast of the *Banda Aceh* region of Indonesia that measured 8.9 on the Richter scale, a series of tsunamis devastated vast coastal regions in 11 countries as far away as East Africa. The earthquake was the most powerful to have occurred in four decades, and generated waves that reached heights as tall as 60 feet on coastal shorelines. The devastation from this event, in regards to the geographical range and number of people affected within the brief timeframe is virtually unprecedented in modern history.

Due to an almost complete lack of tsunami warning systems, no advance notice of the presence or severity of these impending waves was possible for the local populations, many of whom included foreign tourists. As a result, most people had no opportunity to move to higher ground—an action that surely would have prevented injuries and the loss of so many lives. Though the exact number of people killed will never be known, it is assumed to be greater than 150,000 and possibly more than 200,000. There were over 500,000 people who were injured, and ten times that many left homeless.

The reconstruction period for this disaster is expected to last for many years. Countries from around the world provided rescue personnel, equipment, and billions of dollars in relief funding. For information on the U.S. involvement in this event, see Figure 2-5 or visit www.usaid.gov.

Figure 2-5 USG Humanitarian Assistance Indian Ocean Earthquake and Tsunamis.

Source: United States Agency for International Development (http://www.usaid.gov/locations/asia_near_east/tsunami/02.11.05-Tsunami_USAID_Program_Map.pdf)

2004 INDIAN OCEAN TSUNAMI Q&A

Question: Why was the recent earthquake in Sumatra and the resulting tsunami so destructive?

Answer: The reason is actually a combination of factors, including the following: It was generated by an extremely large earthquake; it occurred within the Indian Ocean, which is essentially a basin surrounded by very heavily populated areas; there was no warning system in place in the Indian Ocean basin; and the event occurred on a Sunday morning of what was, for some, a holiday weekend when many were at the beach.

Question: Could the tsunami disaster that has occurred in Indonesia and the Indian Ocean region happen in the United States?

Answer: Yes, although the probability of tsunami is significantly less than other coastal hazards such as hurricanes and storms. However, even though they are rare, as shown in the recent event, the consequences are large enough that they can pose a significant risk. Tsunamis can occur along any coastline, although they occur mostly along the Pacific coastline because of the more frequent seismic hazard. Since they occur so infrequently, the probability is considered too remote to address this hazard in normal building code requirements.

One significant difference between the Indian Ocean and the Pacific Ocean is that the National Oceanic and Atmospheric Administration (NOAA) has a tsunami warning system in place and partner agencies such as FEMA are working with states and local communities to help establish local warning systems and evacuation plans and to raise tsunami awareness among residents and visitors.

Question: What is the greatest tsunami risk to the United States?

Answer: Probably the greatest risk to the United States is believed to be a tsunami that would be generated by an earthquake along the Cascadia subduction zone off the coast of Washington, Oregon, and northern California. Similar to the northern coast of Sumatra, a Cascadia earthquake would be very large, would result in a tsunami, and would give only a few minutes of warning time to the residents along the Pacific Northwest coastline, in many cases not enough time to allow for evacuation, especially during vacation season. This fault last generated an estimated magnitude 9.0–9.5 earthquake and tsunami on January 26, 1700. Although there is Native American folklore and geologic evidence, such as sand deposits, to prove the impact of the tsunami, the actual date has been confirmed from Japanese tsunami records. Although an Atlantic coast tsunami would certainly cause tremendous amounts of damage, the probability of such an event is smaller than a Pacific event.

Question: Has there been a tsunami that has caused fatalities in the United States?

Answer: Yes, several.

On April 1, 1946, a magnitude 7.8 earthquake near Uminak Island in Alaska's Aleutian Islands destroyed a steel reinforced concrete U.S. Coast Guard

continues

lighthouse on Uminak Island, killing all five occupants. The tsunami hit Hawaii five hours later, destroying the Hilo waterfront and killing a total of 165 people, including children at a school on Laupahoehoe Point. It was because of this event that the United States established the Pacific Tsunami Warning Center, now part of NOAA.

On May 22, 1960, a magnitude 9.5 subduction zone earthquake off the coast of Chile resulted in a tsunami that affected the entire Pacific Rim, including Hilo, Hawaii, where it killed 61 people.

On March 28, 1964, the magnitude 9.2 Anchorage earthquake generated a tsunami that caused damage in southeast Alaska, Vancouver Island, Washington, California, and Hawaii. Hardest hit by the tsunami was Crescent City, California, where the tsunami reached 30 feet and destroyed half of the waterfront district. A total of 120 people were killed by the tsunami.

Question: Is a tsunami possible in the Atlantic Ocean?

Answer: Yes. In 1755, an earthquake off the coast of Lisbon, Portugal, reportedly killed thousands along the coast of Portugal, Spain, and North Africa. More recently, a moderate tsunami struck the northwest coast of Puerto Rico in 1918 as a result of an offshore earthquake along the North Atlantic and Caribbean Plate boundary. Also, an earthquake on November 18, 1929, in the Grand Banks of Newfoundland, generated a tsunami that caused considerable damage and loss of life at Placentia Bay, Newfoundland, and resulted in waves that were observed down much of the east coast of the United States.

Although there is the potential for seismic activity in the Caribbean, the Atlantic Ocean generally does not have the type or number of earthquake faults capable of generating a tsunami with the frequency and severity of those in the Pacific. However, there are other potential hazards that could also trigger a tsunami, including volcanic activity along the mid-Atlantic ridge and slumping from pockets of methane hydrate recently found off the coast of South Carolina. Though the probability of such an Atlantic Ocean tsunami is considered rare, a tsunami striking the east coast of the United States or almost anywhere else along the Atlantic Ocean shoreline would result in significant damage and loss of life.

Question: Do we have any quantified tsunami risk assessment information?

Answer: Part of the National Oceanic and Atmospheric Administration (NOAA)-led National Tsunami Hazard Mitigation Program (NTHMP) (www.pmel.noaa.gov/tsunami-hazard) includes a program of developing tsunami inundation maps that show the extent of inundation for the affected area for the Pacific Northwest, Alaska, and Hawaii. These NOAA tsunami inundation maps are now being used at the state and community level to plan for tsunami response and evacuations.

In addition, the FEMA National Flood Insurance Program (NFIP) (www.fema.gov/fima/nfip.shtm) considered tsunami wave heights during the development of its Flood Insurance Rate Maps in areas of Hawaii and the west coast where tsunamis were considered a significantly probable flood threat. In addition, FEMA recently funded a NOAA pilot project under its NFIP Flood Map Modernization Program to develop improved maps and tsunami probabilities, using Seaside, Oregon, as a pilot project.

However, we do not have any reliable risk assessment data, such as information that would be available through HAZUS, FEMA's standardized loss estimation software program. Some interest has been expressed in developing a HAZUS tsunami model that could be based on these inundation maps, but funding has not been available.

Question: Can planning for a large disaster event such as a tsunami make a difference?

Answer: Although a tsunami can generate forces that can overwhelm the best-constructed buildings, planning for such an event can make a difference. A comparison between the 1993 tsunami of Aonae, Japan, and the 1998 tsunami of Warapu, Papua New Guinea, demonstrates how planning can make a difference. Although both events were triggered by earthquakes of similar magnitudes and impacted areas of roughly similar population, the first event killed 15 percent of the population, the second event killed 40 percent of the population. The primary difference was that Japan has a strong program for tsunami public education, awareness, and a warning system that allowed people to get to high ground, whereas Warapu did not. FEMA is aware that public education, awareness, and a warning system can make a real difference in community disaster resistance, and supports continued improvement of community tsunami preparedness, plans, and activities.

Question: Is there anything individuals can do to reduce their vulnerability to the tsunami hazard?

Answer: Residents and visitors to coastal communities should take the time to learn the local evacuation routes and safe areas (visitors' centers often have tsunami evacuation maps and information), and be prepared with emergency supplies that will help them deal with any emergency. Strong ground shaking near the ocean may be the only clue to the arrival of a tsunami within minutes. If shaking is felt, or if you see the ocean suddenly begin to recede, you should go to high ground immediately and wait for further instructions from local officials about when it is safe to return. Tsunami waves can last for hours. Also, subsequent sets of waves are usually the most dangerous, as they can often be higher and contain debris generated from the initial waves.

Question: Is there a federal program that addresses the tsunami hazard?

Answer: Yes, the National Tsunami Hazard Mitigation Program (NTHMP) is a federal/state program formed to address the tsunami hazard, improve tsunami warning, develop tsunami inundation mapping, and mitigate its effects. The program is led by the National Oceanic and Atmospheric Administration (NOAA), which is part of the Department of Commerce, and includes FEMA, which is part of the Department of Homeland Security, along with the U.S.

continues

Geological Survey (USGS) and the National Science Foundation (NSF) as the participating Federal agencies. The NTHMP also includes state emergency management and geoscience agencies from five states (Alaska, California, Hawaii, Oregon, and Washington). NOAA, FEMA, USGS, and the five states make up the steering committee for the NTHMP.

FEMA and the states are involved primarily in the emergency management and mapping issues, NOAA with tsunami modeling and warning system issues, and USGS with seismic system issues. Together, the agencies have developed many products and activities for West Coast communities that have increased their readiness for both long distance and local tsunamis. Future work will continue to improve the level of readiness.

Since tsunami is an earthquake-generated hazard, it is also referenced in the National Earthquake Hazard Reduction Program (NEHRP), which was established by Congress to reduce the risk posed by earthquakes. FEMA is responsible for implementation of the NEHRP, and we have sought to work with NOAA to coordinate activities between the two programs.

Additional information on the NTHMP can be found on the NOAA Web site, (www.pmel.noaa.gov/tsunami-hazard/), and on the FEMA/NEHRP Web site.

Question: Are there any early lessons we can gain from this disaster?
Answer: This disaster demonstrates the importance of tsunami mapping and preparedness activities and the need for tsunami awareness in the United States. The NOAA-led NTHMP is a federal and state program that has several components to address the tsunami hazard, primarily in our Pacific coastal states. NOAA's primary focus has been on developing a tsunami warning system, which is an important component of an overall tsunami program. FEMA has been working in partnership with other federal and state emergency management and science agencies to improve the level of tsunami hazard awareness, planning, and preparedness.

Question: Are there any examples or demonstration projects of tsunami identification and mitigation on a local community level?
Answer: FEMA, through the National Flood Insurance Program (NFIP), NOAA, and USGS are cofunding a $540,000 pilot project to develop risk identification products that will help communities understand their actual level of risk from tsunami in a way that could be conveyed on our existing flood maps. The goal of the project is to develop techniques that can be used to determine the probability and magnitude of tsunami in other communities along the west coast of the United States. The location of the pilot project is Seaside, Oregon. FEMA's NFIP is involved because FEMA is responsible for mapping areas subject to flooding in order to properly rate flood insurance policies and provide risk assessment information to states and local communities.

Question: Is there a program that communities can participate in to reduce their risk from tsunami?

Answer: FEMA supports and promotes NOAA's TsunamiReady Program (www.prh.noaa.gov/ptwc/tsunamiready/tsunamiready.htm) because it includes the same important emergency planning elements that FEMA promotes in all predisaster preparedness activities. Currently, there are 11 TsunamiReady communities located in the Pacific Northwest. The criteria for being recognized as a TsunamiReady Community includes establishing an Emergency Operating Center, warning systems, a community preparedness program, identification of their hazard zone, and establishing evacuation routes and safe areas. Also required is the establishment of plans and drills for schools in the hazard zone, by which the community protects its most precious commodity—its children, its hope, and its lifeblood for the future. This kind of planning, preparedness, and mitigation changes the impact that earthquakes and tsunamis have on communities, and results in a community that is safer and more disaster resistant.

Question: Is it possible to build a structure that would be capable of resisting the extreme forces of a tsunami?

Answer: This question takes on a greater significance because there are several coastal communities along our nation's West Coast that are vulnerable to tsunami triggered by an earthquake on the Cascadia Subduction Zone. An earthquake along this fault could potentially generate a tsunami within minutes, similar to what happened on the north end of Sumatra. Given that many of these coastal communities are located in areas that would be impossible to evacuate in time, which could result in a significant loss of life, FEMA and its mitigation partners at the federal, state, and local levels are looking for alternatives. The only feasible alternative would be vertical evacuation, providing such a structure could be constructed to resist tsunami loads.

For the average structure, generally it would not be economically feasible to construct to withstand the extreme loads of a tsunami. However, we believe it would be possible that a specially designed structure could be built to withstand at least specific tsunami loads without collapse for the purposes of providing community shelter for vertical evacuation. Similarly, the same criteria could possibly be used if the structure was to house a large occupancy load (such as some larger seaside resorts).

Question: Are there any current FEMA design guidance documents that provide design criteria on tsunami?

Answer: FEMA's most recent study of coastal seismic and tsunami loads was done in association with the FEMA Coastal Construction Manual (FEMA-55). This manual was developed to provide design and construction guidance for structures built in coastal areas throughout the United States. The Coastal Construction Manual (CCM) addresses seismic loads for coastal structures and provides information on the tsunami hazard and associated loads. The conclusion of the CCM's authors is that tsunami loads are far too great and that, in general, it is not feasible or practical to design normal structures to withstand these loads. It should be noted that the study was for conventional construction, and did not

continues

take into account the possibility of special design and construction details that would be possible for critical facilities.

Question: Is there any work currently under way to develop tsunami design criteria for shelters or critical facilities?

Answer: Yes, there is a joint National Oceanic and Atmospheric Administration (NOAA)/FEMA-funded effort currently underway to do just that. Given the significant level of risk that exists for the residents of the certain coastal communities in the Pacific Northwest, Alaska, and Hawaii, the cofunded FEMA/NOAA work for the development of guidance for the design of structures that could be used for vertical evacuation will be a significant step toward improving the protection of the residents of these communities.

The first phase of this effort is being managed by the state of Washington under a $100,000 grant from NOAA under the NTHMP. In Phase 1, data regarding tsunamis and their potential forces on structures was collected. The Phase 1 work was preceded by a workshop held in 2003 and attended by engineers from the different affected states. A report on this workshop has been issued by the NTHMP. The overall Phase 1 work is complete and the report is being finalized.

The second phase will determine whether it is possible to design and build a structure to withstand specific tsunami loads, and if so, to develop a technical design and construction guidance document for special facilities that would allow for vertical evacuation from tsunami conditions. This work would continue and build on the work started in Phase 1. Funding for this two-year $400,000 effort will be equally divided between FEMA, the National Earthquake Hazards Reduction Program (NEHRP), and NOAA, through the NTHMP.

The Phase 2 work will be done with input from the engineering and design communities and the states to research and produce the construction design guidance for a tsunami shelter structure capable of withstanding both the severe ground shaking expected during a design earthquake and specific velocities and water pressure from a tsunami that would impact structures. This is a significant challenge since current design practice takes into account earthquake or coastal storm surge but does not address stronger forces that a tsunami would generate. The project will work with Oregon State University's improved tsunami testing basin, recently funded by the National Science Foundation's Network for Earthquake Engineering Simulation (NEES). The project is being done under contract to the Applied Technology Council, and is just getting under way.

A third phase is planned, where information for states and local communities on how this tsunami design guidance can be utilized will be developed. This information will especially be critical for low-lying communities that lack evacuation access to high ground following a local earthquake and that may have to rely on vertical evacuation in existing buildings. Funding is anticipated to be $100,000, also equally divided between NOAA and FEMA.

Source: www.fema.gov

Volcanic Eruptions

A volcano is a mountain that opens downward to a reservoir of molten rock below the surface of the earth. Unlike most mountains, which are pushed up from below, volcanoes are built up by an accumulation of their own eruptive products—lava, ash flows, and airborne ash and dust. When pressure from gases and the molten rock becomes strong enough to cause an explosion, eruptions occur. Gases and rock shoot up through the opening and spill over, or fill the air with lava fragments. Volcanic products are used as building or road-building materials, as abrasive and cleaning agents, and as raw materials for many chemical and industrial uses. Lava ash makes soil rich in mineral nutrients.

Volcanic ash can affect people hundreds of miles away from the cone of a volcano. Several of the deaths from the Mount St. Helens volcano in 1980 were attributed to inhalation of ash. Volcanic ash can contaminate water supplies, cause electrical storms, and collapse roofs. An erupting volcano can also trigger tsunamis, flash floods, earthquakes, rock falls, and mudflows.

Sideways-directed volcanic explosions, known as "lateral blasts," can shoot large pieces of rock at very high speeds for several miles. These explosions can kill by impact, burial, or heat. They have been known to knock down entire forests. Most deaths attributed to the Mount St. Helens volcano were a result of lateral blast and trees that were blown down.

Severe Winter Storms

Severe winter storms consist of extreme cold and heavy concentrations of snowfall or ice. A blizzard combines heavy snowfall, high winds, extreme cold, and ice storms. In the United States, the origins of the weather patterns are from four sources:

- In the Northwestern states, cyclonic weather systems form the North Pacific Ocean or the Aleutian Island region sweep massive low-pressure systems with heavy snow and blizzards.
- In the Midwestern and Upper Plains states, Canadian and Arctic cold fronts push ice and snow deep into the interior region and, in some instances, all the way down to Florida.
- In the Northeast, lake-effect snowstorms develop from the passage of cold air over the relatively warm surfaces of the Great Lakes, causing heavy snowfall and blizzard conditions.
- The Eastern and Northeastern states are affected by extra-tropical cyclonic weather systems in the Atlantic Ocean and Gulf of Mexico that produce snow, ice storms, and occasional blizzards.

Droughts

Drought is defined as a water shortage caused by a deficiency of rainfall and differs from other natural hazards in three ways: (1) A drought's onset and end are difficult to determine because the effects accumulate slowly and may linger even after the apparent termination of an episode; (2) the absence of a precise and universally accepted definition adds to the confusion about whether a drought exists, and if it does, the degree of severity; and (3) drought effects are less obvious and spread over a larger geographic area.

Extreme Heat

Extreme heat is defined as temperatures that hover 10 degrees or more above the average high temperature for the region and last for several weeks. Humid or muggy conditions, which add to the discomfort of high temperatures, occur when a "dome" of high atmospheric pressure traps hazy, damp air near the ground. Excessively dry and hot conditions can provoke dust storms and low visibility. Droughts occur when a long period passes without substantial rainfall. A heat wave combined with a drought is a very dangerous situation.

Coastal Erosion

Coastal erosion is measured as the rate of change in the position or horizontal displacement of a shoreline over a period of time. It generally is associated with storm surges, hurricanes, windstorms, and flooding hazards, and may be exacerbated by human activities such as boat wakes, shoreline hardening, and dredging.

Thunderstorms

Thunderstorms can bring heavy rains (which can cause flash flooding), strong winds, hail, lightning, and tornadoes. Thunderstorms are generated by atmospheric imbalance and turbulence caused by the combination of conditions: (1) unstable warm air rising rapidly into the atmosphere; (2) sufficient moisture to form clouds and rain; and (3) upward lift of air currents caused by colliding weather fronts (cold and warm), sea breezes, or mountains.

Thunderstorms may occur singly, in clusters, or in lines. Thus it is possible for several thunderstorms to affect one location in the course of a few hours. Some of the most severe weather occurs when a single thunderstorm affects one location for an extended period.

Lightning is a major threat during a thunderstorm. In the United States, between 75 and 100 Americans are hit and killed by lightning each year. A thunderstorm is classified as severe if its winds reach or exceed 58 mph, it produces a tornado, or it drops surface hail at least 0.75 inch in diameter.

Significant airplane disasters often are associated with thunderstorms and lightning. It is a myth that lightning never strikes twice in the same place. In fact, lightning will strike several times in the same place in the course of one discharge. A bolt of lightning reaches a temperature approaching 50,000 degrees Fahrenheit in a split second.

Hailstorms

Hailstorms are an outgrowth of a severe thunderstorm in which balls or irregularly shaped lumps of ice greater than 0.75 inch in diameter fall with rain. Hailstorms occur more frequently during late spring and early summer, when the jet stream migrates northward across the Great Plains. Hailstorms cause nearly $1 billion in property and crop damage annually.

Snow Avalanches

A snow avalanche is sliding snow or an ice mass that moves at high velocities. It can shear trees, completely cover entire communities and highway routes, and level buildings. Natural and human-induced snow avalanches most often result from structural weaknesses within the snowpack. The potential for a snow avalanche increases with significant temperature influences.

The primary threat is loss of life of backcountry skiers, climbers, and snowmobilers as a result of suffocation when buried in an avalanche. Around 10,000 avalanches are reported each year. Since 1790, an average of 144 persons have been trapped in avalanches annually: On average, 14 were injured and 14 were died. The estimated annual average damage to structures is $500,000.

Land Subsidence

Land subsidence is the loss of surface elevation caused by the removal of subsurface support; it ranges from broad, regional lowering of the land surface to localized collapse. The primary cause of most subsidence is human activities: underground mining of coal, groundwater or petroleum withdrawal, and drainage of organic soils. The average annual damage from all types of subsidence is conservatively estimated to be at least $125 million (see Table 2-5).

Expansive Soils

Soils and soft rock that tend to swell or shrink because of changes in moisture content are commonly known as expansive soils. Changes in soil volume present a hazard primarily to structures that are built on top of expansive soils. The most extensive

Table 2-5   Top Ten Natural Disasters Ranked by FEMA Relief Costs, 1989–2001

Event	Year	FEMA Funding
Northridge earthquake (CA)	1994	$6.967B
Hurricane Georges (AL, FL, LA, MS, PR, VI)	1998	$2.255B
Hurricane Andrew (FL, LA)	1992	$1.814B
Tropical Storm Allison (FL, LA, MS, PA, TX)	2001	$1.375B
Hurricane Hugo (NC, SC, PR, VI)	1989	$1.307B
Midwest floods (IL, IA, KS, MN, MO, NE, ND, SD, WI)	1993	$1.140B
Hurricane Floyd (CT, DE, FL, ME, MD, NH, NJ, NY, NC, PA, SC, VT, VA)	1999	$1.054B
Loma Prieta earthquake (CA)	1989	$865.8M
Red River Valley floods (MN, ND, SD)	1997	$741.2M
Miami floods (FL)	2000	$623.1M

Source: www.fema.gov

damage occurs to highways and streets. Two major groups of rocks that are prone to expansiveness and that occur more commonly in the West than East are aluminum silicate minerals (i.e., ash, glass, and rocks of volcanic origin) and sedimentary rock (i.e., clay minerals, shale).

Dam Failures

Dam failures are potentially the worst flood events. A dam failure is usually the result of neglect, poor design, or structural damage caused by a major event such as an earthquake. When a dam fails, a gigantic quantity of water is suddenly let loose downstream, destroying anything in its path.

TECHNOLOGICAL HAZARDS

Fires

Fires can be triggered or exacerbated by lightning, high winds, earthquakes, volcanoes, and floods. Lightning is the most significant natural contributor to fires affecting the built environment. Buildings with rooftop storage tanks for flammable liquids are particularly susceptible.

Table 2-6   U.S. Fire Losses, 1991–2003

Year	Fires	Deaths	Injuries	Losses (in $M)
1991	2,041,500	4,465	29,375	$10,906
1992	1,964,500	4,730	28,700	$9,276
1993	1,952,500	4,635	30,475	$9,279
1994	2,054,500	4,275	27,250	$8,630
1995	1,965,500	4,585	25,775	$9,182
1996	1,975,500	4,990	25,550	$9,406
1997	1,795,000	4,050	23,750	$8,525
1998	1,755,000	4,035	23,100	$8,629
1999	1,823,000	3,570	21,875	$10,024
2000	1,708,000	4,045	22,350	$11,207
2001	1,734,500	6,196/3,745*	21,100/20,300**	$44,023/10,583***
2002	1,687,500	3,380	18,425	$10,337
2003	1,584,500	3,925	18,125	$12,307/10,267****

Source: www.usfa.fema.gov

*This number, 3,745, does not include the deaths associated with the September 11, 2001 terrorist attacks. Including those events, there were 6,196 fire-related deaths in 2001.

**This number, 20,300, does not include the injuries associated with the September 11, 2001 terrorist attacks. Including those events, there were 21,100 fire-related injuries in 2001.

***This number, $10,583M does not include the deaths associated with the September 11, 2001 terrorist attacks. Including those events, there was $44,023M in fire-related losses in 2001.

****This number, 10,267, does not include the Southern California Wildfires ("Cedar" and "Old" Wildfires). Including those events, there was $12,307M in fire-related losses in 2003.

Hazardous Materials Incidents

Hazardous materials are chemical substances, which if released or misused, can pose a threat to the environment or health. These chemicals are used in industry, agriculture, medicine, research, and consumer goods. Hazardous materials come in the form of explosives, flammable and combustible substances, poisons, and radioactive materials. These substances are most often released as a result of transportation accidents or because of chemical accidents in plants. Hazardous materials in various forms can cause death, serious injury, long-lasting health effects, and damage to buildings, homes, and other property. Many products containing hazardous chemicals are routinely used and stored in homes. These products are also shipped daily on the nation's highways, railroads, waterways, and pipelines. Varying quantities of hazardous materials are manufactured, used, or stored at an estimated 4.5 million facilities in the United States—from major industrial plants to local dry cleaning establishments or gardening supply stores.

CHLORINE FACT SHEET

What chlorine is

- Chlorine is an element used in industry and found in some household products.
- Chlorine is sometimes in the form of a poisonous gas. Chlorine gas can be pressurized and cooled to change it into a liquid so that it can be shipped and stored. When liquid chlorine is released, it quickly turns into a gas that stays close to the ground and spreads rapidly.
- Chlorine gas can be recognized by its pungent, irritating odor, which is like the odor of bleach. The strong smell may provide an adequate warning to people that they have been exposed.
- Chlorine gas appears to be yellow-green in color.
- Chlorine itself is not flammable, but it can react explosively or form explosive compounds with other chemicals such as turpentine and ammonia.

Where chlorine is found and how it is used

- Chlorine was used during World War I as a choking (pulmonary) agent.
- Chlorine is one of the most commonly manufactured chemicals in the United States. Its most important use is as a bleach in the manufacture of paper and cloth, but it is also used to make pesticides (insect killers), rubber, and solvents.
- Chlorine is used in drinking water and swimming pool water to kill harmful bacteria. It is also as used as part of the sanitation process for industrial waste and sewage.
- Household chlorine bleach can release chlorine gas if it is mixed with other cleaning agents.

How people can be exposed to chlorine

- People's risk for exposure depends on how close they are to the place where the chlorine was released.
- If chlorine gas is released into the air, people may be exposed through skin contact or eye contact. They may also be exposed by breathing air that contains chlorine.
- If chlorine liquid is released into water, people may be exposed by touching or drinking water that contains chlorine.
- If chlorine liquid comes into contact with food, people may be exposed by eating the contaminated food.
- Chlorine gas is heavier than air, so it would settle in low-lying areas.

How chlorine works

- The extent of poisoning caused by chlorine depends on the amount of chlorine a person is exposed to, how the person was exposed, and the length of time of the exposure.
- When chlorine gas comes into contact with moist tissues such as the eyes, throat, and lungs, an acid is produced that can damage these tissues.

Immediate signs and symptoms of chlorine exposure

- During or immediately after exposure to dangerous concentrations of chlorine, the following signs and symptoms may develop:
  - Coughing
  - Chest tightness
  - Burning sensation in the nose, throat, and eyes
  - Watery eyes

continues

- Blurred vision
- Nausea and vomiting
- Burning pain, redness, and blisters on the skin if exposed to gas, skin injury similar to frostbite if exposed to liquid chlorine
- Difficulty breathing or shortness of breath (may appear immediately if high concentrations of chlorine gas are inhaled, or may be delayed if low concentrations of chlorine gas are inhaled)
- Fluid in the lungs (pulmonary edema) within two to four hours
- Showing these signs or symptoms does not necessarily mean that a person has been exposed to chlorine.

What the long-term health effects are

- Long-term complications from chlorine exposure are not found in people who survive a sudden exposure unless they suffer complications such as pneumonia during therapy. Chronic bronchitis may develop in people who develop pneumonia during therapy.

How people can protect themselves, and what they should do if they are exposed to chlorine

- Leave the area where the chlorine was released and get to fresh air. Quickly moving to an area where fresh air is available is highly effective in reducing exposure to chlorine.
  - If the chlorine release was outdoors, move away from the area where the chlorine was released. Go to the highest ground possible, because chlorine is heavier than air and will sink to low-lying areas.
  - If the chlorine release was indoors, get out of the building.
- If you think you may have been exposed, remove your clothing, rapidly wash your entire body with soap and water, and get medical care as quickly as possible.
- *Removing and disposing of clothing:*
  - Quickly take off clothing that has liquid chlorine on it. Any clothing that has to be pulled over the head should be cut off the body instead of pulled over the head. If possible, seal the clothing in a plastic bag. Then seal the first plastic bag in a second plastic bag. Removing and sealing the clothing in this way will help protect you and other people from any chemicals that might be on your clothes.
  - If you placed your clothes in plastic bags, inform either the local or state health department or emergency personnel upon their arrival. Do not handle the plastic bags.
  - If you are helping other people remove their clothing, try to avoid touching any contaminated areas, and remove the clothing as quickly as possible.
- *Washing the body:*
  - As quickly as possible, wash your entire body with large amounts of soap and water. Washing with soap and water will help protect people from any chemicals on their bodies.
  - If your eyes are burning or your vision is blurred, rinse your eyes with plain water for 10 to 15 minutes. If you wear contacts, remove them before rinsing your eyes, and place them in the bags with the contaminated clothing. Do not put the contacts back in your eyes. You should dispose of them even if you do not wear disposable contacts. If you wear eyeglasses, wash them with soap and water. You can put the eyeglasses back on after you wash them.
- If you have ingested (swallowed) chlorine, do not induce vomiting or drink fluids.
- Seek medical attention right away. Dial 911 and explain what has happened.

How chlorine exposure is treated

- No antidote exists for chlorine exposure. Treatment consists of removing the chlorine from the body as soon as possible and providing supportive medical care in a hospital setting.

How people can get more information about chlorine

People can contact one of the following:

- Regional poison control center (1-800-222-1222)
- Centers for Disease Control and Prevention
  - Public Response Hotline (CDC)
    - English (888) 246-2675
    - Español (888) 246-2857
    - TTY (866) 874-2646
  - Emergency Preparedness and Response Web site
  - E-mail inquiries: cdcresponse@ashastd.org
  - Mail inquiries:
    Public Inquiry c/o BPRP
    Bioterrorism Preparedness and Response Planning
    Centers for Disease Control and Prevention
    Mailstop C-18
    1600 Clifton Road
    Atlanta, GA 30333
- Agency for Toxic Substances and Disease Registry (ATSDR) (1-888-422-8737)
  - E-mail inquiries: atsdric@cdc.gov
  - Mail inquiries:
    Agency for Toxic Substances and Disease Registry
    Division of Toxicology
    1600 Clifton Road NE, Mailstop E-29
    Atlanta, GA 30333
- Centers for Disease Control and Prevention (CDC), National Institute for Occupational Safety and Health (NIOSH), *Pocket Guide to Chemical Hazards*

Source: www.cdc.gov

Nuclear Accidents

The potential danger from an accident at a nuclear power plant is exposure to radiation. This exposure could come from the release of radioactive material from the plant into the environment, usually characterized by a plume (cloudlike) formation. The area that the radioactive release may affect is determined by the amount released from the plant, wind direction and speed, and weather conditions (e.g., rain, snow) that would quickly drive the radioactive material to the ground, hence causing increased deposition of radio nuclides. Radioactive materials are composed of atoms that are unstable. An unstable atom gives off its excess energy until it becomes stable. The energy emitted is radiation. The process by which an atom changes from an unstable state to a more stable state by emitting radiation is called *radioactive decay* or *radioactivity*.

Since 1980, each utility that owns a commercial nuclear power plant in the United States has been required to have both an on-site and off-site emergency response plan as a condition of obtaining and maintaining a license to operate that plant. On-site emergency response plans are approved by the Nuclear Regulatory Commission (NRC). Off-site plans (which are closely coordinated with the utility's on-site emergency response plan) are evaluated by FEMA and provided to the NRC, who must consider the FEMA findings when issuing or maintaining a license.

Radioactive materials, if handled improperly, or radiation that is accidentally released into the environment can be dangerous because of the harmful effects of certain types of radiation on the body. The longer a person is exposed to radiation and the closer the person is to the radiation, the greater the risk. Although radiation cannot be detected by the senses (e.g., sight, smell), it is easily detected by scientists with sophisticated instruments that can detect even the smallest levels of radiation.

Terrorism

Terrorism is the use of force or violence against persons or property in violation of the criminal laws of the United States for purposes of intimidation, coercion, or ransom. Terrorists often use threats to create fear among the public, to try to convince citizens that their government is powerless to prevent terrorism, and to get immediate publicity for their causes.

Before the September 11, 2001 attacks on New York and the Pentagon, most terrorist incidents in the United States have been bombing attacks, involving detonated and undetonated explosive devices, tear gas, and pipe and fire bombs. The effects of terrorism can vary significantly from loss of life and injuries to property damage and disruptions in services such as electricity, water supply, public transportation, and communications.

One way governments attempt to reduce people's vulnerability to terrorist incidents is by increasing security at airports and other public facilities. The U.S. government also works with other countries to limit the sources of support for terrorism.

The Federal Bureau of Investigations (FBI) categorizes terrorism in the United States as one of two types: domestic terrorism or international terrorism. Domestic terrorism involves groups or individuals whose terrorism activities are directed at elements of government or population without foreign direction. International terrorism involves groups or individuals whose terrorist activities are foreign-based and/or directed by countries or groups outside the United States or whose activities transcend national boundaries.

Weapons of Mass Destruction

The U.S. Military defines weapons of mass destruction (WMD) as the broad family of weapons, including conventional, biological, chemical, nuclear, or other advanced weapons, that are characterized by their broad-sweeping intended effects, such as inflicting mass casualties and/or physical destruction. There are many different ways that WMDs are categorized. One of the more common categorizations, which include Chemical, Biological, Nuclear, and Radiological agents, is referred to by the acronym CBRN. Although these weapons are considered WMDs because of their potential for creating such widespread destruction, it should be noted that they can also be distributed in such a way as to harm or kill only one or a very few individuals, but still maintain that potential—and, as such, still be considered weapons of mass destruction.

Chemical warfare agents are poisonous vapors, aerosols, liquids, or solids that have toxic effects on people, animals, or plants (see Table 2-8). They can be released by bombs, sprayed from aircraft, boats or vehicles, or used as a liquid to create a hazard to people and the environment. Some chemical agents may be odorless and tasteless. They can have an immediate effect (a few seconds to a few minutes) or a delayed effect (several hours to several days). Though potentially lethal, chemical agents are difficult to deliver in lethal concentrations. Outdoors, the agents often dissipate rapidly. Chemical agents are also difficult to produce.

There are six types of agents:

- Pulmonary, or "choking" agents
- Blood agents
- Vesicants or blister agents
- Nerve agents
- Incapacitating agents
- Riot-control agents, or "irritants"

LIST OF CHEMICAL AGENTS
Compiled by the Centers for Disease Control

Abrin
Adamsite (DM)
Agent 15
Ammonia
Arsenic
Arsine (SA)
Benzene
Bromobenzylcyanide (CA) **NEW! Aug 1, 2003**
BZ
Cannabinoids
Chlorine (CL)
Chloroacetophenone (CN) **NEW! Aug 1, 2003**
Chlorobenzylidenemalononitrile (CS) **NEW! Aug 1, 2003**
Chloropicrin (PS) **NEW! Aug 1, 2003**
Cyanide
Cyanogen Chloride (CK)
Cyclohcxyl Sarin (GF)
Dibenzoxazepine (CR) **NEW! Aug 1, 2003**
Diphenylchloroarsine (DA)
Diphenylcyanoarsine (DC)
Diphosgene (DP)
Distilled Mustard (HD)
Ethyldichloroarsine (ED)
Ethylene Glycol
Fentanyls and Other Opioids
Hydrofluoric Acid
Hydrogen Chloride
Hydrogen Cyanide (AC)
Lewisite (L, L-1, L-2, L-3)
LSD
Mercury
Methyldichloroarsine (MD)
Mustard Gas (H) (Sulfur Mustard)
Mustard/Lewisite (HL)
Mustard/T
Nitrogen Mustard (HN-1, HN-2, HN-3)
Nitrogen Oxide (NO)
Paraquat
Perflurorisobutylene (PHIB)
Phenodichloroarsine (PD)

continues

Phenothiazines
Phosgene (CG)
Phosgene Oxime (CX)
Phosphine
Potassium Cyanide (KCN)
Red Phosphorous (RP)
Ricin
Sarin (GB)
Sesqui Mustard
Sodium Azide
Sodium Cyanide (NaCN)
Soman (GD)
Stibine
Strychnine
Sulfur Mustard (H) (Mustard Gas)
Super Warfarin
Sulfur Trioxide-Chlorosulfonic Acid (FS)
Tabun (GA)
Teflon and Perflurorisobutylene (PHIB)
Thallium
Titanium Tetrachloride (FM)
Unidentified Chemical
VX
White Phosphorus
Zinc Oxide (HC)

Source: http://www.bt.cdc.gov/agent/agentlistchem.asp

Biological agents are organisms or toxins either naturally occurring or genetically engineered, that can kill or incapacitate people, livestock, and crops. There are three basic groups of biological agents that would likely be used as weapons, which include:

- Bacteria
- Viruses
- Toxins

Most biological agents are difficult to grow and maintain. Although many of these agents decay rapidly when exposed to sunlight and other environmental factors, others such as anthrax spores (see Table 2-10) can be very resilient and survive for decades or longer. Biological agents can be dispersed by aerosolization (spraying them in the air), by human-to-human or animal-to-human infection, and through food and water contamination. Human-to-human transmission has been the primary source of infection in past epidemics that involved pathogens capable of use as a biological weapon, including smallpox, plague, and the Lassa virus.

LIST OF BIOLOGICAL AGENTS
Compiled by the Centers for Disease Control

Anthrax (*Bacillus anthracis*)

Bacillus anthracis (anthrax)

Botulism (*Clostridium botulinum* toxin)

Brucella species (brucellosis)

Brucellosis (*Brucella* species)

Burkholderia mallei (glanders)

Burkholderia pseudomallei (melioidosis)

Chlamydia psittaci (psittacosis)

Cholera (Vibrio cholerae)

Clostridium botulinum toxin (botulism)

Clostridium perfringens (Epsilon toxin)

Coxiella burnetii (Q fever)

E. coli O157:H7 (*Escherichia coli*)

Emerging infectious diseases such as Nipah virus and hantavirus

Epsilon toxin of *Clostridium perfringens*

Escherichia coli O157:H7 (*E. coli*)

Food safety threats (e.g., *Salmonella* species, *Escherichia coli* O157:H7, *Shigella*)

Francisella tularensis (tularcmia)

Glanders (*Burkholderia mallei*)

Melioidosis (*Burkholderia pseudomallei*)

Plague (*Yersinia pestis*)

Psittacosis (*Chlamydia psittaci*)

Q fever (*Coxiella burnetii*)

Ricin toxin from *Ricinus communis* (castor beans)

Rickettsia prowazekii (typhus fever)

Salmonella species (salmonellosis)

Salmonella Typhi (typhoid fever)

Salmonellosis (Salmonella species)

Shigella (shigellosis)

Shigellosis (Shigella)

Smallpox (variola major)

Staphylococcal enterotoxin B

Tularemia (*Francisella tularensis*)

Typhoid fever (*Salmonella* Typhi)

Typhus fever (*Rickettsia prowazekii*)

Variola major (smallpox)

Vibrio cholerae (cholera)

Viral encephalitis (alphaviruses; e.g., Venezuelan equine encephalitis, eastern equine encephalitis, western equine encephalitis)

Viral hemorrhagic fevers (filoviruses, e.g., Ebola, Marburg; and arenaviruses, e.g., Lassa, Machupo)

Water safety threats (e.g., *Vibrio cholerae, Cryptosporidium parvum*)

Yersinia pestis (plague)

Source: http://www.bt.cdc.gov/agent/agentlist.asp

ANTHRAX FACT SHEET
ANTHRAX: WHAT YOU NEED TO KNOW

What Is Anthrax?

Anthrax is a serious disease caused by *Bacillus anthracis*, a bacterium that forms spores. A bacterium is a very small organism made up of one cell. Many bacteria can cause disease. A spore is a cell that is dormant (asleep) but may come to life with the right conditions.

There are three types of anthrax:

- Skin (cutaneous)
- Lungs (inhalation)
- Digestive (gastrointestinal)

How Do You Get It?

Anthrax is not known to spread from one person to another.

Anthrax from animals. Humans can become infected with anthrax by handling products from infected animals or by breathing in anthrax spores from infected animal products (like wool, for example). People also can become infected with gastrointestinal anthrax by eating undercooked meat from infected animals.

Anthrax as a weapon. Anthrax also can be used as a weapon. This happened in the United States in 2001. Anthrax was deliberately spread through the postal system by sending letters with powder containing anthrax. This caused 22 cases of anthrax infection.

How Dangerous Is Anthrax?

The Centers for Disease Control and Prevention classify agents with recognized bioterrorism potential into three priority areas (A, B, and C). Anthrax is classified a Category A agent. Category A agents are those that:

- pose the greatest possible threat for a bad effect on public health
- may spread across a large area or need public awareness
- need a great deal of planning to protect the public's health

In most cases, early treatment with antibiotics can cure cutaneous anthrax. Even if untreated, 80 percent of people who become infected with cutaneous anthrax do not die. Gastrointestinal anthrax is more serious because between one-fourth and more than half of cases lead to death. Inhalation anthrax is much more severe. In 2001, about half of the cases of inhalation anthrax ended in death.

What Are the Symptoms?

The symptoms (warning signs) of anthrax are different depending on the type of the disease:

- Cutaneous: The first symptom is a small sore that develops into a blister. The blister then develops into a skin ulcer with a black area in the center. The sore, blister, and ulcer do not hurt.
- Gastrointestinal: The first symptoms are nausea, loss of appetite, bloody diarrhea, and fever, followed by bad stomach pain.
- Inhalation: The first symptoms of inhalation anthrax are like cold or flu symptoms and can include a sore throat, mild fever and muscle aches. Later symptoms include cough, chest discomfort, shortness of breath, tiredness and muscle aches. (Caution: Do not assume that just because a person has cold or flu symptoms that they have inhalation anthrax.)

How Soon Do Infected People Get Sick?

Symptoms can appear within seven days of coming in contact with the bacterium for all three types of anthrax. For inhalation anthrax, symptoms can appear within a week or can take up to 42 days to appear.

How Is Anthrax Treated?

Antibiotics are used to treat all three types of anthrax. Early identification and treatment are important.

Prevention after exposure. Treatment is different for a person who is exposed to anthrax, but is not yet sick. Health-care providers will use antibiotics (such as ciprofloxacin, doxycycline, or penicillin) combined with the anthrax vaccine to prevent anthrax infection.

Treatment after infection. Treatment is usually a 60-day course of antibiotics. Success depends on the type of anthrax and how soon treatment begins.

Can Anthrax Be Prevented?

There is a vaccine to prevent anthrax, but it is not yet available for the general public. Anyone who may be exposed to anthrax, including certain members of the U.S. armed forces, laboratory workers, and workers who may enter or reenter contaminated areas, may get the vaccine. Also, in the event of an attack using anthrax as a weapon, people exposed would get the vaccine.

What Should I Do if I Think I Have Anthrax?

If you are showing symptoms of anthrax infection, call your health-care provider right away.

What Should I Do if I Think I Have Been Exposed to Anthrax?

Contact local law enforcement immediately if you think that you may have been exposed to anthrax. This includes being exposed to a suspicious package or envelope that contains powder.

What Is CDC Doing to Prepare for a Possible Anthrax Attack?

CDC is working with state and local health authorities to prepare for an anthrax attack. Activities include:

- Developing plans and procedures to respond to an attack using anthrax.
- Training and equipping emergency response teams to help state and local governments control infection, gather samples, and perform tests. Educating health-care providers, media, and the general public about what to do in the event of an attack.
- Working closely with health departments, veterinarians, and laboratories to watch for suspected cases of anthrax. Developing a national electronic database to track potential cases of anthrax.
- Ensuring that there are enough safe laboratories for quickly testing of suspected anthrax cases.
- Working with hospitals, laboratories, emergency response teams, and health-care providers to make sure they have the supplies they need in case of an attack.

Source: www.cdc.gov

A **radiation** threat, commonly referred to as a "dirty bomb" or "radiological dispersion device (RDD)," is the use of common explosives to spread radioactive materials over a targeted area. Radiological weapons are distinct from nuclear blasts. In a radiological attack, the force of the explosion and radioactive contamination will be much more localized. Although the blast will be immediately obvious, the presence of radiation will not be clearly defined until trained personnel with specialized equipment arrive and monitor environmental conditions. The radioactive material will be harmful to those exposed, and may be very difficult to remove or contain. The terror (fear) effect of a radiological attack, however, is expected to be more of a threat than the actual physical consequences that result.

RADIATION FACT SHEET
FREQUENTLY ASKED QUESTIONS (FAQS) ABOUT A RADIATION EMERGENCY

What Is Radiation?

- Radiation is a form of energy that is present all around us.
- Different types of radiation exist, some of which have more energy than others.
- Amounts of radiation released into the environment are measured in units called curies. However, the dose of radiation that a person receives is measured in units called rem.

For more information about radiation, check the following Web sites: www.epa.gov/radiation, www.orau.gov/reacts/define.htm.

How Can Exposure Occur?

- People are exposed to small amounts of radiation every day, both from naturally occurring sources (such as elements in the soil or cosmic rays from the sun), and man-made sources. Man-made sources include some electronic equipment (such as microwave ovens and television sets), medical sources (such as x-rays, certain diagnostic tests, and treatments), and from nuclear weapons testing.
- The amount of radiation from natural or man-made sources to which people are exposed is usually small; a radiation emergency (such as a nuclear power plant accident or a terrorist event) could expose people to small or large doses of radiation, depending on the situation.
- Scientists estimate that the average person in the United States receives a dose of about one-third of a rem per year. About 80% of human exposure comes from natural sources and the remaining 20% comes from man-made radiation sources—mainly medical x-rays.
- Internal exposure refers to radioactive material that is taken into the body through breathing, eating, or drinking.
- External exposure refers to an exposure to a radioactive source outside of our bodies.
- Contamination refers to particles of radioactive material that are deposited anywhere that they are not supposed to be, such as on an object or on a person's skin.

For more information about radiation, check the following Web sites: www.epa.gov/radiation, www.orau.gov/reacts/define.htm.

What Happens When People Are Exposed to Radiation?

- Radiation can affect the body in a number of ways, and the adverse health effects of exposure may not be apparent for many years.
- These adverse health effects can range from mild effects, such as skin reddening, to serious effects such as cancer and death, depending on the amount of radiation absorbed by the body (the dose), the type of radiation, the route of exposure, and the length of time a person was exposed.
- Exposure to very large doses of radiation may cause death within a few days or months.
- Exposure to lower doses of radiation may lead to an increased risk of developing cancer or other adverse health effects later in life.

For more information about health effects from radiation exposure, check the following Web sites:

- www.epa.gov/radiation
- www.orau.gov/reacts/injury.htm
- www.bt.cdc.gov/radiation/healthfacts.asp

What Types of Terrorist Events Might Involve Radiation?

- Possible terrorist events could involve introducing radioactive material into the food or water supply, using explosives (like dynamite) to scatter radioactive materials (called a "dirty bomb"), bombing or destroying a nuclear facility, or exploding a small nuclear device.
- Although introducing radioactive material into the food or water supply most likely would cause great concern or fear, it probably would not cause much contamination or increase the danger of adverse health effects.
- Although a dirty bomb could cause serious injuries from the explosion, it most likely would not have enough radioactive material in a form that would cause serious radiation sickness among large numbers of people.

continues

However, people who were exposed to radiation scattered by the bomb could have a greater risk of developing cancer later in life, depending on their dose.

- A meltdown or explosion at a nuclear facility could cause a large amount of radioactive material to be released. People at the facility would probably be contaminated with radioactive material and possibly be injured if there was an explosion. Those people who received a large dose might develop acute radiation syndrome. People in the surrounding area could be exposed or contaminated.

- Clearly, an exploded nuclear device could result in a lot of property damage. People would be killed or injured from the blast and might be contaminated by radioactive material. Many people could have symptoms of acute radiation syndrome. After a nuclear explosion, radioactive fallout would extend over a large region far from the point of impact, potentially increasing people's risk of developing cancer over time.

For more information about radiation terrorist events, check the following Web sites:

- www.bt.cdc.gov/radiation/terrorismqa.asp
- www.orau.gov/reacts
- www.nrt.org
- www.energy.gov
- www.nrc.gov
- www.epa.gov

What Preparations Can I Make for a Radiation Emergency?

- Your community should have a plan in place in case of a radiation emergency. Check with community leaders to learn more about the plan and possible evacuation routes.
- Check with your child's school, the nursing home of a family member, and your employer to see what their plans are for dealing with a radiation emergency.
- Develop your own family emergency plan so that every family member knows what to do.
- At home, put together an emergency kit that would be appropriate for any emergency. The kit should include the following items:
  - A flashlight with extra batteries
  - A portable radio with extra batteries
  - Bottled water
  - Canned and packaged food
  - A hand-operated can opener
  - A first-aid kit and essential prescription medications
  - Personal items such as paper towels, garbage bags, and toilet paper

For more information about preparing for a radiation emergency event, check the following Web sites:

- ww.fema.gov
- www.redcross.org/services/disaster/beprepared/
- www.epa.gov/swercepp/
- www.ojp.usdoj.gov/bja

How Can I Protect Myself during a Radiation Emergency?

- After a release of radioactive materials, local authorities will monitor the levels of radiation and determine what protective actions to take.

- The most appropriate action will depend on the situation. Tune to the local emergency response network or news station for information and instructions during any emergency.
- If a radiation emergency involves the release of large amounts of radioactive materials, you may be advised to "shelter in place," which means to stay in your home or office; or you may be advised to move to another location.
- If you are advised to shelter in place, you should do the following:
  - Close and lock all doors and windows.
  - Turn off fans, air conditioners, and forced-air heating units that bring in fresh air from the outside. Only use units to recirculate air that is already in the building.
  - Close fireplace dampers.
  - If possible, bring pets inside.
  - Move to an inner room or basement.
  - Keep your radio tuned to the emergency response network or local news to find out what else you need to do.
- If you are advised to evacuate, follow the directions that your local officials provide. Leave the area as quickly and orderly as possible. In addition,
  - Take a flashlight, portable radio, batteries, first-aid kit, supply of sealed food and water, hand-operated can opener, essential medicines, and cash and credit cards.
  - Take pets only if you are using your own vehicle and going to a place you know will accept animals. Emergency vehicles and shelters usually will not accept animals.

For more information about emergency response, check the following Web sites:

- www.fema.gov
- www.redcross.org/services/disaster/beprepared/
- www.epa.gov/swercepp/
- www.ojp.usdoj.gov/bja

Should I Take Potassium Iodide during a Radiation Emergency?

- Potassium iodide (KI) should only be taken in a radiation emergency that involves the release of radioactive iodine, such as an accident at a nuclear power plant or the explosion of a nuclear bomb. A "dirty bomb" most likely will not contain radioactive iodine.
- A person who is internally exposed to radioactive iodine may experience thyroid disease later in life. The thyroid gland will absorb radioactive iodine and may develop cancer or abnormal growths later on. KI will saturate the thyroid gland with iodine, decreasing the amount of harmful radioactive iodine that can be absorbed.
- KI protects only the thyroid gland and does not provide protection from any other radiation exposure.
- Some people are allergic to iodine and should not take KI. Check with your doctor about any concerns you have about potassium iodide.

For more information about KI, check the following Web sites:

- www.bt.cdc.gov/radiation/ki.asp
- www.fda.gov/cder/drugprepare/KI_Q&A.htm
- www.fda.gov/cder/guidance/4825fnl.htm

Source: www.cdc.gov

A **nuclear** blast is an explosion with intense light and heat, a damaging pressure wave, and widespread radioactive material that can contaminate the air, water, and ground surfaces for miles around. The detonation of a nuclear weapon involves the release of great amounts of destructive energy resulting from an intentional initiation of a chain fission or fusion nuclear reaction. A highly refined, weapons-grade nuclear fuel is required for a reaction of this kind. Although experts may predict at this time that a nuclear attack is less likely than other types, terrorism by its nature is unpredictable.

Sources: www.ready.gov, www.dhs.gov

RISK ASSESSMENT

Most practitioners and academics refer to the term *risk assessment* as a process or methodology that can be used for evaluating risk. In this context, risk is defined as (1) the probability and frequency of a hazard occurring, (2) the level of exposure of people and property to the hazard, and (3) the effects or costs, both direct and indirect, of this exposure. There are various approaches to developing a risk assessment methodology, ranging from qualitative to quantitative, as well as several computer-based models for natural hazard risk assessment, currently in use in the United States and Japan.

The validity and use of any risk assessment is determined by the quality and availability of data. Because these two factors are still unknown and will not be determined until the in-country risk templates have been compiled, the determination of the most effective approach will not be made until the data has been collected and reviewed; however, a general discussion of the suggested approach will be undertaken.

As mentioned previously, various accepted methodologies could be applied. These include the risk matrix approach that is qualitative and is designed to support risk management planning and decision making. The Composite Exposure Indicator (CEI) approach is based on the effects of a single or multiple hazards on a series of indicator variables focused primarily on infrastructure, such as roads, pipelines, hospitals, public water supply, and so on. The CEI is a measure of exposure of 14 variables that produces a number that is then correlated to the population affected. Numerous approaches result in vulnerability analyses that have been applied to earthquake and hurricane (coastal) hazards. The differences between these approaches often relate to *how* direct costs or *if* indirect costs are measured. Internal to the World Bank, several individuals have developed methodologies to assess environmental risks, health risks, and other hazards. Common to most of these methodologies is a series of essential elements or steps that must be undertaken. In general these steps are as follows:

1. *Identify and characterize the hazard*. What are the characteristics of the hazard (e.g., high-velocity winds, ground shaking)? What causes the hazard event, and how does it trigger or relate to other hazards?

2. *Evaluate each hazard for the severity and frequency.* What is the probability of a hazard event happening annually, every 10 years, once a century? What factors enhance or deter the probabilities? What measurements or scales can be applied to determine severity? Could other factors influence severity and frequency (e.g., El Niño, global warming)?

3. *Estimate the risk.* Identify and quantify what will be affected by the hazard event. This step imposes the human and built environment that could be affected, damaged, or disrupted by a hazard event. Included in the analysis would be the general building stock (commercial and residential), inventories of lifelines, and essential, critical facilities. Population and development concentrations would be included.

4. *Determine the potential societal and economic (direct) effects and the indirect effects or costs.* In estimating direct economic losses, data that would be included is the cost of repair or replacement of damaged structures or lifelines, nonstructural damage, loss of contents and business inventory, and related loss of function costs. Agricultural (crop) losses figure prominently in this category. Other costs could be income loss, relocation costs, and rental losses that occur as a consequence of the event.

Social costs are predominantly categorized as casualties, injuries, displaced households, and the cost of sheltering. Indirect effects and costs are more difficult to calculate and the data more difficult to obtain. Examples of indirect economic effects can include increase in unemployment, business interruption and loss of production, reduction in demand and consumer spending, and tax base losses. Indirect losses are more easily calculated at the local and regional levels because the information needed relative to population, employment, and tax base and the nature of the economy and businesses is more easily identified.

The costs to federal, state, and local governments, individuals, and businesses of responding to disaster events are often not incorporated into the cost-effect equation, but in many cases these costs have a significant effect on agencies' budgets and should be considered.

Two other steps should be included in looking at a risk assessment methodology:

5. *Determine the acceptable level of risk.* An analysis is undertaken of the information or data assembled in steps 1 to 4 to establish an acceptable level of risk. This means simply: What level of damage or impact will be tolerated? Societal effects and the less tangible, direct, and indirect costs make this evaluation a more difficult part of the process. Compounding this difficulty are the public perception of risk and the political consequences of taking or not taking action to address the risks.

6. *Identify risk-reduction opportunities.* This critical step takes the risk assessment methodology beyond process to decision making and action. At this point, cost-effective actions that will reduce or mitigate unacceptable risks should be identified and implemented. A variety of structural and nonstructural alternatives can be combined with technology, legislation, and other solutions to design a risk-reduction implementation plan consistent with the degree of risks.

TECHNOLOGY

The nation's ability to identify hazards and quantify risk has significantly improved in the last 10 years. Technological advances have refined the ability to identify and understand the nature of hazards and develop better risk assessment methods. Recent technological advances include the use of satellite imagery and radar to map ever-changing floodplains and areas of coastal erosions, the FEMA-developed HAZUS loss estimation model that provides us with loss estimates from various earthquake scenarios, and the technology that has created safe rooms for homes in tornado-prone areas. The research and scientific agencies of the federal government and the university community continue to develop new approaches to measuring, mapping, and predicting natural hazards. With the reality of September 11, technology is focusing on new methods to detect, prevent, or provide an antidote for the various biological and chemical agents that could be used in a terrorist event.

CONCLUSION

In the process by which hazard risks are managed, often called Hazards Risk Management, the identification of hazards is the key factor that determines what preparative and preventive measures will be taken by the community. In other words, a community needs to know their risks to manage them.

Through the monitoring of hazards, emergencies, and disaster throughout the world, and research conducted into the mechanism by which natural, technological, and intentional hazards operate, a greater understanding of risk is being achieved. Without this valuable information that is collected, societies would be much less able to manage the consequences of the low incidence, high catastrophe events, such as tsunamis or weapons of mass destruction, that have traditionally gone unaddressed or done so in a haphazard manner. In sum, information is power, and with information about hazards societies will have the power to act effectively to reduce or eliminate their risk.

Of course, with increased knowledge comes increased responsibility. The provision of hazard information and management tools to states and communities is but one necessary step in the risk reduction process. Success of these efforts requires that they assume responsibility and take appropriate action. Emergency management provides the impetus for incorporating these considerations into the planning and governing of our communities.

Hazards will persist. Some, particularly technological hazards, may be reduced by our efforts, but our ability to control or eliminate natural hazards is questionable. Recent efforts to undo some of the former channelization and flood control projects undertaken by the U.S. Army Corps of Engineers, once thought to be an effective measure to eliminate flood risk, are vivid examples of our inability to control nature. However, there is still a strong argument for an increased emphasis on improved science in hazard identification and increased financial support for hazards mapping, both of which have been effective components in community hazards risk management efforts.

As our knowledge about hazards continues to expand, the economic and social logic of applying long-term solutions for reducing the risks posed by these hazards through mitigation and preparedness will gain momentum. The cost-to-benefit ratios of mitigation and preparedness efforts will become more attractive to local political bodies, and, eventually, disaster losses will begin to fall substantially. However, each and all of these local successes will be wholly dependent upon the leadership potential and motivational abilities of an emergency management professional, who will be the driving force behind any such positive momentum that exists.

3. The Disciplines of Emergency Management: Mitigation

INTRODUCTION

Disasters are a reality of living in the natural world. Despite humans' attempts to control nature, dating back to the early Egyptians and continuing to this century's massive flood control efforts, natural hazards continue.

Over the last decade, the social and economic costs of disasters to the United States and throughout the world have grown significantly. From the period 1990 to 1999, FEMA spent more than $25.4 billion to provide disaster assistance in the United States. During the 1990s, the economic toll of natural disasters topped $608 billion worldwide, more than the previous four decades combined. The causes of this growth are myriad. Climatological changes such as El Niño, global warming, and sea level rise are one factor. Add to these changes the effects of societal actions such as increased development, deforestation and clear-cutting, migration of population to coastal areas, and filling in of floodplains, and a recipe for disaster results.

The discipline of mitigation provides the means for reducing these impacts. *Mitigation* is defined as a sustained action to reduce or eliminate risk to people and property from hazards and their effects. This discussion of mitigation focuses on natural hazards mitigation efforts and programs in the United States. Techniques for mitigation of technological hazards will be referenced, but the body of knowledge and applications in this area are still evolving; however, many of the successful natural hazards techniques such as building codes do have applicability to technological hazards.

The function of mitigation differs from the other emergency management disciplines because it looks at long-term solutions to reducing risk as opposed to preparedness for hazards, the immediate response to a hazard, or the short-term recovery from a hazard event. Mitigation is usually not considered part of the emergency phase of a disaster as in response, or as part of emergency planning as in preparedness. The definition lines do get a little blurred regarding recovery. As discussed in Chapter 5, applying mitigation strategies should be a part of recovery from disaster; however, even in this context, these are actions that will reduce the impacts, or risks, over time.

The recovery function of emergency management still represents one of the best opportunities for mitigation, and until recently, this phase in a disaster plan provided the most substantial funding for mitigation activities. Recently there has been a trend toward greater federal spending on predisaster mitigation, which is discussed later in this chapter.

Another difference sets mitigation apart from the other disciplines of emergency management. Implementing mitigation programs and activities requires the participation and support of a broad spectrum of players outside of the traditional emergency management circle. Mitigation involves, among others, land-use planners, construction and building officials, both public and private, business owners, insurance companies, community leaders, and politicians.

The skills and tools for accomplishing mitigation (i.e., planning expertise, political acumen, marketing and public relations, and consensus building) are different from the operational, first-responder skills that more often characterize emergency management professionals. In fact, historically, emergency management professionals have been reluctant to take a lead role in promoting mitigation. A state director of emergency management once said words to the effect: "I will never lose my job for failing to do mitigation, but I could lose my job if I mess up a response."

With the exception of the fire community, who were early leaders in the effort to mitigate fire risks through support for building codes, code enforcement, and public education, the emergency management community has remained focused on their response and recovery obligations; however, this trend is changing for several reasons. Leadership at the federal level, larger disasters, substantial increases in funding, and more value and professionalism in emergency management have resulted in greater acknowledgement of the importance of mitigation.

This chapter discusses the tools of mitigation, the impediments to mitigation, federal programs that support mitigation, and several case studies that demonstrate how these tools have been applied to successfully reduce various risks.

TOOLS FOR MITIGATION

Over the years, the United States has made great strides in reducing the number of deaths that occur in natural disasters. Through building codes, warning systems, and public education, the number of deaths and casualties from natural disasters in the last century has significantly declined; however, economic effects and property damages have escalated. Many people believe that these costs are preventable and that the tools exist to dramatically reduce these costs.

Technological disasters such as the Oklahoma City bombing and the terrorist attacks of September 11, 2001, are not as easy to analyze. There is much speculation about how improved intelligence and security could reduce the human effects of these disasters. From a property perspective, many people believe that some reduction in impacts could be achieved through application of traditional mitigation techniques such as improved building construction for blast effects. Other technological disasters such as the Valdez oil spill, the Three Mile Island emergency, and so on could have been prevented through better inspections, training, education, and exercises. These measures reflect good preparedness activities more than mitigation.

In any case, further research and analyses are needed to answer the questions posed by the effects of terrorist events and similar technological hazards.

Most practitioners agree that the primary intent of mitigation is to ensure that fewer communities and individuals become victims of disasters. The goal of mitigation is to create economically secure, socially stable, better built, and more environmentally sound communities that are out of harm's way.

The following widely accepted mitigation tools are used to reduce risk:

- Hazard identification and mapping
- Design and construction applications
- Land-use planning
- Financial incentives
- Insurance
- Structural controls

Hazard Identification and Mapping

This is the most obvious tool for mitigation. You can't mitigate a hazard if you don't know what it is or whom it affects. The most essential part of any mitigation strategy or plan is an analysis of what the hazards are in a particular area. The resources for hazards identification are numerous. The federal government has extensive programs that map virtually every hazard, and these products are available to communities. FEMA's National Flood Insurance Program (NFIP) provides detailed flood maps and studies, and the U.S. Geological Survey (USGS) provides extensive earthquake and landslide studies and maps. Many state agencies have refined the products for hazard identification. For example, special soil stability studies and geological investigations, which are required in some parts of California, further refine this analysis.

Geographic information systems (GIS) have become ubiquitous and staples for all local planning organizations. What is often missing from the available tools is the ability to superimpose the human and built environment onto the hazards, thereby providing a quantified level of risk. FEMA has developed one such tool called HAZUS. HAZUS is a nationally applicable methodology for estimating losses from earthquakes at the community or regional level. FEMA is currently expanding HAZUS to cover hurricane or wind losses and floods.

Design and Construction Applications

The design and construction process provides one of the most cost-effective means of addressing risk. This process is governed by building codes, architecture and design criteria, and soils and landscaping considerations. Code criteria that support risk reduction usually apply only to new construction, substantial renovation, or renovation to change the type or use of the building. Enactment of building codes are the responsibility of the states, and most state codes are derivatives of one of the three model codes, which reflect geographical differences across the United States. Some states delegate code adoption responsibility to more local governmental authorities. Because of cost, codes that require rehabilitation of existing potentially hazardous structures have been rarely implemented. The Los Angeles seismic retrofit ordinance is a rare example. The case study of the Virgin Islands at the end of this chapter illustrates the importance of building codes to mitigation.

The construction process offers other opportunities. For example, using fire-retardant building materials such as slate instead of wood for roofing is important in areas of wildland/urban interface such as Oakland, California. Constructing houses on pilings allows for uninterrupted flow of high-velocity waves in coastal areas.

Landscaping is particularly critical in areas of potential wildfires because vegetation close to structures can become fuel for a fire. Clearing, grading, and siting all have potential impacts to soil stability and erosion and can be included as part of a design or building permit review process.

The federal government has made a significant investment in developing technical guidance for improving the building and construction of structures in hazard areas, particularly earthquake, wind, and flood-prone areas. There has been some discussion of developing a National Code to support mitigation efforts. Because the Constitutional responsibility for public health and safety resides with the states, a National Code developed by the federal government is not politically feasible or practical.

Land-Use Planning

Mitigation programs are most successful when undertaken at the local level, where most decisions about development are made. The strategies for land-use planning offer many options for effecting mitigation, including acquisition, easements, storm water management, annexation, environmental review, and floodplain management plans. It also encompasses a myriad of zoning options such as density controls, special uses permits, historic preservation, coastal zone management, and subdivision controls.

Land-use planning was one of the earliest tools used to encourage mitigation. In 1968 Congress passed the National Flood Insurance Act that established the NFIP. This act required local governments to pass a floodplain management ordinance in return for federally backed, low-cost flood insurance being available to the community. This act started one of the largest federal mapping efforts because the government promised local governments that they would provide them with the technical tools to determine where the floodplains were in their communities so they could steer development away from these areas. A more complete discussion of the NFIP can be found later in this chapter.

Moving structures out of harm's way through property acquisition is clearly the most effective land-use planning tool, but it is also the most costly. Following the Midwest floods of 1993, FEMA worked with Congress to make property acquisition more feasible by providing a substantial increase in funding for acquisition after a disaster. The case study on Missouri at the end of this chapter provides documentation on how well an acquisition strategy can work.

There are many other examples of how land-use planning and ordinances can promote risk reduction. The North Carolina coastal setback ordinance seeks to preserve the fragile and eroding coastlines of its barrier islands. The Alquist-Priola Act in California limits development near known earthquake faults.

Financial Incentives

This is one of the emerging areas for promoting mitigation. Among the approaches being used by localities to reduce risk are creation of special tax assessments, passage of tax increases or bonds to pay for mitigation, relocation assistance, and targeting of federal community development or renewal grant funds for mitigation.

The economic effects of repetitive flooding led the citizens of Napa, California and Tulsa, Oklahoma to pass small tax increases to pay for flood-mitigation activities. In both cases, the tax had minimal effect on the community citizens but had a major effect in reducing the potential economic losses from future floods. Berkeley, California has passed more than 10 different bond issues to support seismic retrofit of public buildings, schools, and private residences.

Funding from the Community Development Block Grant (CDBG), a HUD program, has been used extensively to support local efforts at property acquisition and relocation. These funds have been used to meet the nonfederal match on other federal funding, which has often been a stumbling block to local mitigation. Other federal programs of the Small Business Administration (SBA) and the Economic Development Administration provide financial incentives for mitigation.

Other emerging areas of financial tools include special assessment districts, impact fees, and transfer of development rights. All these tools provide either incentives or penalties to developers as a means of promoting good risk-reduction development practices.

Insurance

Some people would argue with the inclusion of insurance as a mitigation tool. Their reasoning is that insurance by itself really only provides for a transfer of the risk from the individual or community to the insurance company. Although this is true, the NFIP is the prime example of how, if properly designed, the insurance mechanism can be a tool for mitigation. The NFIP is considered to be one of the most successful mitigation programs ever created.

The NFIP was created by Congress in response to the damages from multiple, severe hurricanes and inland flooding and the rising costs of disaster assistance after these floods. At that time, flood insurance was not readily available or affordable through the private insurance market. Because many of the people being affected by this flooding were low-income residents, Congress agreed to subsidize the cost of the insurance so the premiums would be affordable. The idea was to reduce the costs to the government of disaster assistance through insurance. The designers of this program, with great insight, thought the government should get something for their subsidy. So in exchange for the low-cost insurance, they required that communities pass an ordinance directing future development away from the floodplain.

The NFIP was designed as a voluntary program and, as such, did not prosper during its early years, even though flooding disaster continued. Then in 1973, after

Hurricane Agnes, the legislation was modified significantly. The purchase of federal flood insurance became mandatory on all federally backed loans. In other words, anyone buying a property with a Veterans Administration (VA) or Federal Housing Administration (FHA) loan had to purchase the insurance. Citizen pressure to buy the insurance caused communities to pass ordinances and join the NFIP. The NFIP helped the communities by providing them with a variety of flood hazard maps to define their flood boundaries and set insurance rates.

The 1993 Midwest floods triggered another major reform to the NFIP. This act strengthened the compliance procedures. It told communities that if they didn't join the program, they would be eligible for disaster assistance only one time. Any further request would be denied. As a positive incentive, the act established a Flood Mitigation Assistance (FMA) fund for flood planning, flood mitigation grants, and additional policy coverage for meeting the tougher compliance requirements such as building elevation.

Over the years, the NFIP has created other incentive programs such as the Community Rating System. This program rewards those communities that go beyond the minimum floodplain ordinance requirements with reduced insurance premiums. The NFIP represents one of the best public/private partnerships. Through the Write Your Own program, private insurers are given incentives to market and sell flood insurance.

Today more than 20,000 communities in the NFIP have mitigation programs in place. Other attempts have been made to duplicate this program for wind and earthquake hazards, but these have not received the support necessary to pass in the Congress. If another major earthquake occurs, the issue of creating a federally supported earthquake or all-hazards insurance will resurface.

Structural Controls

Structural controls are controversial as a mitigation tool. Structural controls usually have been used to protect existing development. In doing so, they can have both positive and negative effects on the areas they are not protecting. In addition, as the name implies, they are used to control the hazard, not reduce it. Invariably, as was seen so graphically in the Midwest floods, the structures lose control and nature wins; however, in some circumstances, structural controls are the only alternative.

The most common form of structural control is the levee. The U.S. Army Corps of Engineers has designed and built levees as flood control structures across the United States. Levees are part of the aging infrastructure of America. As mitigation tools, they have obvious limitations. They can be overtopped or breached, as in the 1993 Midwest floods, they give residents a false sense of safety that often promotes increased development, and they can exacerbate the hazard in other locations. After the 1993 floods, a major rethinking of dependency on levees has occurred. Efforts are being made to acquire structures built behind the levees, new design criteria are being considered, and other more wetland-friendly policies are being adopted. For a city like New Orleans, however, which is built below sea level and where relocation is impractical, levees can be used effectively to protect flood-prone areas.

Other structural controls are intended to protect along coastal areas. Seawalls, bulkheads, breakwaters, groins, and jetties are intended to stabilize the beach or reduce the impacts of wave action. These structures are equally controversial because they protect in one place and increase the damage in another. The shore of New Jersey is a prime example of the failure of seawalls as a solution to shoreline erosion problems. Cape May, New Jersey, where cars used to be raced on the beach, lost all of its beachfront. An ongoing beach replenishment project is the only thing that has brought some of it back.

IMPEDIMENTS TO MITIGATION

If so many tools can be applied, why haven't risk-reduction and mitigation programs been more widely applied? There are several factors, including denial of the risk, political will, costs and lack of funding, and the taking issue. Despite the best technical knowledge, historic occurrence, public education, and media attention, many individuals don't want to recognize that they or their communities are vulnerable. Recognition requires action and it could have economic consequences as businesses decide to locate elsewhere if they find the community is at risk. Some people are willing to try to beat the odds, but if a disaster strikes, they know the government will help them out. Gradually, attitudes are changing. Potential liability issues are making communities more aware, media attention to disasters has brought public pressure, and the government has provided both incentives for, and penalties for not, taking action.

As previously mentioned, mitigation provides a long-term benefit. The U.S. political system tends to focus on short-term rewards. Developers are large players in the political process and often are concerned that mitigation means additional costs. Mitigation strategies and actions require political vision and will. As Tip O'Neill, former Speaker of the U.S. House of Representatives, said, "All politics is local." Well, so is mitigation. Local elected officials are the individuals who have to promote, market, and endorse adopting risk reduction as a goal. For many elected officials, the development pressures are too much, funding is lacking, and other priorities dominate their agendas; however, with the increasing attention to the economic, social, and political costs of not dealing with their risks, more elected officials are recognizing that they can't afford to *not* take action.

Mitigation costs money. Most mitigation of new structures or development can be passed on to the builder or buyer without much notice. Programs to retrofit existing structures or acquisition and relocation projects are expensive and almost always beyond the capacity of the local government. Funding for mitigation comes primarily from federal programs that need to be matched with state or local dollars. As state and local budgets constrict, their ability to match is reduced. Strong arguments can be made that it is in the best financial interest of the federal government to support mitigation. These arguments and a series of large disasters resulted in substantial increases in federal funding, including new monies for predisaster mitigation, but the fact remains that mitigation needs far outweigh mitigation funding.

Many mitigation actions involve privately owned property. A major legal issue surrounding this is the taking issue. The Fifth Amendment to the U.S. Constitution prohibits the taking of property without just compensation. What constitutes a taking, under what circumstances, and what is just compensation have been the focus of numerous legal cases. Several have dealt with the use of property in the floodplain and the use of oceanfront property on a barrier island. The decisions have been mixed, and taking will continue to be an issue in implementing mitigation programs and policies.

FEDERAL MITIGATION PROGRAMS

FEMA is responsible for most of the programs of the federal government that support mitigation; this section focuses on these programs. As noted earlier, the SBA, Economic Development Administration (EDA), and HUD have policies that support mitigation. The PATH program at HUD supports incorporating mitigation into public housing. The Environmental Protection Agency (EPA) has several programs in floodplain management and in 2002 is initiating a new pilot program for national watersheds. The National Earthquake Hazards Reduction Program, which is described in a following section, includes several other federal agencies; however, the predominant federal agency involved in disaster mitigation is FEMA. FEMA's programs include the NFIP (described earlier in the chapter), the Hazard Mitigation Grant Program (HMGP), the Pre-Disaster Mitigation Program (PDM), the National Earthquake Hazard Reduction Program (NEHRP), the National Hurricane Program, the National Dam Safety Program, and the Fire Prevention and Assistance Grant Program.

The Hazard Mitigation Grant Program

HMGP is the largest source of funding for state and local mitigation activities. This program provides grants to state and local governments to implement long-term hazard mitigation programs after a major disaster has been declared by the President. HMGP projects must reduce the risk, and the benefits of the project must exceed the costs.

Examples of activities supported by HMGP include the following:

- Acquisition of property on a voluntary basis and commitment to open use of the property
- Retrofitting of structures and lifelines
- Elevation of structures
- Vegetation management programs
- Building code enforcement
- Localized flood-control projects
- Public education and awareness

This program was enacted by Congress in 1988 as part of the Robert T. Stafford Act that was a major reworking of federal disaster policy. Besides creating the HMGP, it established a cost sharing of disaster assistance by the states. At the time, the formula for state HMGP funding was 15 percent of the public assistance costs, and it had a 50 percent federal, 50 percent state cost share.

From the period 1988 to 1993, many states did not take advantage of the HMGP funding because it was difficult to meet the matching requirements, even though the 15 percent cap was often not very much. After the devastation of the 1993 Midwest floods, Congressman Volkmer from Missouri championed a change to the legislation that would significantly increase the states' ability to mitigate. Congress amended the legislation to allow for a 75 percent federal, 25 percent state match, and dramatically increased the amount of funding to 15 percent of the total disaster costs. The rationale for these changes was to work aggressively to move people and structures out of the floodplain. As the Missouri case study at the end of this chapter documents, the rationale was sound.

HMGP has allowed states to hire staff to work on mitigation and requires development of a State Hazard Mitigation Plan as a condition of funding. This program brought about a change in the emergency management community at the state and local levels. With adequate funding, states and localities began to hire staff designated to work on mitigation.

HMGP has its detractors and, in 2002, the Federal Office of Management and Budget (OMB) proposed that this program be eliminated in favor of a new predisaster competitive grant program. It is unlikely that Congress will eliminate this program.

Pre-Disaster Mitigation Program (PDM)

Through the Disaster Mitigation Act of 2000, Congress approved creation of a national PDM to provide mitigation funding not dependent on a disaster declaration. The genesis of PDM was an initiative of the Clinton administration called Project Impact: Building Disaster-Resistant Communities. Project Impact grew out of the devastating disasters of the 1990s. Many of the communities hit by these disasters took months and even years to recover emotionally and financially. James Lee Witt, then Director of FEMA, questioned the wisdom of spending more than $2.5 billion per year on disaster relief and not a penny to reduce disasters before they happen. The mitigation tools and techniques were available, so why not work to prevent individuals and communities from becoming victims of disasters? With a small amount of seed money, FEMA launched Project Impact in 1997 in seven pilot communities.

The concept behind the initiative was simple. The mitigation activities had to be designed and tailored to the hazards in that community, and all sectors of the community had to become involved in order for it to be effective and sustainable. Project Impact brought the business community into the emergency management umbrella. Communities were asked to achieve the following four goals:

1. Build a community partnership.
2. Assess the risks.
3. Prioritize risk-reduction actions.
4. Build support by communicating your actions.

By 2001, more than 200 communities were participating in Project Impact, and Congress had appropriated $25 million to the initiative. Seattle, Washington was one of the original pilot communities. In 2002, when a 6.8 earthquake struck Seattle, the mayor attributed the success of their Project Impact activities for the minimal damages and prompt recovery. The Tulsa case study provides an example of a Project Impact community.

In 2002, the Bush administration decided to drop the Project Impact name and concept in exchange for a competitive grant program as their approach to PDM. They have requested $300 million and proposed that this program replace both Project Impact and the HMGP.

The National Earthquake Hazard Reduction Program

The National Earthquake Hazard Reduction Program (NEHRP) is a federal government effort focused upon reducing the risks to life and property from future earthquakes in the United States. Congress established the program in 1977 (Public Law 95-124) as a long-term, nationwide program to reduce the risks to life and property in the United States resulting from earthquakes. This is accomplished through the establishment and maintenance of an effective earthquake hazards reduction program.

The NEHRP works to improve understanding, characterization, and prediction of hazards and vulnerabilities; improve model building codes and land use practices; reduce risk through post-earthquake investigations and education; develop and improve design and construction techniques; improve mitigation capacity; and accelerate application of research results. The NEHRP provides funding to states to establish programs that promote public education and awareness, planning, loss estimation studies, and some minimal mitigation activities. FEMA also supports state and local governments by providing HAZUS, a computer risk modeling tool for communities to use for estimating potential losses from natural hazards.

The primary NEHRP program agencies are:

- Federal Emergency Management Agency (FEMA)
- National Institute of Standards and Technology (NIST)
- National Science Foundation (NSF)
- United States Geological Survey (USGS)

Since its inception, Congress has reviewed and reauthorized NEHRP every two or three years. Congress recently completed a thorough two-year review of NEHRP, resulting in enactment of the NEHRP Reauthorization Act of 2004 (P.L. 108-360), which the President signed into law on October 25, 2004. Public Law 108-360 designates NIST as the lead agency for NEHRP, transferring that responsibility from FEMA which filled that role since the program's inception. The NIST director chairs the NEHRP Interagency Coordinating Committee, which comprises the directors of the primary program agencies, the White House Office of Science and Technology Policy (OSTP), and the Office of Management and Budget (OMB). In addition, the law assigns NIST significant new R&D responsibilities to close the research-to-implementation gap and accelerate the use of new earthquake risk mitigation technologies based on the earth sciences and engineering knowledge developed through NEHRP efforts.

The specific roles of each of the agencies within NEHRP are summarized here:

- FEMA is responsible for emergency response and management, estimation of loss potential, and implementation of mitigation actions.
- NIST conducts applied earthquake engineering research to provide the technical basis for building codes, standards, and practices, and provides the NEHRP lead agency function.
- NSF conducts basic research in seismology, earthquake engineering, and social, behavioral, and economic sciences, and operates the Network for Earthquake Engineering Simulation (which includes the tsunami wave basin research facility and supporting tsunami research).
- USGS operates the seismic networks, develops seismic hazard maps, coordinates post-earthquake investigations, and conducts applied earth sciences research (which includes tsunami research and risk assessment).
- NSF and USGS jointly support the Global Seismographic Network (GSN), the main facility for pinpointing earthquakes in real time.

The NEHRP Reauthorization act of 2004 authorized $900 million to be spent during the period from 2004 to 2009. The law also authorizes the spending of $72.5 million, over a three-year period, for the creation of a National Windstorm Impact Reduction Program that will be modeled according to the NEHRP model, for the purpose of studying the impact of wind-related hazards on structures and the mitigation of these consequences.

Source Information: www.bfrl.nist.gov

The National Hurricane Program

This FEMA program supports activities at the federal, state, and local levels that focus on the physical effects of hurricanes, improved response capabilities, and new mitigation techniques for the built environment. The program has done significant work in storm surge modeling and evacuation planning, design and construction of properties in hurricane-prone areas, and public education and awareness programs for schools and communities. The amount of funding that FEMA receives for this program is in the range of $3 million annually, which is clearly not commensurate with the risk.

The National Dam Safety Program

The National Dam Safety Program Act of 1996 formally established the National Dam Safety Program and named the Director of FEMA as its coordinator. Initiatives under the Act include funding to the states to establish and maintain dam safety programs; training for state dam safety staff and inspectors; technical and archival research in dam safety; education of the public in the hazards of dam failure and related matters; the establishment of the National Dam Safety Review Board; and support for the Interagency Committee on Dam Safety. This act, which is part of the Water Resources Development Act of 1996, expired in fiscal year 2002.

The Fire Prevention and Assistance Act

This program was created in 2001 to address the needs of the nation's paid and volunteer fire departments and to support prevention activities. Congress had long-standing concerns about status of this first-responder community. New threats from potential biochemical terrorism, increasing wildfire requirements, and a stagnant search-and-rescue capability provided the rationale for funding this program. This multimillion-dollar grant program provides competitive grants to fire companies throughout the United States. In the wake of the September 11 events, the appropriations for this program tripled in 2002.

FEMA'S ASSISTANCE TO FIREFIGHTERS GRANT PROGRAM

The purpose of the program is to award one-year grants directly to fire departments of a state to enhance their abilities with respect to fire and fire-related hazards. This program seeks to identify departments that lack the basic tools and resources necessary to protect the health and safety of the public and their firefighting personnel. The primary goal is to provide assistance to meet these needs.

Assistance to Firefighters Grant Program
Fiscal Year 2002 Award Recipients (through August 12, 2002)

Category	Number of Awards		Amount of Awards	
	2002	2003	2002	2003
Fire Operations & Firefighter Safety	4,731	7,014	$281,091,066	$502,157,331
Fire Prevention	215	294	$10,926,998	$14,070,509
Firefighting Vehicles	315	1,374	$39,277,630	$185,113,255
Emergency Medical Services	53	68	$3,069,736	$4,547,325
Total	**5,314**	**8,753**	**$334,365,430**	**$705,888,420**

Source: www.usfa.fema.gov

CONCLUSION

Disasters occur in every state. The direct costs of these events are staggering, but the indirect effects to the economy and the social fabric of communities is even worse. Mitigation works. The case studies included in this chapter are just a few examples of successful, sustained programs that are reducing risk and making communities safer. Mitigation programs exist at all levels of government, and there is a growing interest in the private sector for taking mitigation actions to reduce their risk exposure. To many people, even in a time when terrorism preoccupies the emergency management psyche, mitigation is—and should be—the future direction of emergency management.

CASE STUDIES

TULSA SAFE ROOM PROGRAM

Tulsa, Oklahoma lies in the heart of Tornado Alley. Tornadoes with major damage have hit Tulsa on average of every four or five years. Most recently, the May 3, 1999, tornadoes killed 44 people and decimated communities throughout Oklahoma. As a result of these storms, the President declared a major disaster. Oklahoma was provided the opportunity to take advantage of new construction technology to mitigate the effects of tornadoes.

The concept of "safe room" construction was developed and pilot-tested in 1998 by the Wind Engineering Research Center of Texas Tech University with financial support from FEMA. Safe rooms are anchored and armored rooms that provide shelter during tornadoes, even above ground. Tulsa proposed to FEMA that it use its HMGP funding provided through the President's declaration to provide grants to homeowners to build safe rooms in their homes (see Figure 3-1).

Figure 3-1   November 23, 2001, Tulsa, Oklahoma (Disaster Alley in the Eastland Mall). A safe room wall section is shown here. The insulated concrete form is cut away to show reinforcing steel. The cavity is filled with concrete. Photo by Kent Baxter/FEMA News Photo.

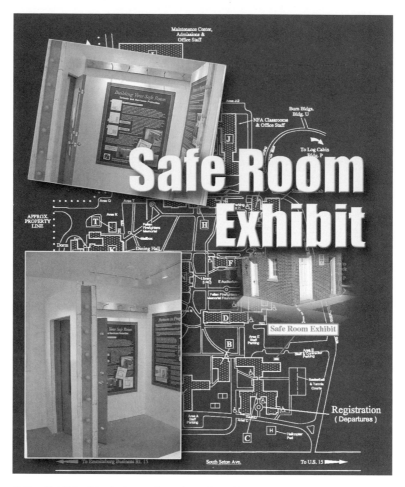

Figure 3-2   Exhibit of techniques for a tornado safe room. FEMA Photo.

Under their Project Impact designation, Tulsa formed a coalition of partners, including FEMA, Oklahoma State Emergency Management, Home Builders of Greater Tulsa, Tulsa Public Works, State Farm Insurance, and other community partners. This coalition then agreed on building and construction standards, permitting, certification and compliance procedures, and public education and awareness programs. This coalition set as their goal to build a tornado safe room in every newly constructed and existing home by the year 2020. This program was supported through a variety of public and private funding, but the major key to its success was the partnership of the building and construction community (see Figure 3-2).

Tulsa builders embraced the safe room concept and quickly made it a positive marketing tool for their business. The city continued to encourage growth of the program by providing certain financial incentives. Eleven major Tulsa builders launched the first safe room subdivision in a new upscale residential area of Tulsa. It is believed to be the first safe room subdivision in Oklahoma, and perhaps the first in the nation, financed entirely by private builders.

The program continues to expand not just within Tulsa and Oklahoma, but to other states and communities in "tornado alley" as well. Within Tulsa, wheelchair-accessible safe rooms have been designed and built. The next step is building safe rooms in public buildings and schools. The technology exists, but the societal questions of size, access, and quantity of space and related issues are still being worked on.

The Tulsa safe room project provides an excellent example of taking advantage of the opportunity afforded in the postdisaster climate. Its success provides an even better example of how building coalitions, particularly with the private sector, ensures sustainability of the mitigation program.

THE CASTAIC UNION SCHOOL DISTRICT

The Castaic Union School District, located in Southern California, is a case study that demonstrates the threat from multiple hazards. After the 1994 Northridge Earthquake, Castaic Union School District conducted a study of the earthquake-related risks that threatened their elementary and middle schools and administration buildings. The assessment revealed that earthquake-related structural damage was not the only risk the school district faced.

The district maintained and operated 63 buildings (77,000 square feet of usable space) in Northern Los Angeles County that consisted of a mix of permanent and portable structures with construction dates as far back as 1917. These structures service approximately 1,200 students and 115 staff members. The San Andreas and San Gabriel fault systems, two of the most active faults in the country, pass through the area in which the district is located. In addition, the USGS has concluded that significant new earthquake activity may occur along both the San Andreas and San Gabriel systems.

These factors led the Castaic Union School District to conclude in their study that the probability of a large earthquake affecting their facilities was high. They also learned, however, that the risk went well beyond possible damages caused by ground shaking. Along with the expected seismic damage, the study revealed two additional threats: flooding from the Castaic Dam and fire or explosion from a rupture in nearby oil pipelines.

The district's risk assessment study indicated that the school buildings were located within the inundation area of the Castaic Dam (located only 1.7 miles upstream). If the dam were to fail, the school buildings and their occupants would be inundated with catastrophic flooding. The 2,200-acre reservoir above the dam could release nearly 105 billion gallons of water, inundating the area below the dam with 50 feet of water. In 1992, the California Department of Water Resources (DWR) reexamined the seismic performance of the dam. Based on the analyses, the DWR considers the dam to meet all current safety requirements and to be able to resist failure caused by the maximum credible earthquake; however, the district's risk assessment concluded the probability the Castaic Dam will fail is never zero.

Along with the threat posed by the Castaic Dam, the study also revealed that the buildings were at high risk of damage from both fire and explosion if nearby pipelines failed. Two high-pressure crude oil pipelines currently cross the campus (a 1925 gas-welded pipeline and a 1964 modern arc-welded steel pipeline), both of which could rupture during ground shaking or ground displacement in earthquakes. An analysis of the lines and the fault conditions near the district indicated a 35 percent chance of failure somewhere in the Castaic area as a result of any large earthquake.

This information caused alarm about the safety of the district's facilities. In the event of a pipeline failure, a fire or explosion could result from the ignition of the released oil, putting both facilities and people at great risk. Additionally, the ability to prevent a nearby fire from spreading would be limited by the decreased reliability of water lines and hydrants, as well as the increased demands on emergency fire services after an earthquake.

Using the results of the district's risk analysis, it was determined that the potential economic costs from either a dam failure or oil pipeline break following an earthquake were enormous. The first potential cost to the school district would be incurred from both building and content damage. Replacement of the school buildings would cost an estimated $7.7 million. Second, if such an earthquake occurred, alternate school facilities would have to be located and rented at an estimated cost of more than $500,000 per year. Third, the community would have to absorb the costs of losing the educational services provided by the district in the time period between the actual loss of the facilities and the relocation to temporary facilities. The school district calculated the cost of the lost public services based on the operating expenses required to provide the services. The daily cost of lost educational services was estimated at $28,601.

In addition to these direct and indirect financial losses, the risk of earthquake-related casualties in the district's facilities was determined to be significant. In an earthquake-induced dam failure, the predicted speed of inundation on the campus caused the risk of casualties to be very high. When calculating this risk, a casualty rate of 250 individuals was determined based on the average hourly rate of campus usage in a typical week. In the event of a dam failure during school hours, the loss of life could be as high as 1,200 students and 115 faculty members. In an earthquake-induced potential pipeline failure, the district calculated a casualty rate of nine individuals and injury rate of 45 individuals. Once again, the actual number of casualties increases dramatically if the earthquake and pipeline failure occurs during school hours.

Through the cost-benefit analysis, the district determined that the most feasible method to reduce their risks would be to condemn the structures on the old, high-risk site and relocate the campus to a low-risk area. Given the nature and severity of the potential hazards, mitigation options other than relocation were judged infeasible.

Once the decision was made to relocate, the district went to work to identify an alternate site for the school facilities. The selected location for the campus was completely out of the dam inundation area and far removed from the high-pressure oil pipelines. Thus the risk posed by the dam and oil pipelines hazards would be eliminated. Although the campus would still be within an active earthquake fault area, the new campus buildings would be constructed to fully conform to 1995 building code provisions, thus making them more resistant to seismic damage than the buildings being replaced.

The district then agreed to turn the land over to the Newhall County Water District as soon as the relocation effort was under way. The old school property is located above two active wells, which the water district can use to supply their customers in Castaic. In doing so, they changed the property deed to restrict human habitation and development and to return the site to natural open space.

The Castaic School District financed the relocation effort through a combination of grant money from FEMA and the sale of bonds. The district applied for and received a $7.2 million grant through FEMA's Hazard Mitigation Grant Program for the market value of the property, including the existing structures and infrastructure. The district used this funding, plus $20 million generated by school bonds, to rebuild the elementary school, district office, and middle school, and to relocate the elementary school students into temporary buildings during the construction of the new facilities. The new middle school opened in the fall of 1996 and the new elementary school opened in August 1997.

VIRGIN ISLANDS BUILDING CODE

On September 18, 1989, Hurricane Hugo, a Category 4 storm, passed over the Virgin Islands with sustained winds of 130 mph, leaving near-total devastation in its wake. Losses of $1.5 billion included damage or destruction of 95 percent of the buildings and 90 percent of the power supply. Almost all public buildings, including hospitals, schools, and shelters, sustained major damage or were destroyed. The tourist industry was in a shambles. All communications with Puerto Rico and the mainland were severed. A Presidential disaster declaration was announced.

The Government of the Virgin Islands, with support from FEMA, began an immediate effort to identify measures to mitigate damage from future storms. Projects identified included upgrading the building codes and building practices, training building inspectors, initiating projects to harden the power grid, and establishing public education programs to show residents how to perform simple mitigation measures and their value.

With technical assistance from FEMA, a new building code was written and implemented. The code required anchoring systems, hurricane clips, shutters, and other measures to hold buildings together and reduce flying debris. Piers, water production, distribution, and oil storage facilities were strengthened. A massive public education program was launched.

When Hurricane Marilyn hit, the public buildings performed well, but most single-family homes lost their roofs. Once again, the building codes were amended to strengthen the quality of residential construction. The Governor's Office initiated a comprehensive program to repair damaged roofs. The Home Protection Roofing Program provided more than 350 homeowners with roofs to withstand a Category 2 storm.

Hurricane Georges, occurring in September 1998, packing winds of more than 100 mph, put these measures to the test. The results were excellent. Public and

Figure 3-3   Guam Memorial Hospital before and after rebuilding and adding mitigation techniques for high wind. FEMA Photo.

private efforts had retrofitted or rebuilt most of the structures on the island by September 1998. Damage to homes was limited to less than 2 percent of the islands. All hotels survived with little or no damage. Power was interrupted to 15 percent of the island but was fully restored within two weeks. Schools and other public structures were undamaged and provided safe havens for the residents to ride out the storm. Officials attribute the reduction in damages not just to the stronger code but also to the intensive education effort for building officials, contractors, and building owners about proper building practices and other mitigation strategies (see Figure 3-3).

ARNOLD, MISSOURI

The City of Arnold, Missouri is located about 20 miles southwest of St. Louis at the confluence of the Meramec and Mississippi rivers. The geography of Arnold causes it to be impacted from backwaters from the Mississippi and direct flooding of the Meramec and its tributaries. The floodplains of both these rivers had experienced extensive development. Because of these concerns, Arnold had adopted a floodplain management program in 1991; however, it had no storm water management program.

The Midwest floods of the spring and summer of 1993 resulted in record flood losses and damages totaling between $12 to $16 billion. Nine states, 532 counties, and more than 55,000 homes were flooded. The 1993 floods had a devastating effect on the 18,000 residents of Arnold. Approximately 250 structures were under water, and more than 528 households applied for disaster assistance, which amounted to more than $2 million. The city had to operate more than 60 sandbag sites to hold off the waters. Parts of the town were under water for up to two weeks.

Figure 3-4   An example of relocation of homes out of the floodplain. FEMA Photo.

When the water receded, the City of Arnold started an aggressive program to voluntarily buy out properties in the floodplain. It proposed the purchase of single-family homes, commercial structures, and mobile homes. It developed a plan to turn the purchased land into an open space greenway along the west banks of the Meramec and Mississippi rivers. They initiated a public education campaign for the purchase of flood insurance because only 208 of the 908 floodplain properties had flood insurance.

Although they were unable to implement their 1991 flood plain management plan, their commitment to mitigation paid off. Arnold received significant HMGP funding for their buyout because of their commitments. By combining HMGP, community development block grant (CDBG), and other HUD funding, it proceeded with its buyout program. Initial estimates put the program costs at $3.5 million, but in the end it would cost $7.3 million.

In the midst of this effort, Arnold experienced another major flood in 1995. The 1995 flood was the fourth largest flood in Arnold's history, but this time the results were dramatically different. Only four sandbag sites were needed, only 26 households applied for assistance, and the damage costs were less than $40,000.

Arnold continued its buyout program into 1996, working to obtain funding to remove the last 34 properties. The city continues to make other structural changes, including bridge elevations to restore the floodplain to its natural state and to provide a buffer for any future flooding.

4. The Disciplines of Emergency Management: Response

INTRODUCTION

When a disaster event such as a flood, earthquake, or hurricane occurs, the first responders to this event are always local police, fire, and emergency medical personnel. Their job is to rescue and attend to those injured, suppress fires, secure and police the disaster area, and begin the process of restoring order. They are supported in this effort by local emergency management personnel and community government officials.

If the size of the disaster event is so large that the capabilities of local responders are overwhelmed and the costs of the damage inflicted exceeds the capacity of the local government, the mayor or county executive will turn to the governor and state government for assistance in responding to the event and in helping the community to recover. The governor will turn to the state's emergency management agency and possibly the state National Guard and other state resources to provide this assistance to the stricken community.

If the governor decides, based on information generated by community and state officials, that the size of the disaster event exceeds the state's capacity to respond, the governor will make a formal request to the President for a presidential major disaster declaration. This request is prepared by state officials in cooperation with regional staff from FEMA. The governor's request is analyzed first by the FEMA regional office and then forwarded to FEMA headquarters in Washington, D.C. FEMA headquarters staff review and evaluate the governor's request and forward their analysis and recommendation to the President. The President considers FEMA's recommendation and then makes a decision to grant the declaration or turn it down.

If the President grants a major disaster declaration, FEMA activates the National Response Plan (NRP) and proceeds to direct 32 federal departments and agencies, including the American Red Cross, in support of state and local efforts to respond to and recover from the disaster event. The Presidential declaration also makes available several disaster assistance programs through FEMA and other federal agencies designed to assist individuals and communities to begin the process of rebuilding their homes, their community infrastructure, and their lives.

When a major disaster strikes in the United States, the aforementioned chronology describes how the most sophisticated and advanced emergency management system in the world responds and begins the recovery process. This system is built on coordination and cooperation among a significant number of federal, state, and local government agencies, volunteer organizations, and, more recently, the business community.

In the 1990s the emergency management system in the United States was tested repeatedly by major disaster events such as the 1993 Midwest floods, the 1994 Northridge, California earthquake, and a series of devastating hurricanes and tornadoes. In each instance, the system worked to bring the full resources of the federal, state, and local governments to produce the most comprehensive and effective response possible. The system also leveraged the capabilities and resources of America's cadre of volunteer organizations to provide immediate food and shelter. In recent years, government officials and agencies at all levels have begun to reach out to the business community to both leverage their response capabilities and to work closer with them in the recovery effort.

The September 11 terrorist attacks have caused all levels of government to reevaluate response procedures and protocols. The unusual loss of so many first responders to this disaster event has resulted in numerous after-action evaluations that likely will lead to changes in the procedures and protocols for first responders in the future. Additionally, the possibility of future terrorism attacks has focused attention on how best to protect first responders from harm in future attacks. These issues are discussed in detail in Chapter 9.

This chapter describes how local, state, and federal government officials and their partners respond to disasters in this country. The chapter includes sections discussing local response, state response, volunteer groups response, the Incident Command System, the FRP, and communications among responding agencies.

LOCAL RESPONSE

Minor disasters occur daily in communities around the United States. Local fire, police, and emergency medical personnel respond to these events usually in a systematic and well-planned course of action. Firefighters, police officers, and emergency medical technicians respond to the scene. Their job is to secure the scene and maintain order, rescue and treat those injured, contain and suppress fire or hazardous conditions, and retrieve the dead.

The types of minor disasters responded to at the community level include hazardous materials transportation and storage incidents, fires, and localized flooding. Local officials are also the first responders to major disaster events such as large floods, hurricanes, and major earthquakes, but in these instances their efforts are supported, upon request by community leaders, by state government and, by request of the governor and approval of the President, by the federal government.

The actions of local first responders are driven by procedures and protocols developed by the responding agency (i.e., fire, police, and emergency medical). Most communities in the United States have developed communitywide emergency plans that incorporate these procedures and protocols. These community emergency plans also identify roles and responsibilities for all responding agencies and personnel for a wide range of disaster scenarios. These plans also include copies of the statutory authorities that provide the legal backing for emergency operations in the community.

In the aftermath of the September 11 terrorist events, many communities are reviewing and reworking their community emergency plans to include procedures and protocols for responding to all forms of terrorist attacks, including bioterrorism and weapons of mass destruction.

First Responder Roles and Responsibilities

The roles and responsibilities of first responders are often detailed in the community emergency plan. A review of the Madison County, North Carolina, All-Hazard Plan provides a typical example of the contents of community emergency plans and the designation of roles and responsibilities among local first responders.

CONTENTS OF MADISON COUNTY (NC) ALL-HAZARD PLAN

- Instructions for use
- Basic plan
- Glossary
- Acronyms and abbreviations
- Laws and ordinances
- Madison County Emergency Management Ordinance
- Madison County State of Emergency Ordinance
- Proclamation of State of Emergency
- Proclamation of Terminating
- Mutual aid
- Madison County Operation Plan (assignment of responsibilities)
  - Chairperson, County Commissioners
  - County Manager
  - Finance
  - Emergency Managment Coordinator
  - Radiological Officer
  - Damage Assessment Officer (tax assessor)
  - Sheriff
  - Towns
  - County Fire Marshal and Fire Chiefs
  - Incident Commander
  - EMS Coordinator
  - Social Services Director
  - Amateur Radio Emergency Service
  - Health Director
  - Medical Center Diaster Coordinator
  - Medical Examiner
  - Mental Health Coordinator

continues

- • Superintendent of Schools
- • American Red Cross
- • Public Works
- • Salvation Army
- • Direction and control
- • Communications
- • Notification and warning
- • Emergency public information
- • Law enforcement
- • Fire and rescue
- • Public work/landfill
- • Health and medical services
- • Evacuation and transportation
- • Shelter and mass care, including Red Cross
- • Damage assessment/recovery
- • Radiological protection
- • Resource management
- • Nuclear threat/hazard
- • Hazardous Materials Southern Railway (ATT) EOC
- • Hurricans and flooding
- • Transportation accidents
- • Mass casualties
- • Winter storms
- • Tornadoes
- • Civil disorders
- • Dam failure
- • Major incidents at public schools
- • I-40 detour traffic
- • Search and rescue plan
- • 911 failure
- • Power failure/countywide
- • Formation of LEPC
- • Contingency plan
- • (ATT) EOC—Federal Response Plan—Southern Railroad/HAZ Plan

Source: Madison County All-Hazard Plan

Local Emergency Managers

It is usually the responsibility of the designated local emergency manager to develop
and maintain the community emergency plans. This individual often holds one or
more other positions in local government such as fire or police chief and serves only
part-time as the community's emergency manager. The profession of local emergency management has been maturing since the 1980s. There are now more oppor-

tunities for individuals to receive formal training in emergency management in the United States. Currently, more than 80 junior college, undergraduate, and graduate programs offer courses and degrees in emergency management and related fields. Additionally, FEMA's Emergency Management Institute (EMI) located in Emmitsburg, Maryland offers emergency management courses on campus and through distance learning programs. EMI has also worked closely with junior colleges, colleges and universities, and graduate schools to develop coursework and curriculums in emergency management. More information on EMI and other emergency management education programs can be found in Chapter 6.

THE CERTIFIED EMERGENCY MANAGER PROGRAM

The International Association of Emergency Managers (IAEM) created the Certified Emergency Manager© (CEM) Program to raise and maintain professional standards. It is an internationally recognized program that certifies achievements within the emergency management profession.

CEM certification is a peer review process administered through the International Association of Emergency Managers. You do not have to be an IAEM member to be certified, although IAEM membership does offer you a number of benefits that can assist you through the certification process. Certification is maintained in five-year cycles.

The CEM Program is served by a CEM Commission that is composed of emergency management professionals, including representatives from allied fields, education, the military, and private industry. Development of the CEM Program was supported by FEMA, the National Emergency Management Association (NEMA), and a host of allied organizations.

Source: IAEM, www.iaem.org

ROLES AND RESPONSIBILITIES OF THE EMERGENCY MANAGEMENT COORDINATOR IN MADISON COUNTY ALL-HAZARD PLAN

Emergency Management Coordinator

a. Perform assigned duties according to state statutes and local ordinances.
b. Responsible for planning in accordance with federal and state guidelines and coordinating of emergency operations within the jurisdiction.
c. Maintain current inventories of public information resources.
d. Ensure regular drills and exercises are conducted to test the functions of the EOP annually.
e. Identify resources county and private and maintain current inventories of county-owned resources, including sources and quantities, and develop mutual aid agreements to control these resources.

continues

f. Request funding for maintaining equipment for radiation hazard evaluation and exposure control.

g. Establish and equip the County Emergency Operating Center (EOC) to include primary and backup radio communications (fixed and mobile), and provide for operations on a continuous basis as required.

h. Ensure adequate training for the emergency management organization.

i. Ensure means are available within the jurisdiction to gather necessary information (i.e., fuel storage facilities, major distributors, and end-user status), during the energy emergency status.

j. Provide emergency information materials for the public including non-English-speaking groups.

k. Prepare written statements of agreements with the media to provide for dissemination of essential emergency information and warning to the public, including the appropriate protective actions to be taken.

l. Coordinate exercises and tests of emergency systems within the jurisdiction.

m. Maintain liaison with utility companies to arrange for backup water, power, and telephone service during emergencies.

n. Maintain working relationships with the media and a current list of radio stations, television stations, and newspapers to be used for public information releases.

o. Alert and activate, as required by the County Emergency Management Organization, when informed of an emergency within the county.

p. Receive requests for assistance from municipalities within the county and direct aid to areas where needed.

q. Coordinate disaster assessment teams conducting field surveys.

r. Conduct a public information campaign to disseminate disaster assistance information as necessary.

s. Maintain listing of medical facilities.

t. Collect data and prepare damage assessment reports.

u. Provide for the storage, maintenance, and replenishment/replacement of essential equipment and materials (e.g., medical supplies, food and water, radiological instruments).

x. Develop a schedule for testing, maintaining, and repairing EOC and other emergency equipment.

y. Develop and maintain the EOC Standard Operating Guides, including an activation checklist and notification/recall roster.

z. Establish and maintain coordination with other jurisdictional EOCs as appropriate.

aa. Provide for adequate coordination of recovery activities among private, state, and federal agencies/organizations.

bb. Develop procedures to warn areas not covered by existing warning systems.

cc. Coordinate warning resources with neighboring counties.

dd. Develop and maintain a public information and education program.

ee. Assist the public information officer (PIO) in disseminating public information and education program.

ff. Identify and develop procedures for potential evacuation areas in accordance with the county's hazard analysis.

gg. Identify population groups requiring special assistance during evacuation (e.g., senior citizens, the very ill and disabled, nursing homes, prison population) and assure that they have evacuation procedures in place.

hh. Establish Disaster Assistance Centers if appropriate.

ii. Initiate the return of the population as soon as conditions are safe at the direction of the Chairman, Board of County Commissioners.

jj. Initiate the crisis upgrading and marking of shelters.

kk. Identify and survey congregate care shelter facilities that have lodging and mass feeding capabilities.

ll. Develop procedures to activate and deactivate shelters and ensure that ARC and DSS develop shelter SOGs.

mm. Establish public information and education programs on sheltering.

nn. Assist with designating facilities and arranging for the shelter needs of institutionalized or special needs groups.

oo. Designate shelter facilities in the reception area with the shortest commuting distance to the hazardous area for essential workers and their families.

pp. Appoint a Damage Assessment Officer to coordinate overall damage assessment operations.

qq. Recruit damage assessment team members.

rr. Secure resources to support and assist with damage assessment activities (e.g., maps, tax data, cameras, identification, report forms).

ss. Establish a Utilities Liaison to coordinate information flow between the EOC and affected utilities.

tt. Assist with identification and notification of applicants that may be eligible for Public Assistance programs.

uu. Develop a flood warning system for areas in the county subject to frequent flooding.

vv. Appoint a Radiological Officer or perform duties of that office.

ww. Acquire and provide radiological monitoring equipment.

xx. Coordinate overall radiological protection activities.

yy. Coordinate resource use under emergency conditions and provide a system to protect these resources.

zz. Support the LEPC in maintaining liaison with facility emergency coordinators to ensure availability of current information concerning hazards and response to an incident.

aaa. Ensure a critique of incident responses to assess and update procedures as needed.

continues

bbb. Serve as the Community Emergency Coordinator as identified in SARA, Title Jill.
ccc. Assist the area staff and the energy policy council in obtaining the essential data for implementation of contingency plans.
ddd. Assure coordination of planning efforts among jurisdictions (e.g., municipalities, counties, facilities), including the development of notification/warning, response, and remediation procedures for covered facilities.
eee. Ensure serviceability of radiological monitoring instruments.
 fff. Alert all emergency support services to the dangers associated with technological hazards and fire during emergency operations.
ggg. Advise decision makers on the hazards associated with hazardous materials.

Source: Madison County All Hazard Plan

More and more communities have designated emergency managers responsible for guiding response and recovery operations. Training and education programs in emergency management are expanding dramatically, resulting in a growing number of professionally trained and certified local emergency managers. The maturing of this profession can only lead to more effective and efficient local responses to future disaster events.

STATE RESPONSE

Each of the 50 states and six territories that constitute the United States maintains a state government office of emergency management. The names of the office vary from state to state. For example, in California it is called the Office of Emergency Services (OES), in Tennessee it is the Tennessee Emergency Management Agency (TEMA), in North Carolina it is the Department of Emergency Management (DEM), and in Florida it is the Florida Division of Emergency Management. A full list of State Emergency Management Organizations is presented in Appendix C.

Also, where the emergency management office resides in state government varies from state to state. In California, OES is located in the Office of the Governor; in Tennessee, TEMA reports to the Adjunct General; and in Florida, the emergency management function is located in the Office of Community Affairs. National Guard Adjutant Generals manage state emergency management offices in more than half the 56 states and territories. The remaining state emergency management offices are led by civilian employees.

Funding for state emergency management offices comes principally from FEMA and state budgets. For years, FEMA has provided up to $175 million annually to states to fund state and local government emergency management activities. This money is used by state emergency management agencies to hire staff, conduct training and exercises, and purchase equipment. A segment of this funding is targeted for local emergency management operations as designated by the state. State budgets also provide funding for emergency management operations, but this funding historically has been inconsistent, especially in those states with minimal annual disaster activity.

The principal resource available to governors in responding to a disaster event in their state is the National Guard. The resources of the National Guard that can be used in disaster response include personnel, communications systems and equipment, air and road transport, heavy construction and earth-moving equipment, mass care and feeding equipment, and emergency supplies such as beds, blankets, and medical supplies.

Response capabilities and capacities are strongest in those states and territories that experience high levels of annual disaster activity. North Carolina is one of those states with high risk of hurricanes and floods. How the North Carolina Department of Emergency Management describes its response process on its Web site provides an example of state response functions.

RESPONSE BY THE NORTH CAROLINA DEPARTMENT OF EMERGENCY MANAGEMENT RESPONSE

The division's emergency response functions are coordinated in a proactive manner from the State Emergency Operations Center located in Raleigh. Proactive response strategies used by the division include:

- Area Commands that are strategically located in an impacted region to assist with local response efforts using state resources
- Central warehousing operations managed by the state that allow for immediate delivery of bottled water, ready-to-eat meals, blankets, tarps, and the like; field deployment teams manned by division and other state agency personnel that assist severely impacted counties coordinate and prioritize response activity
- Incident action planning that identifies response priorities and resource requirements 12 to 24 hours in advance

The State Emergency Response Team (SERT), which consists of top-level management representatives of each state agency involved in response activities, provides the technical expertise and coordinates the delivery of the emergency resources used to support local emergency operations.

When resource needs are beyond the capabilities of state agencies, mutual aid from other unimpacted local governments and states may be secured using the Statewide Mutual Aid agreement or Emergency Management Assistance Compact. Federal assistance may also be requested through the Federal Emergency Response Team, which collocates with the SERT during major disasters.

Source: North Carolina Department of Emergency Management, www.dcm. dcc.state.nc.us

VOLUNTEER GROUP RESPONSE

Volunteer groups are on the front line of any disaster response. National groups such as the American Red Cross and the Salvation Army roster and maintain local chapters of volunteers who are trained in emergency response. These organizations work with local, state, and federal authorities to address the immediate needs of disaster victims. These organizations provide shelter, food, and clothing to disaster victims who have lost their homes to disasters large and small.

In addition to the Red Cross and the Salvation Army, numerous volunteer groups across the country provide aid and comfort to disaster victims. The National Volunteer Organizations Against Disasters (NVOAD) consists of 34 national member organizations, 52 state and territorial VOADs, and a growing number of local VOADs involved in disaster response and recovery operations around the country and abroad. Formed in 1970, NVOAD helps member groups at a disaster location to coordinate and communicate in order to provide the most efficient and effective response. A list of the NVOAD member organizations is provided.

LIST OF NVOAD MEMBER ORGANIZATIONS

Adventist Community Services, www.adventist.communityservices.org
American Radio Relay League, www.arrl.org
American Red Cross, www.redcross.org
American's Second Harvest, www.secondharvest.org
Ananda Marga Universal Relief Team, www.amurt.org
Catholic Charities USA, www.catholiccharitiesusa.org
Christian Disaster Response, www.cdresponse.org
Church of the Brethren, www.brethren.org
Church World Services, www.cwserp.org
Episcopal Relief and Development, www.er-d.org
Friends Disaster Service
Humane Society of the United States, www.hsus.org
International Relief Friendship Foundation, IRFFint@aol.com
International Aid, www.gospelcom.net/ia
Lutheran Disaster Response, www.elca.org/dcs/disaster
Mennonite Disaster Services, www.mds.mennonite.net
National Emergency Response Team, www.nert-usa.org
National Organization for Victim Assistance, www.try-nova.org
Nazarene Disaster Response, www.nazarenedisasterresponse.org
Northwest Medical Teams International, www.nwmti.org
The Phoenix Society for Burn Survivors, www.phoenix-society.org
The Points of Light Foundation, www.pointsoflight.org
Presbyterian Disaster Assistance, www.pcusa.org/pcusa/wmd/pda/index.html
REACT International, www.reactintl.org

The Salvation Army, www.salvationarmyusa.org
Society of St. Vincent de Paul, http://home.aol.com/svdpus
Southern Baptist Disaster Relief, www.namb.net/dr/pages/beginnings.asp
United Jewish Communities, www.ujcna.org
United Methodist Committee on Relief,
    http://gbgm-umc.org/umcor/emergency.stm
United States Service Command, www.ussc-hq.org
Volunteers of America, www.voa.org
World Vision, www.worldvision.org

Source: National Volunteer Organizations Against Disasters (NVOAD)

GERMAN SALVATIONISTS PROVIDE AID TO FLOOD VICTIMS

August 15, 2002. Salvationists in Dresden, Germany have been working tirelessly to help people affected by the flooding that has brought chaos to the city and much of the surrounding area. The River Elbe is already at its highest point since the mid-19th century and water levels are still rising. More than 3,000 people have so far been forced to evacuate their homes.

When the flooding started, Salvationists from the Salvation Army center in Dresden immediately offered support to the emergency workers, as no official supplies were then being provided to fire and ambulance personnel. However, the main focus of attention quickly shifted to the victims of the flooding.

The Salvation Army corps (church) building in Dresden is located on high ground and, unlike many buildings in the city, still has power, so cooking and food preparation are possible. More than 2,000 meals have been provided so far. Local hotels and a bakery are assisting with preparation and two Salvation Army mobile kitchens are being used to deliver food.

There is great concern for the many elderly people who are unable to leave their properties but who now have no power for cooking or heating. In addition to providing hot soup, consideration is being given to assisting these elderly people to find alternative, temporary accommodation. Many offers of help have come in, including from a nurse who put herself forward to provide assistance after the hospital she was working in was evacuated.

Donations of clothing and supplies, up to now, have had to be turned down because of a lack of storage space. The Salvation Army's International Emergency Services Office has arranged for US$25,000 to be sent out from International Headquarters and these funds will be used, among other things, to hire a suitable warehouse where donations can be stored.

Source: Salvation Army

INCIDENT COMMAND SYSTEM

A difficult issue in any response operation is determining who is in charge of the overall response effort. The Incident Command System (ICS) was developed after the 1970 fires in southern California. Duplication of efforts, lack of coordination, and communication hindered all agencies responding to the expanding fires. The main function of ICS is to establish a set of planning and management systems that would help the agencies responding to a disaster to work together in a coordinated and systematic approach. The step-by-step process enables the numerous responding agencies to effectively use resources and personnel to respond to those in need.

There are multiple functions in the ICS system. They include common use of terminology, integrated communications, a unified command structure, resource management, and action planning. A planned set of directives includes assigning one coordinator to manage the infrastructure of the response, assigning personnel, deploying equipment, obtaining resources, and working with the numerous agencies that respond to the disaster scene. In most instances the local fire chief or fire commissioner is the Incident Commander.

For the ICS to be effective, it must provide for effective operations at three levels of incident character: (1) single jurisdiction and/or single agency, (2) single jurisdiction with multiple agency support, and (3) multijurisdictional and/or multiagency support. The organizational structure must be adaptable to a wide variety of emergencies (i.e., fire, flood, earthquake, and rescue). The ICS includes agency autonomy, management by objectives, unity integrity, functional clarity, and effective span of control. The logistics, coordination, and ability of the multiple agencies to work together must adhere to the ICS so that efficient leadership is maintained during the disaster. One of the most significant problems before the ICS was that agencies who would respond to major disasters would assign their own commander and there would be power struggles, miscommunication, and duplication of efforts (Irwin, 1980).

There are five major management systems within the ICS. They include Command, Operations, Planning, Logistics, and Finance.

- The *Command Section* includes developing, directing, and maintaining communication and collaboration with the multiple agencies on site, working with the local officials, the public, and the media to provide up-to-date information regarding the disaster.
- The *Operations Section* handles the tactical operations, coordinates the command objectives, and organizes and directs all resources to the disaster site.
- The *Planning Section* provides the necessary information to the command center to develop the action plan to accomplish the objectives. This section also collects and evaluates information as it is made available.
- The *Logistics Section* provides personnel, equipment, and support for the Command Center. They handle the coordination of all services that are involved in the response, from locating rescue equipment to coordinating the response for volunteer organizations such as the Salvation Army and the Red Cross.

- The *Finance Section* is responsible for accounting for funds used during the response and recovery aspect of the disaster. The Finance Section monitors costs related to the incident and provides accounting procurement time recording cost analyses.

In today's world, the public, private, and political values at risk in major emergencies demand the most efficient methods of response and management. Meeting this demand when multiple and diverse agencies are involved becomes a difficult task. The Unified Command concept of ICS offers a process that all participating agencies can use to improve overall management, whether their jurisdiction is of a geographical or functional nature (Irwin, 1980).

The Unified Command is best used when there is a multiagency response. Because of the nature of the disaster, multiple government agencies need to work together to monitor the response and manage the large number of personnel who respond to the scene (see Figure 4-1). It allows for the integration of the agencies to operate under one overall response management.

PROCEDURES FOR INCIDENT COMMAND SYSTEM

For an ICC to be effective, procedures need to be followed closely:

- A command post needs to be established.
- Proper equipment, such as computers, radios, and telephone lines, need to be installed and in working order.
- A media/press area needs to be established.
- Topographic maps need to be located and posted. After tornados, street signs or other identifying landmarks are destroyed and rescue personnel are unable to use traditional road maps.
- Locate/prepare a missing persons list.
- Monitor the movement and location of triage areas and transportation of victims.
- Have the ability to maintain continuous communication with local hospitals to monitor the number of victims received.
- Establish and grid the search area.
- Based on the type of disaster, such as flooding, responders may have to use boats to search for and rescue victims.
- Determine what resources are available within the local area and what resources are being deployed.
- As the response system expands, reevaluate tasks that need to be performed and develop new tasks.

Source: Irwin, 2002

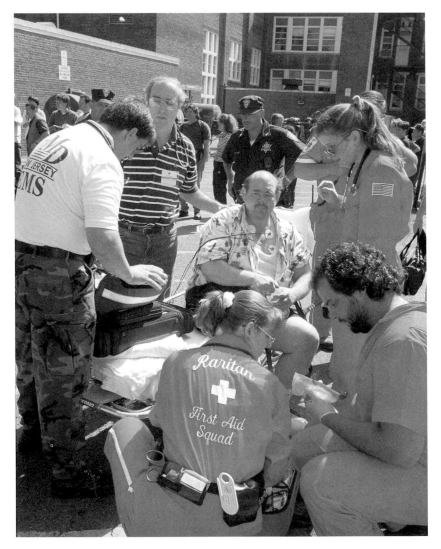

Figure 4-1   New York, NY, October 30, 2001. FEMA/NY State Disaster Field Office personnel meet to coordinate federal, state, and local disaster assistance programs. Photo by Andrea Booher/FEMA News Photo Photo by Dave Gatley/FEMA News Photo.

The Incident Commander (IC) prepares to delegate responsibilities as needed, to maintain focus on the overall situation. The IC needs to assign positions, such as debriefers, coordinators, and unit leaders, to mange the command center. As the response and recovery process proceeds, the IC needs to have an ongoing dialogue with staff and officials to monitor and manage the response. The IC needs to evaluate the continuing needs of the responders and determine if additional resources are needed. In the after-action reports, discussion and evaluation of the disaster determines the success based on the initial competence and effectiveness of the Incident Commander and the Center.

THE FEDERAL RESPONSE

Once the governor has determined that a disaster event has overwhelmed the capacity of state and local governments to effectively respond and to subsequently fund the recovery effort, the governor forwards a letter to the President requesting a presidential disaster declaration. This is the first step toward involving federal officials, agencies and departments, and resources in a disaster event. If the event is declared a major disaster by the President, 32 federal departments and agencies, including the American Red Cross, work together to support the efforts of state and local officials.

The Department of Homeland Security, through FEMA, is responsible for coordinating all federal activities in support of state and local response and recovery efforts in a presidentially declared disaster. In such an instance, FEMA activates the National Response Plan (NRP). FEMA also manages several programs that provide disaster assistance to individuals and affected communities. These programs are discussed in detail in Chapter 5.

Presidential Disaster Declaration Process

The presidential disaster declaration makes available the resources of the federal government to the disaster area. Although a formal declaration does not have to be signed for the federal government to respond, the governor must make a formal request for assistance and specify in the request the specific needs of the disaster area. The presidential major disaster declaration process is provided as follows:

Federal, state, local, tribal, private-sector, and nongovernmental organizations report threats, incidents, and potential incidents using established communications and reporting channels. The HSOC receives threat and operational information regarding incidents or potential incidents and makes an initial determination to initiate the coordination of federal information-sharing and incident management activities.

PRESIDENTIAL MAJOR DISASTER DECLARATION PROCESS

A disaster declaration should include the following guidelines:

- Contact is made between the affected state and the FEMA regional office. This contact may take place before or immediately following the disaster.
- If it appears the situation is beyond state and local capacity, the state requests FEMA to conduct a joint Preliminary Damage Assessment (PDA). Participants in the PDA will include FEMA, state, and local government representatives and other federal agencies.
- Based on the PDA findings, the governor submits a request to the President through the FEMA regional director for either a major disaster or an emergency declaration and identifying the counties impacted.

---

continues

- The FEMA regional office submits a summary of the event and a recommendation based on the results of the PDA to FEMA headquarters, along with the governor's request.
- Upon receipt of these documents, headquarters senior staff convenes to discuss the request and determine the recommendation to be made to the President.
- FEMA's recommendation is forwarded to the White House for review.
- The President declares a major disaster or an emergency.

Source: Federal Response Plan, April 1999

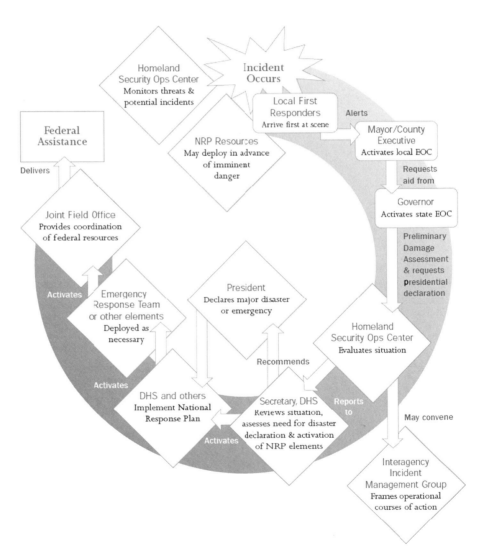

Figure 4-2   Flow of initial national-level incident management actions.
Source: DHS National Response Plan.

The decision to make a disaster declaration is completely at the discretion of the President. There are no set criteria to follow and no government regulations to guide which events are declared by the President and which events are not. FEMA has developed several factors it considers in making its recommendation to the President, including individual property losses per capita, level of damage to existing community infrastructure, and insurance coverage. In the end, however, the decision to make the declaration is the President's alone.

A presidential disaster declaration can be made in as short a time as a few hours, as was the case in the 1994 Northridge earthquake and the 1995 Oklahoma City bombing. Sometimes it takes weeks for damages to be assessed and the capability of state and local jurisdictions to fund response and recovery efforts to be evaluated. If the governor's request is turned down by the President, the governor has a right to appeal and can be successful, especially if new damage data become available and are included in the appeal.

Presidential declarations are routinely sought for such events as large floods, hurricanes, earthquakes, and big tornadoes. In recent years, governors have become more inventive and have requested presidential disaster declarations for snow removal, drought, West Nile virus, and economic losses caused by failing industries such as the Northwest salmon spawning decline.

Since 1976, there have been 906 presidential disaster declarations, averaging 34 declarations per year (see Table 4-1). As an example of disaster declaration activity in a single year, in 2004 there were 68 major disaster declarations in 37 states and 6 territories including 18 hurricanes and tropical storms and 35 severe storms (see Table 4-2).

Table 4-1 Total Major Disaster Declarations, 1976–2004

Year	Total Disaster Declarations	Year	Total Disaster Declarations
1976	30	1992	45
1977	22	1993	32
1978	25	1994	36
1979	42	1995	32
1980	23	1996	75
1981	15	1997	44
1982	24	1998	65
1983	21	1999	50
1984	34	2000	45
1985	27	2001	45
1986	28	2002	49
1987	23	2003	56
1988	11	2004	68
1989	31	Total	1079
1990	38	Average	37.2
1991	43		

Source: www.fema.gov

Table 4-2 FEMA Major Disaster Activity, January 1, 2004 to December 31, 2004

Date	State	Incident
1/13	American Samoa	High winds, high surf and heavy rainfall associated with tropical cyclone Heta
1/13	California	Earthquake
1/26	Ohio	Severe storms, flooding, mudslides, and landslides
2/5	Maine	Severe storms, flooding, snow melt, and ice jams
2/13	South Carolina	Severe ice storm
2/19	Oregon	Severe winter storms
4/10	Micronesia	Typhoon Sudal
4/21	Massachusetts	Flooding
4/23	Illinois	Severe storms and tornadoes
4/29	New Mexico	Severe storms and flooding
5/5	North Dakota	Severe storms, flooding, and ground saturation
5/7	Arkansas	Severe storms, flooding and landslides
5/25	Iowa	Severe storms, tornadoes, and flooding
5/25	Nebraska	Severe storms, tornadoes and flooding
6/3	Indiana	Severe storms, tornadoes, and flooding
6/3	Ohio	Severe storms and flooding
6/7	West Virginia	Severe storms, flooding, and landslides
6/8	Louisiana	Severe storms and flooding
6/10	Kentucky	Severe storms, tornadoes, flooding, and mudslides
6/11	Missouri	Severe storms, tornadoes, and flooding
6/15	Virginia	Severe storms, tornadoes, and flooding
6/18	Wisconsin	Severe storms and flooding
6/30	California	Flooding as a result of a levee break
6/30	Arkansas	Severe storms and flooding
6/30	Michigan	Severe storms, tornadoes, and flooding
7/16	New Jersey	Severe storms and flooding
7/20	South Dakota	Severe storms and flooding
7/29	Guam	High winds, flooding, and mudslides as a result of tropical storm Tingting
7/29	Northern Mariana Islands	Flooding, high surf, high winds, and wind-driven rain associated with typhoon Tingting
8/3	Kansas	Severe storms, flooding, and tornadoes
8/3	New York	Severe storms and flooding
8/6	Pennsylvania	Severe storms and flooding
8/6	Kentucky	Severe storms and flooding
8/6	West Virginia	Severe storms, flooding, and landslides
8/13	Florida	Hurricane Charley and tropical storm Bonnie
8/26	Northern Mariana Islands	Flooding, high surf, storm surge, and high winds as a result of super typhoon Chaba
8/26	Nevada	Wildland fire
9/1	South Carolina	Hurricane Charley
9/1	Indiana	Tornadoes and flooding
9/3	Virginia	Severe storms, flooding, and tornadoes associated with tropical depression Gaston
9/4	Florida	Hurricane Frances
9/10	South Carolina	Tropical storm Frances
9/15	Mississippi	Hurricane Ivan
9/15	Alabama	Hurricane Ivan
9/15	Louisiana	Hurricane Ivan
9/15	North Carolina	Tropical storm Gaston

Table 4-2 *Continued*

Date	State	Incident
9/16	Florida	Hurricane Ivan
9/17	Puerto Rico	Tropical storm Jeanne and resulting landslides and mudslides
9/18	Georgia	Hurricane Ivan
9/18	North Carolina	Hurricane Ivan
9/19	Pennsylvania	Tropical depression Ivan
9/19	Ohio	Severe storms and flooding
9/19	Pennsylvania	Severe storms and flooding associated with tropical depression Frances
9/20	West Virginia	Severe storms, flooding, and landslides
9/23	Vermont	Severe storms and flooding
9/24	Georgia	Tropical storm Frances
9/26	Florida	Hurricane Jeanne
9/30	Kansas	Severe storms, flooding, and tornadoes
10/01	New York	Tropical depression Ivan
10/01	New York	Severe storms and flooding
10/01	New Jersey	Tropical depression Ivan
10/07	Minnesota	Severe storms and flooding
10/07	Tennessee	Severe storms and flooding
10/07	US Virgin Islands	Tropical storm Jeanne
10/07	South Carolina	Tropical storm Frances
10/18	Virginia	Severe storms and flooding from the remnants of Hurricane Jeanne
11/15	Delaware	Severe storms, tornadoes, and flooding from the remnants of Hurricane Jeanne
11/15	Alaska	Severe winter storm, tidal surges, and flooding

Source: www.fema.gov

Federal Response Plan (FRP) and National Response Plan (NRP)

In 1992 the Federal Emergency Management Agency (FEMA) developed the Federal Response Plan (FRP). FEMA defined the FRP as a

> Signed agreement among 27 Federal departments and agencies, including the American Red Cross, that: Provides the mechanism for coordinating delivery of Federal assistance and resources to augment efforts of State and local governments overwhelmed by a major disaster or emergency, Supports implementation of the Robert T. Stafford Disaster Relief and Emergency Assistance Act, as amended (42 U.S.C. 5121, et seq.), as well as individual agency statutory authorities and Supplements other Federal emergency operations plans developed to address specific hazards.

The fundamental goal of the FRP was to maximize available Federal resources in support of response and recovery actions taken by state and local emergency officials.

TYPES OF FEDERAL ASSISTANCE AVAILABLE

The Federal Response Plan (FRP) made available the following types of assistance:

To deliver immediate relief:
- Initial response resources, including food, water, emergency generators
- Emergency services to clear debris, open critical transportation routes, provide mass sheltering and feeding

To speed return to normal and reduce damage from future occurrences:
- Loans and grants to repair or replace damaged housing and personal property
- Grants to repair or replace roads and public buildings, incorporating to the extent practical hazard-reduction structural and nonstructural measures
- Technical assistance to identify and implement mitigation opportunities to reduce future losses
- Other assistance, including crisis counseling, tax relief, legal services, job placement

Source: Federal Response Plan

Following the absorption of FEMA into the Department of Homeland Security, on February 18, 2003, President Bush signed Presidential Directive 5 (HSPD-5) "to enhance the ability of the United States to manage domestic incidents by establishing a single, comprehensive national incident management system." This action authorized the design and development of a National Response Plan (NRP) to "align Federal coordination structures, capabilities, and resources into a unified, all-discipline, and all-hazards approach to domestic incident management." The Department of Homeland Security writes,

> The NRP is an all-discipline, all-hazards plan that establishes a single, comprehensive framework for the management of domestic incidents. It provides the structure and mechanisms for the coordination of Federal support to State, local, and tribal incident managers and for exercising direct Federal authorities and responsibilities. The NRP assists in the important homeland security mission of preventing terrorist attacks within the United States; reducing the vulnerability to all natural and manmade hazards; and minimizing the damage and assisting in the recovery from any type of incident that occurs. (From the NRP Letter of Agreement)

The NRP was designed according to the template of the National Incident Management System (NIMS; released March 1, 2004), so as to ensure that a consistent doctrinal framework exists for the management of incidents at all jurisdictional levels, regardless of the incident cause, size, or complexity. NIMS was created to integrate effective practices in emergency preparedness and response into a comprehensive national framework for incident management. NIMS enables responders at all levels to work together more effectively and efficiently to manage domestic incidents no matter what the cause, size, or complexity, including catastrophic acts of terrorism and disasters.

DHS lists the benefits of the NIMS system to be:

- Standardized organizational structures, processes and procedures
- Standards for planning, training, and exercising, and personnel qualification standards
- Equipment acquisition and certification standards
- Interoperable communications processes, procedures, and systems
- Information management systems
- Supporting technologies—voice and data communications systems, information systems, data display systems, and specialized technologies

Consistent with the model provided in the NIMS, the NRP can be partially or fully implemented in the context of a threat, anticipation of a significant event, or the response to a significant event. Selective implementation through the activation of one or more of the system's components allows for flexibility in meeting the unique operational and information-sharing requirements of the situation at hand and enabling effective interaction between various federal and non-federal entities.

The NRP provides the framework for federal interaction with state, local, and tribal governments; the private sector; and NGOs in the context of domestic incident prevention, preparedness, response, and recovery activities. It describes capabilities and resources and establishes responsibilities, operational processes, and protocols to help protect the nation from terrorist attacks and other natural and manmade hazards; save lives; protect public health, safety, property, and the environment; and reduce adverse psychological consequences and disruptions. Finally, the NRP serves as the foundation for the development of detailed supplemental plans and procedures to effectively and efficiently implement federal incident management activities and assistance in the context of specific types of incidents.

The NRP establishes mechanisms to:

- Maximize the integration of incident-related prevention, preparedness, response, and recovery activities
- Improve coordination and integration of federal, state, local, tribal, regional, private-sector, and nongovernmental organization partners
- Maximize efficient utilization of resources needed for effective incident management and Critical Infrastructure/Key Resources (CI/KR) protection and restoration
- Improve incident management communications and increase situational awareness across jurisdictions and between the public and private sectors
- Facilitate emergency mutual aid and federal emergency support to state, local, and tribal governments
- Facilitate federal-to-federal interaction and emergency support
- Provide a proactive and integrated federal response to catastrophic events
- Address linkages to other federal incident management and emergency response plans developed for specific types of incidents or hazards

The NRP covers the full range of complex and constantly changing requirements in anticipation of, or in response to, threats or acts of terrorism, major disasters, and other emergencies. The NRP also provides the basis to initiate long-term community recovery and mitigation activities. The NRP establishes interagency and multi-jurisdictional mechanisms for federal government involvement in, and DHS coordination of, domestic incident management operations. This includes coordinating structures and processes for incidents requiring:

- Federal support to state, local, and tribal governments
- Federal-to-federal support
- The exercise of direct federal authorities and responsibilities, as appropriate under the law
- Public and private-sector domestic incident management integration

The NRP distinguishes between incidents that require DHS coordination, termed Incidents of National Significance, and the majority of incidents occurring each year that are handled by responsible jurisdictions or agencies through other established authorities and existing plans.

In addition, the NRP:

- Recognizes and incorporates the various jurisdictional and functional authorities of federal departments and agencies; state, local, and tribal governments; and private-sector organizations in domestic incident management.
- Details the specific domestic incident management roles and responsibilities of the Secretary of Homeland Security, Attorney General, Secretary of Defense, Secretary of State, and other departments and agencies involved in domestic incident management as defined in HSPD-5 and other relevant statutes and directives.
- Establishes the multiagency organizational structures and processes required to implement the authorities, roles, and responsibilities of the Secretary of Homeland Security as the principal federal official for domestic incident management.

The NRP is applicable to all federal departments and agencies that may be requested to provide assistance or conduct operations in the context of actual or potential Incidents of National Significance. This includes the American Red Cross, which functions as an Emergency Support Function (ESF) primary organization in coordinating the use of mass care resources in a presidentially declared disaster or emergency. The NRP is applicable to incidents that may occur at sites under the control of the legislative or judicial branches of the federal government.

Based on the criteria established in HSPD-5, Incidents of National Significance are those high-impact events that require a coordinated and effective response by an appropriate combination of federal, state, local, tribal, private-sector, and nongovernmental entities in order to save lives, minimize damage, and provide the basis for long-term community recovery and mitigation activities.

The NRP bases the definition of Incidents of National Significance on situations related to the following four criteria set forth in HSPD-5:

1. A federal department or agency acting under its own authority has requested the assistance of the Secretary of Homeland Security.

2. The resources of state and local authorities are overwhelmed and federal assistance has been requested by the appropriate state and local authorities. Examples include:
   a. Major disasters or emergencies as defined under the Stafford Act
   b. Catastrophic incidents
3. More than one federal department or agency has become substantially involved in responding to an incident. Examples include:
   a. Credible threats, indications or warnings of imminent terrorist attack, or acts of terrorism directed domestically against the people, property, environment, or political or legal institutions of the United States or its territories or possessions
   b. Threats or incidents related to high-profile, large-scale events that present high-probability targets such as National Special Security Events (NSSEs) and other special events as determined by the Secretary of Homeland Security, in coordination with other federal departments and agencies
6. The Secretary of Homeland Security has been directed to assume responsibility for managing a domestic incident by the President.

Additionally, since Incidents of National Significance typically result in impacts far beyond the immediate or initial incident area, the NRP provides a framework to enable the management of cascading impacts and multiple incidents as well as the prevention of, and preparation for, subsequent events. Examples of incident management actions from a national perspective include:

- Increasing nationwide public awareness
- Assessing trends that point to potential terrorist activity
- Elevating the national Homeland Security Advisory System (HSAS) alert condition and coordinating protective measures across jurisdictions
- Increasing countermeasures such as inspections, surveillance, security, counterintelligence, and infrastructure protection
- Conducting public health surveillance and assessment processes and, where appropriate, conducting a wide range of prevention measures to include, but not be limited to, immunizations
- Providing immediate and long-term public health and medical response assets
- Coordinating federal support to state, local, and tribal authorities in the aftermath of an incident
- Providing strategies for coordination of federal resources required to handle subsequent events
- Restoring public confidence after a terrorist attack
- Enabling immediate recovery activities, as well as addressing long-term consequences in the impacted area

Signatory Partners

There are 32 Signatory Partners in the NRP. Each of these partners serves as a primary agency or support agency in one or more of the 15 Emergency Support Functions (ESF) in the NRP. FEMA defines primary and support agencies as follows:

- **Primary Agencies.** A federal agency designated as an ESF primary agency serves as a federal executive agent under the Federal Coordinating Officer (or Federal Resource Coordinator for non-Stafford Act incidents) to accomplish the ESF mission. When an ESF is activated in response to an Incident of National Significance, the primary agency is responsible for:
  - Orchestrating federal support within their functional area for an affected state
  - Providing staff for the operations functions at fixed and field facilities
  - Notifying and requesting assistance from support agencies
  - Managing mission assignments and coordinating with support agencies, as well as appropriate state agencies
  - Working with appropriate private-sector organizations to maximize use of all available resources
  - Supporting and keeping other ESFs and organizational elements informed of ESF operational priorities and activities
  - Executing contracts and procuring goods and services as needed
  - Ensuring financial and property accountability for ESF activities
  - Planning for short-term and long-term incident management and recovery operations
  - Maintaining trained personnel to support interagency emergency response and support teams
- **Support Agencies.** When an ESF is activated in response to an Incident of National Significance, support agencies are responsible for:
  - Conducting operations, when requested by DHS or the designated ESF primary agency, using their own authorities, subject-matter experts, capabilities, or resources
  - Participating in planning for short-term and long-term incident management and recovery operations and the development of supporting operational plans, SOPs, checklists, or other job aids, in concert with existing first-responder standards
  - Assisting in the conduct of situational assessments
  - Furnishing available personnel, equipment, or other resource support as requested by DHS or the ESF primary agency
  - Providing input to periodic readiness assessments
  - Participating in training and exercises aimed at continuous improvement of prevention, response, and recovery capabilities
  - Identifying new equipment or capabilities required to prevent or respond to new or emerging threats and hazards, or to improve the ability to address existing threats
  - Nominating new technologies to DHS for review and evaluation that have the potential to improve performance within or across functional areas
  - Providing information or intelligence regarding their agency's area of expertise

The signatory partners are provided in the following list:

NATIONAL RESPONSE PLAN SIGNATORY PARTNERS

Department of Agriculture
Department of Commerce
Department of Defense
Department of Education
Department of Energy
Department of Health and Human Services
Department of Homeland Security
Department of Housing and Urban Development
Department of the Interior
Department of Justice
Department of Labor
Department of State
Department of Transportation
Department of the Treasury
Department of Veterans Affairs
Central Intelligence Agency
Environmental Protection Agency
Federal Bureau of Investigation
Federal Communications Commission
General Services Administration
National Aeronautic and Space Administration
National Transportation Safety Board
Nuclear Regulatory Commission
Office of Personnel Management
Small Business Administration
Social Security Administration
Tennessee Valley Authority
U.S. Agency for International Development
U.S. Postal Service
American Red Cross
Corporation for National and Community Service
National Voluntary Organizations Active in Disaster

Source: www.dhs.gov

Emergency Support Functions (ESFs)

The NRP applies a functional approach that groups the capabilities of federal departments and agencies and the American Red Cross into Emergency Support Functions (ESFs) to provide the planning, support, resources, program implementation, and emergency services that are most likely to be needed during Incidents of National Significance. The federal response to actual or potential Incidents of National Significance typically is provided through the full or partial activation of the ESF structure as necessary.

The ESFs serve as the coordination mechanism to provide assistance to state, local, and tribal governments or to federal departments and agencies conducting missions of primary federal responsibility. ESFs may be selectively activated for both Stafford Act and non-Stafford Act incidents where federal departments or agencies request DHS assistance or under other circumstances as defined in HSPD-5.

Each ESF is composed of primary and support agencies. The NRP identifies primary agencies on the basis of authorities, resources, and capabilities. Support agencies are assigned based on resources and capabilities in a given functional area. The ESF structure provides a structure within which to mobilize the components necessary to best address the requirements of each incident. For example, a large-scale natural disaster or massive terrorist event may require the activation of all ESFs. A localized flood or tornado might require activation only of a select number of ESFs.

The scope of each ESF is summarized as follows.

ESF #1—Transportation

- Federal and civil transportation support
- Transportation safety
- Restoration/recovery of transportation infrastructure
- Movement restrictions
- Damage and impact assessment

ESF #2—Communications

- Coordination with telecommunications industry
- Restoration/repair of telecommunications infrastructure
- Protection, restoration, and sustainability of national cyber and information technology resources

ESF #3—Public Works and Engineering

- Infrastructure protection and emergency repair
- Infrastructure restoration
- Engineering services, construction management
- Critical infrastructure liaison

ESF #4—Firefighting

- Firefighting activities on federal lands
- Resource support to rural and urban firefighting operations

ESF #5—Emergency Management

- Coordination of incident management efforts
- Issuance of mission assignments
- Resource and human capital
- Incident action planning
- Financial management

ESF #6—Mass Care, Housing, and Human Services

* Mass care
* Disaster housing
* Human services

ESF #7—Resource Support

* Resource support (facility space, office equipment and supplies, contracting services, etc.)

ESF #8—Public Health and Medical Services

* Public health
* Medical
* Mental health services
* Mortuary services

ESF #9—Urban Search and Rescue

* Life-saving assistance
* Urban search and rescue

ESF #10—Oil and Hazardous Materials Response

* Oil and hazardous materials (chemical, biological, radiological, etc.) response
* Environmental safety and short- and long-term cleanup

ESF #11—Agriculture and Natural Resources

* Nutrition assistance
* Animal and plant disease/pest response
* Food safety and security
* Natural and cultural resources and historic properties protection and restoration

ESF #12—Energy

* Energy infrastructure assessment, repair, and restoration
* Energy industry utilities coordination
* Energy forecast

ESF #13—Public Safety and Security

* Facility and resource security
* Security planning and technical and resource assistance
* Public safety/security support
* Support to access, traffic, and crowd control

continues

ESF #14—Long-Term Community Recovery and Mitigation

- Social and economic community impact assessment
- Long-term community recovery assistance to states, local governments, and the private sector
- Mitigation analysis and program implementation

ESF #15—External Affairs

- Emergency public information and protective action guidance
- Media and community relations
- Congressional and international affairs
- Tribal and insular affairs

Source: www.dhs.gov

Roles and Responsibilities

The NRP also defines the roles and responsibilities of public, private, and nonprofit parties involved in incident management at the local, state, and national levels.

Governor. As a state's chief executive, the governor is responsible for the public safety and welfare of the people of that state or territory. The governor:

- Is responsible for coordinating state resources to address the full spectrum of actions to prevent, prepare for, respond to, and recover from incidents in an all-hazards context to include terrorism, natural disasters, accidents, and other contingencies.
- Under certain emergency conditions, typically has police powers to make, amend, and rescind orders and regulations.
- Provides leadership and plays a key role in communicating to the public and in helping people, businesses, and organizations cope with the consequences of any type of declared emergency within state jurisdiction.
- Encourages participation in mutual aid and implements authorities for the state to enter into mutual aid agreements with other states, tribes, and territories to facilitate resource-sharing.
- Is the Commander-in-Chief of state military forces (National Guard when in State Active Duty or Title 32 Status and the authorized state militias).
- Requests federal assistance when it becomes clear that state or tribal capabilities will be insufficient or have been exceeded or exhausted.

Local Chief Executive Officer. A mayor or city or county manager, as a jurisdiction's chief executive, is responsible for the public safety and welfare of the people of that jurisdiction. The local Chief Executive Officer:

- Is responsible for coordinating local resources to address the full spectrum of actions to prevent, prepare for, respond to, and recover from incidents involving all hazards including terrorism, natural disasters, accidents, and other contingencies.

- Dependent upon state and local law, has extraordinary powers to suspend local laws and ordinances, such as to establish a curfew, direct evacuations, and, in coordination with the local health authority, to order a quarantine.
- Provides leadership and plays a key role in communicating to the public, and in helping people, businesses, and organizations cope with the consequences of any type of domestic incident within the jurisdiction.
- Negotiates and enters into mutual aid agreements with other jurisdictions to facilitate resource-sharing. Requests state and, if necessary, federal assistance through the governor of the state when the jurisdiction's capabilities have been exceeded or exhausted.

Tribal Chief Executive Officer. The tribal Chief Executive Officer is responsible for the public safety and welfare of the people of that tribe. The tribal Chief Executive Officer, as authorized by tribal government:

- Is responsible for coordinating tribal resources to address the full spectrum of actions to prevent, prepare for, respond to, and recover from incidents involving all hazards including terrorism, natural disasters, accidents, and other contingencies.
- Has extraordinary powers to suspend tribal laws and ordinances, such as to establish a curfew, direct evacuations, and order a quarantine.
- Provides leadership and plays a key role in communicating to the tribal nation, and in helping people, businesses, and organizations cope with the consequences of any type of domestic incident within the jurisdiction.
- Negotiates and enters into mutual aid agreements with other tribes/jurisdictions to facilitate resource-sharing. Can request state and federal assistance through the governor of the state when the tribe's capabilities have been exceeded or exhausted.
- Can elect to deal directly with the federal government. (Although a state governor must request a presidential disaster declaration on behalf of a tribe under the Stafford Act, federal agencies can work directly with the tribe within existing authorities and resources.)

Secretary of Homeland Security. Pursuant to HSPD-5, the Secretary of Homeland Security:

- Is responsible for coordinating federal operations within the United States to prepare for, respond to, and recover from terrorist attacks, major disasters, and other emergencies.
- Serves as the "principal Federal official" for domestic incident management. The secretary is also responsible for coordinating federal resources utilized in response to or recovery from terrorist attacks, major disasters, or other emergencies if and when any of the following four conditions applies:
  - A federal department or agency acting under its own authority has requested DHS assistance.
  - The resources of state and local authorities are overwhelmed and federal assistance has been requested.

- More than one federal department or agency has become substantially involved in responding to the incident.
- The secretary has been directed to assume incident management responsibilities by the President.

Attorney General. The Attorney General is the chief law enforcement officer in the United States. In accordance with HSPD-5 and other relevant statutes and directives, the Attorney General has lead responsibility for criminal investigations of terrorist acts or terrorist threats:

- By individuals or groups inside the United States
- Directed at U.S. citizens or institutions abroad

Generally acting through the FBI, the Attorney General—in cooperation with other federal departments and agencies engaged in activities to protect national security—coordinates the activities of the other members of the law enforcement community. Nothing in the NRP derogates the Attorney General's status or responsibilities.

Secretary of Defense. DoD has significant resources that may be available to support the federal response to an Incident of National Significance. The Secretary of Defense authorizes Defense Support of Civil Authorities (DSCA) for domestic incidents as directed by the President or when consistent with military readiness operations and appropriate under the circumstances and the law. The Secretary of Defense retains command of military forces under DSCA, as with all other situations and operations. Nothing in the NRP impairs or otherwise affects the authority of the Secretary of Defense over the DoD.

Secretary of State. The Secretary of State is responsible for coordinating international prevention, preparedness, response, and recovery activities relating to domestic incidents, and for the protection of U.S. citizens and U.S. interests overseas.

Nongovernmental Organizations (NGOs). NGOs collaborate with first responders, governments at all levels, and other agencies and organizations providing relief services to sustain life, reduce physical and emotional distress, and promote recovery of disaster victims when assistance is not available from other sources.

Private Sector. DHS and NRP primary and support agencies coordinate with the private sector to effectively share information, form courses of action, and incorporate available resources to prevent, prepare for, respond to, and recover from Incidents of National Significance. The roles, responsibilities, and participation of the private sector during Incidents of National Significance vary based on the nature of the organization and the type and impact of the incident. Private-sector organizations may be involved as:

- **An Impacted Organization or Infrastructure.** Private-sector organizations may be affected by direct or indirect consequences of the incident. Examples of privately owned infrastructure include transportation, telecommunications, private utilities, financial institutions, and hospitals.

- **A Response Resource.** Private-sector organizations may provide response resources (donated or compensated) during an incident, including specialized teams, equipment, and advanced technologies.
- **A Regulated and/or Responsible Party.** Owners/operators of certain regulated facilities or hazardous operations may bear responsibilities under the law for preparing for and preventing incidents from occurring, and responding to an incident once it occurs. For example, federal regulations require owners/operators of Nuclear Regulatory Commission-regulated nuclear facilities to maintain emergency (incident) preparedness plans, procedures, and facilities and to perform assessments, prompt notifications, and training for a response to an incident.
- **A Member of State/Local Emergency Organizations.** Private-sector organizations may serve as an active partner in local and state emergency preparedness and response organizations and activities.

Citizen Involvement. Strong partnerships with citizen groups and organizations provide support for incident management prevention, preparedness, response, recovery, and mitigation. The U.S. Citizen Corps brings these groups together and focuses efforts of individuals through education, training, and volunteer service to help make communities safer, stronger, and better prepared to address the threats of terrorism, crime, public health issues, and disasters of all kinds.

THE NATIONAL RESPONSE PLAN COORDINATING STRUCTURES

- **Incident Command Post (ICP).** The field location at which the primary tactical-level, on-scene incident command functions are performed. The ICP may be collocated with the incident base or other incident facilities and is normally identified by a green rotating or flashing light.
- **Area Command (Unified Area Command).** An organization established (1) to oversee the management of multiple incidents that are each being handled by an ICS organization or (2) to oversee the management of large or multiple incidents to which several Incident Management Teams have been assigned. Area Command has the responsibility to set overall strategy and priorities, allocate critical resources according to priorities, ensure that incidents are properly managed, and ensure that objectives are met and strategies followed. Area Command becomes Unified Area Command when incidents are multijurisdictional. Area Command may be established at an EOC facility or at some location other than an ICP.
- **Local Emergency Operations Center (EOC).** The physical location at which the coordination of information and resources to support local incident management activities normally takes place.
- **State Emergency Operations Center (EOC).** The physical location at which the coordination of information and resources to support state incident management activities normally takes place.

continues

- **Homeland Security Operations Center (HSOC).** The primary national hub for domestic incident management operational coordination and situational awareness. The HSOC is a standing 24/7 interagency organization fusing law enforcement, national intelligence, emergency response, and private-sector reporting. The HSOC facilitates homeland security information-sharing and operational coordination with other federal, state, local, tribal, and nongovernmental EOCs.
- **Interagency Incident Management Group (IIMG).** The IIMG is a federal headquarters-level multiagency coordination entity that facilitates federal domestic incident management for Incidents of National Significance. The Secretary of Homeland Security activates the IIMG based on the nature, severity, magnitude, and complexity of the threat or incident. The Secretary of Homeland Security may activate the IIMG for high-profile, large-scale events that present high probability targets, such as NSSEs, and in heightened threat situations. The IIMG is composed of senior representatives from DHS components, other Federal departments and agencies, and nongovernmental organizations, as required. The IIMG membership is flexible and can be tailored or task-organized to provide the appropriate subject-matter expertise required for the specific threat or incident.
- **National Response Coordination Center (NRCC).** The NRCC is a multiagency center that provides overall federal response coordination for Incidents of National Significance and emergency management program implementation. FEMA maintains the NRCC as a functional component of the HSOC in support of incident management operations. The NRCC monitors potential or developing Incidents of National Significance and supports the efforts of regional and field components. The NRCC resolves federal resource support conflicts and other implementation issues forwarded by the Joint Field Office (JFO). Those issues that cannot be resolved by the NRCC are referred to the IIMG.
- **Regional Response Coordination Center (RRCC).** The RRCC is a standing facility operated by FEMA that is activated to coordinate regional response efforts, establish federal priorities, and implement local federal program support. The RRCC operates until a JFO is established in the field and/or the Principal Federal Officer, Federal Coordinating Officer, or Federal Resource Coordinator can assume their NRP coordination responsibilities. The RRCC establishes communications with the affected state emergency management agency and the National Response Coordination Center (NRCC), coordinates deployment of the Emergency Response Team-Advance Element (ERT-A) to field locations, assesses damage information, develops situation reports, and issues initial mission assignments.
- **Strategic Information and Operations Center (SIOC).** The FBI SIOC is the focal point and operational control center for all federal intelligence, law enforcement, and investigative law enforcement activities related to domestic terrorist incidents or credible threats, including leading attribution investigations. The SIOC serves as an information clearinghouse to help collect,

process, vet, and disseminate information relevant to law enforcement and criminal investigation efforts in a timely manner. The SIOC maintains direct connectivity with the HSOC and IIMG. The SIOC, located at FBI headquarters, supports the FBI's mission in leading efforts of the law enforcement community to detect, prevent, preempt, and disrupt terrorist attacks against the United States. The SIOC houses the National Joint Terrorism Task Force (NJTTF). The mission of the NJTTF is to enhance communications, coordination, and cooperation among federal, state, local, and tribal agencies representing the intelligence, law enforcement, defense, diplomatic, public safety, and homeland security communities by providing a point of fusion for terrorism intelligence and by supporting Joint Terrorism Task Forces (JTTFs) throughout the United States.

- **Joint Field Office (JFO).** The JFO is a temporary federal facility established locally to coordinate operational federal assistance activities to the affected jurisdiction(s) during Incidents of National Significance. The JFO is a multiagency center that provides a central location for coordination of federal, state, local, tribal, nongovernmental, and private-sector organizations with primary responsibility for threat response and incident support. The JFO enables the effective and efficient coordination of federal incident-related prevention, preparedness, response, and recovery actions. The JFO utilizes the scalable organizational structure of the NIMS Incident Command System (ICS). The JFO organization adapts to the magnitude and complexity of the situation at hand, and incorporates the NIMS principles regarding span of control and organizational structure: management, operations, planning, logistics, and finance/administration. Although the JFO uses an ICS structure, the JFO does not manage on-scene operations. Instead, the JFO focuses on providing support to on-scene efforts and conducting broader support operations that may extend beyond the incident site.

- **Joint Operations Center (JOC).** The JOC branch is established by the Senior Federal Law Enforcement Officer (SFLEO) (e.g., the FBI SAC during terrorist incidents) to coordinate and direct law enforcement and criminal investigation activities related to the incident. The JOC Branch ensures management and coordination of federal, state, local, and tribal investigative/law enforcement activities. The emphasis of the JOC is on prevention as well as intelligence collection, investigation, and prosecution of a criminal act. This emphasis includes managing unique tactical issues inherent to a crisis situation (e.g., a hostage situation or terrorist threat). When this branch is included as part of the Joint Field Office (JFO), it is responsible for coordinating the intelligence and information function (as described in NIMS), which includes information and operational security, and the collection, analysis, and distribution of all incident related intelligence. Accordingly, the Intelligence Unit within the JOC Branch serves as the interagency fusion center for all intelligence related to an incident.

FIELD-LEVEL ORGANIZATIONAL STRUCTURES: JFO COORDINATION GROUP

The field-level organizational structures and teams deployed in response to an Incident of National Significance, include the following potential members of the JFO Coordination Group:

- **Principal Federal Official (PFO).** The PFO is personally designated by the Secretary of Homeland Security to facilitate federal support to the established Incident Command System (ICS) Unified Command structure and to coordinate overall federal incident management and assistance activities across the spectrum of prevention, preparedness, response, and recovery. The PFO ensures that incident management efforts are maximized through effective and efficient coordination. The PFO provides a primary point of contact and situational awareness locally for the Secretary of Homeland Security.
- **Federal Coordinating Officer (FCO).** The FCO manages and coordinates federal resource support activities related to Stafford Act disasters and emergencies. The FCO:
  - Assists the Unified Command and/or the Area Command.
  - Works closely with the Principal Federal Official (PFO), Senior Federal Law Enforcement Official (SFLEO), and other Senior Federal Officials (SFOs).

  In Stafford Act situations where a PFO has not been assigned, the FCO provides overall coordination for the Federal components of the JFO and works in partnership with the State Coordinating Officer (SCO) to determine and satisfy state and local assistance requirements.
- **Senior Federal Law Enforcement Official (SFLEO).** The SFLEO is the senior law enforcement official from the agency with primary jurisdictional responsibility as directed by statute, presidential directive, existing federal policies, and/or the Attorney General. The SFLEO directs intelligence/investigative law enforcement operations related to the incident and supports the law enforcement component of the Unified Command on-scene. In the event of a terrorist incident, this official will normally be the FBI Senior Agent-in-Charge (SAC).
- **Federal Resource Coordinator (FRC).** The FRC manages federal resource support activities related to non-Stafford Act Incidents of National Significance when Federal-to-Federal support is requested from DHS by another Federal agency. The FRC is responsible for coordinating the timely delivery of resources to the requesting agency. In non-Stafford Act situations when a federal department or agency acting under its own authority has requested the assistance of the Secretary of Homeland Security to obtain support from other federal departments and agencies, DHS designates an FRC. In these situations, the FRC coordinates support through interagency agreements and memoranda of understanding (MOUs).

- **State/Local/Tribal Official(s).** The JFO Coordination Group also includes state representatives such as:
  - The **State Coordinating Officer (SCO)**, who serves as the state counterpart to the FCO and manages the state's incident management programs and activities.
  - The **Governor's Authorized Representative**, who represents the governor of the impacted state.
  - Possibly, local area representatives with primary statutory authority for incident management.
- **Senior Federal Officials (SFOs).** The JFO Coordination Group may also include representatives of other federal departments or agencies with primary statutory responsibility for certain aspects of incident management. SFOs utilize existing authorities, expertise, and capabilities to assist in management of the incident working in coordination with the PFO, FCO, SFLEO, and other members of the JFO Coordination Group. When appropriate, the JFO Coordination Group may also include U.S. attorneys or other senior officials or their designees from Department of Justice (DOJ) to provide expert legal counsel.

FIELD-LEVEL ORGANIZATIONAL STRUCTURES: JFO COORDINATION STAFF

The JFO structure will normally include a Coordination Staff. The JFO Coordination Group determines the extent of this staffing based on the type and magnitude of the incident. The roles and responsibilities of the JFO Coordination Staff are summarized here:

- **Chief of Staff.** The JFO Coordination Staff may include a Chief of Staff and representatives providing specialized assistance, which may include support in the following areas: safety, legal counsel, equal rights, security, infrastructure liaison, and other liaisons.
- **External Affairs Officer.** The External Affairs Officer provides support to the JFO leadership in all functions involving communications with external audiences. External Affairs includes Public Affairs, Community Relations, Congressional Affairs, State and Local Coordination, Tribal Affairs, and International Affairs, when appropriate. Resources for the various External Affairs Functions are coordinated through ESF #15. The External Affairs Officer also is responsible for overseeing operations of the Federal Joint Information Center (JIC) established to support the JFO. The JIC:
  - Is a physical location where public affairs professionals from organizations involved in incident management activities work together to provide critical emergency information, crisis communications, and public affairs support.

continues

- Serves as a focal point for the coordination and dissemination of information to the public and media concerning incident prevention, preparedness, response, recovery, and mitigation.
- **Defense Coordinating Officer (DCO).** If appointed by DoD, the DCO serves as DoD's single point of contact at the JFO. With few exceptions, requests for Defense Support of Civil Authorities (DSCA) originating at the JFO will be coordinated with and processed through the DCO. The DCO may have a Defense Coordinating Element (DCE) consisting of a staff and military liaison officers in order to facilitate coordination and support to activated Emergency Support Functions (ESFs). Specific responsibilities of the DCO (subject to modification based on the situation) include processing requirements for military support, forwarding mission assignments to the appropriate military organizations through DoD-designated channels, and assigning military liaisons, as appropriate, to activated ESFs.

FIELD-LEVEL ORGANIZATIONAL STRUCTURES: JFO SECTIONS

The role of each JFO Section is presented here:

- **Operations Section.** The Operations Section coordinates operational support to on-scene incident management efforts. Branches may be added or deleted as required, depending on the nature of the incident. The Operations Section also is responsible for coordination with other Federal command posts that may be established to support incident management activities. The Operations Section may include the following elements:
  - The **Response and Recovery Operations Branch** coordinates the request and delivery of federal assistance and support from various special teams. This branch is composed of four groups: Emergency Services, Human Services, Infrastructure Support, and Community Recovery and Mitigation.
  - The **Law Enforcement Investigative Operations Branch/Joint Operations Center (JOC)** is established by the Senior Federal Law Enforcement Official (SFLEO) (e.g., the FBI SAC during terrorist incidents) to coordinate and direct law enforcement and criminal investigation activities related to a terrorist incident. The JOC branch ensures management and coordination of federal, state, local, and tribal investigative/law enforcement activities. The emphasis of the JOC is on prevention as well as intelligence collection, investigation, and prosecution of a criminal act. This emphasis includes managing unique tactical issues inherent to a crisis situation (e.g., a hostage situation or terrorist threat).
  - For National Special Security Events (NSSEs), a third branch, **the Security Operations Branch**, or **Multi-agency Command Center (MACC)**,

may be added to coordinate protection and site security efforts. In these situations, the Operations Section Chief is designated by mutual agreement of the JFO Coordination Group based on the agency with greatest jurisdictional involvement and statutory authority for the current incident priorities. The agency providing the Operations Section Chief may change over time as incident priorities change.

- **Planning Section.** The Planning Section provides current information to the JFO Coordination Group to ensure situational awareness, determine cascading effects, identify national implications, and determine specific areas of interest requiring long-term attention. The Planning Section also provides technical and scientific expertise. The Planning Section is composed of the following units: Situation, Resources, Documentation, Technical Specialists, and Demobilization. The Planning Section may also include an Information and Intelligence Unit (if not assigned elsewhere), and an HSOC representative who aids in the development of reports for the HSOC and IIMG.

- **Logistics Section.** The Logistics Section coordinates logistics support that includes:
  - Control and accountability for federal supplies and equipment
  - Resource ordering
  - Delivery of equipment, supplies, and services to the JFO and other field locations
  - Facility location, setup, space management, building services, and general facility operations
  - Transportation coordination and fleet management services
  - Information and technology systems services, administrative services such as mail management and reproduction, and customer assistance

  The Logistics Section may include Coordination and Planning, Resource Management, Supply, and Information Services Branches.

- **Finance and Administration Section (Comptroller).** The Finance and Administration Section is responsible for the financial management, monitoring, and tracking of all federal costs relating to the incident and the functioning of the JFO while adhering to all federal laws, acts, and regulations. The position of the Financial and Administration Chief will be held exclusively by a comptroller who serves as the Senior Financial Advisor to the team leader (e.g., FCO) and represents the coordinating agency's Chief Financial Officer (CFO) as prescribed by the CFO Act of 1990.

FIELD-LEVEL ORGANIZATIONAL STRUCTURES: RESPONSE TEAMS

Various teams are ready to deploy in response to threats or incidents. These teams include the following:

- **ERT Advance Element (ERT-A).** The ERT-A conducts assessments, and initiates coordination with the state and initial deployment of federal resources. It is headed by a team leader from FEMA and is composed of program and support staff and representatives from selected ESF primary agencies. Each FEMA region maintains an ERT ready to deploy during the early stages of an incident to:
  - The state EOC or to other locations to work directly with the state to obtain information on the impact of the event and to identify specific state requests for federal incident management assistance.
  - The affected area to establish field communications, locate and establish field facilities, and set up support activities.
- **National Emergency Response Team (ERT-N).** The ERT-N deploys for large-scale, high-impact events, or as required. An ERT-N may predeploy based on threat conditions. The Secretary of Homeland Security determines the need for ERT-N deployment, coordinating the plans with the affected region and other federal agencies. The ERT-N includes staff from FEMA headquarters and regional offices as well as other federal agencies.
- **Federal Incident Response Support Team (FIRST).** The FIRST is a forward component of the ERT-A that provides on-scene support to the local Incident Command or Area Command structure in order to facilitate an integrated inter-jurisdictional response. The FIRST is designed to be a quick and readily deployable resource to support the federal response to Incidents of National Significance. The FIRST deploys within two hours of notification, to be on-scene within 12 hours of notification. FEMA maintains and deploys the FIRST. Upon the subsequent deployment of an ERT, the FIRST integrates into the Operations Section of the JFO.
- **Domestic Emergency Support Teams (DEST).** The DEST may be deployed to provide technical support for management of potential or actual terrorist incidents. Based upon a credible threat assessment, the Attorney General, in consultation with the Secretary of Homeland Security, may request authorization through the White House to deploy the DEST. The PFO and a small staff component may deploy with the DEST to facilitate their timely arrival and enhance initial situational awareness. Upon arrival at the JFO or critical incident location, the DEST may act as a stand-alone advisory team to the FBI SAC providing required technical assistance or recommended operational courses of action.

FIELD-LEVEL ORGANIZATIONAL STRUCTURES: RESPONSE TEAMS

- **Other Federal Teams.** There are numerous special teams available to support incident management and disaster response and recovery operations. Examples include:
  - Damage assessment teams
  - The Nuclear Incident Response Team (NIRT)
  - Disaster Medical Assistance Teams (DMATs)
  - HHS Secretary's Emergency Response Team
  - DOL/OSHA's Specialized Response Teams
  - Veterinarian Medical Assistance Teams (VMATs)
  - Disaster Mortuary Operational Response Teams (DMORTs)
  - National Medical Response Teams (NMRTs)
  - Scientific and Technical Advisory and Response Teams (STARTs)
  - Donations Coordination Teams
  - Urban Search and Rescue (US&R) task forces and incident support teams
  - Federal Type 1 and Type 2 Incident Management Teams (IMTs)
  - Domestic Animal and Wildlife Emergency Response Teams and Mitigation Assessment Teams

INCIDENT MANAGEMENT ACTIONS

- **Notification and Assessment.** Federal, state, local, tribal, private-sector, and nongovernmental organizations report threats, incidents, and potential incidents using established communications and reporting channels. The Homeland Security Operations Center (HSOC) receives threat and operational information regarding incidents or potential incidents and makes an initial determination to initiate the coordination of federal information-sharing and incident management activities. When notified of a threat or an incident with possible national-level implications, the HSOC assesses the situation and notifies the Secretary of Homeland Security accordingly.
- **Reporting.** Federal, state, tribal, private-sector, and nongovernmental Emergency Operations Centers (EOCs) report incident information to the HSOC. In most situations, incident information will be reported using existing mechanisms to state or federal operations centers, which in turn will report the information to the HSOC. Information regarding potential terrorist threats normally is reported initially to a local or regional Joint Terrorism Task Force (JTTF) and, subsequently, from the FBI Strategic Information and Operations Center (SIOC) to the HSOC if the FBI deems the threat to be credible.
- **Activation.** For actual or potential Incidents of National Significance, the HSOC reports the situation to the Secretary of Homeland Security and/or senior staff as delegated by the secretary, who then determines the need to activate components of the NRP to conduct further assessment of the situa-

continues

tion, initiate interagency coordination, share information with affected jurisdictions, and/or initiate deployment of resources. Concurrently, the secretary also makes a determination of whether or not an event meets the criteria established for a potential or actual Incident of National Significance as defined in the NRP. When the secretary declares an Incident of National Significance, federal departments and agencies are notified by the HSOC (as operational security considerations permit), and may be called upon to staff the Interagency Incident Management Group (IIMG) and National Response Coordination Center (NRCC). The affected state(s) and tribes also are notified by the HSOC using appropriate operational security protocols. In the preincident mode, such notification may be conducted discreetly, on a need-to-know basis, so as to preserve the operational security and confidentiality of certain law enforcement and investigative operations. The NRCC and RRCC deploy, track, and provide incident-related information until the JFO is established.

- **Response.** Once an incident occurs, the priority shifts to immediate and short-term response activities to preserve life, property, the environment, and the social, economic, and political structure of the community. Actions also are taken to prevent and protect against other potential threats. Examples of response actions include immediate law enforcement, fire, and emergency medical service actions; mass care, public health, and medical services; emergency restoration of critical infrastructure; control of environmental contamination; and responder health and safety protection. During the response to a terrorist event, law enforcement actions to collect and preserve evidence and to apprehend perpetrators are critical. These actions take place simultaneously with response operations necessary to save lives and protect property.

- **Recovery.** Recovery involves actions needed to help individuals and communities return to normal when feasible. The JFO is the central coordination point among Federal, State, local, and tribal agencies and voluntary organizations for delivering recovery assistance programs. Long-term environmental recovery may include cleanup and restoration of public facilities, businesses, and residences; reestablishment of habitats and prevention of subsequent damage to natural resources; protection of cultural or archeological sites; and protection of natural, cultural, and historical resources from intentional damage during other recovery operations.

- **Mitigation.** Hazard mitigation involves reducing or eliminating long-term risk to people and property from hazards and their side effects. The JFO's Community Recovery and Mitigation Branch is responsible for coordinating the delivery of all mitigation programs within the affected area, including hazard mitigation for:
  - Grant programs for loss reduction measures (if available).
  - Delivery of loss reduction building-science expertise.
  - Coordination of federal flood insurance operations.
  - Community education and outreach necessary to foster loss reduction.

- **Demobilization.** When a centralized federal coordination presence is no longer required in the affected area, the JFO Coordination Group implements the demobilization plan to transfer responsibilities and close out the JFO. After the JFO closes, long-term recovery program management and monitoring transition to individual agencies' regional offices and/or headquarters, as appropriate.

- **Remedial Actions and After-Action Reports.** DHS formally convenes interagency meetings called *hotwashes* to identify critical issues requiring headquarters-level attention, lessons learned, and best practices associated with the federal response to Incidents of National Significance. Hotwashes typically are conducted at major transition points over the course of incident management operations, and should include state, local, and tribal participation. Identified issues are validated and promptly assigned to appropriate organizations for remediation. Following an incident, the JFO Coordination Group submits an after-action report to DHS Headquarters detailing operational successes, problems, and key issues affecting incident management. The report includes appropriate feedback from all federal, state, local, tribal, nongovernmental, and private-sector partners participating in the incident.

Source: FEMA Emergency Management Institute

Two of the US&R task forces have agreements with the U.S. Agency for International Development (USAID) to provide search-and-rescue services oversees. These two task forces are Metro-Dade Fire Department in Florida and the Fairfax County Fire & Rescue in Virginia. A full list of US&R task forces is presented in Table 4-3.

Other FEMA Response Resources

FEMA manages a cadre of nearly 4,000 temporary Disaster Assistances Employees (DAEs), who support FEMA response and recovery activities in the field in areas such as logistics, facility management, public affairs, community relations, and customer service. FEMA manages a mobile operations capability that provides communications and logistical support to state and local emergency officials.

THE DAE EXPERIENCE

By Pat Glithero, FEMA Region V

Whenever the Federal Emergency Management Agency (FEMA) responds to a disaster, local and staff officials encounter and work with numerous FEMA staff, many of them women and men of FEMA's Disaster Assistance Employee (DAE) program. DAEs respond as needed to presidential-declared emergencies and disasters across the nation and territories and remain until the Disaster Field Office (DFO) closes. They then return home to resume their lives as officials, businessmen, professionals, retirees, and in a myriad of other jobs.

continues

Table 4-3   FEMA Urban Search-and-Rescue Task Forces

State	Number	Organization
Arizona	AZ-TF1	Phoenix, Arizona
California	CA-TF1	LA City Fire Dept.
	CA-TF2	LA County Fire Dept.
	CA-TF3	Menlo Park Fire Dept.
	CA-TF4	Oakland Fire Dept.
	CA-TF5	Orange Co. Fire Authority
	CA-TF6	Riverside Fire Dept.
	CA-TF7	Sacramento Fire Dept.
	CA-TF8	San Diego Fire Dept.
Colorado	CO-TF1	State of Colorado
Florida	FL-TF1	Metro-Dade Fire Dept.
	FL-TF2	Miami Fire Dept.
Indiana	IN-TF1	Marion County
Maryland	MD-TF1	Montgomery Fire Rescue
Massachusetts	MA-TF1	City of Beverly
Missouri	MO-TF1	Boone County Fire Protection District
Nebraska	NE-TF1	Lincoln Fire Dept.
Nevada	NV-TF1	Clark County Fire Dept.
New Mexico	NM-TF1	State of New Mexico
New York	NY-TF1	NYC Fire and EMS, Police
Ohio	OH-TF1	Miami Valley US&R
Pennsylvania	PA-TF1	Commonwealth of Pennsylvania
Tennessee	TN-TF1	Memphis Fire Dept.
Texas	TX-TF1	State of Texas Urban Search & Rescue
Utah	UT-TF1	Salt Lake Fire Dept.
Virginia	New York	Fairfax Co. Fire & Rescue Dept.
	VA-TF2	Virginia Beach Fire Dept.
Washington	WA-TF1	Puget Sound Task Force

Source: FEMA, www.fema.gov

DAEs join the cadre for many reasons. Some have benefited from FEMA programs in disasters and want to share their gratitude. Many believe strongly in programs designed to help fellow citizens. To many retirees, the DAE program allows a continued work opportunity alongside colleagues with whom they have worked for many years on a full-time basis. Some just appreciate working a job that uses their skills and allows them to feel they make a difference. Working in various geographic parts of the United States at a state or local level is a positive by product of disaster deployment. FEMA offers many opportunities for training classes and learning that otherwise might not be available in the private sector.

At the beginning, or end, of any DFO, though, what means most to many DAEs is the sense of camaraderie and family that DAEs share with each other and other FEMA staff. With a total workforce of less than 2,500 nationwide, FEMA DAEs and staff become part of a community that comes together as needed, does the job, and then parts officially until the next call.

Source: FEMA, www.fema.gov

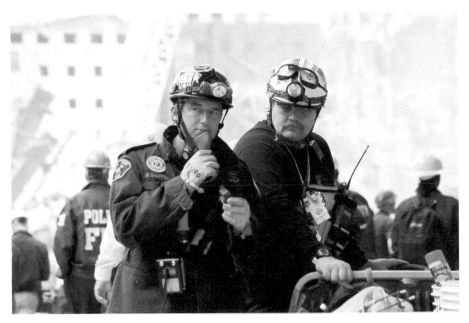

Figure 4-3 New York, New York, September 27, 2001. FEMA, New York Fire Fighters, and the Urban Search and Rescue teams worked very closely throughout the cleanup effort at the World Trade Center. Photo by Bri Rodriguez/FEMA News Photo.

Figure 4-4 Malibu, California, 1996. A California Department of Forestry official watches the wildfire as it burns up a hillside. FEMA News Photo.

FEMA's MOBILE OPERATIONS CAPABILITY

Disasters may require resources beyond the capabilities of the local or state authorities. In response to regional requests for support, FEMA provides mobile telecommunications, operational support, life support, and power generation assets for the on-site management of disaster and all-hazard activities. This support is managed by the Response and Recovery Directorate's Mobile Operations Division (RR-MO).

The Mobile Operations Division has a small headquarters staff and five geographically dispersed Mobile Emergency Response Support (MERS) Detachments and the Mobile Air Transportable Telecommunications System (MATTS) to:

- Meet the needs of the government emergency managers in their efforts to save lives, protect property, and coordinate disaster and all-hazard operations.
- Provide prompt and rapid multimedia communications, information processing, logistics, and operational support to federal, state, and local agencies during catastrophic emergencies and disasters for government response and recovery operations.

The MERS and MATTS support the Disaster Field Facilities. They support the federal, state, and local responders—not the disaster victims.

Available Support

Each of the MERS Detachments can concurrently support a large Disaster Field Office and multiple field operating sites within the disaster area. MERS is equipped with self-sustaining telecommunications, logistics, and operations support elements that can be driven or airlifted to the disaster location. MATTS and some of the MERS assets can be airlifted by C-130 military cargo aircraft.

The MERS and MATTS are available for immediate deployment. As required, equipment and personnel will deploy promptly and provide:

- Multimedia communications and information processing support, especially for the Communications Section, Emergency Support Function (ESF) #2 of the Federal Response Plan (FRP)
- Operational support, especially for the Information and Planning Section, ESF #5 of the FRP
- Liaison to the Federal Coordinating Officer (FCO)
- Logistics and life support for emergency responders
- Automated information and decision support capability
- Security (facility, equipment, and personnel) management and consultation

Most equipment is preloaded or installed on heavy-duty, multiwheel drive trucks. Some equipment is installed in transit cases.

Source: FEMA, www.fema.gov

COMMUNICATIONS AMONG RESPONDING AGENCIES

General

Overlapping responsibilities and unclear delineation makes communications among responding agencies crucial. Responding agencies to a disaster event may include emergency management agencies from all levels of government, nongovernmental organizations, other responding government agencies, such as law enforcement and public health, the medical and scientific communities, and even businesses. Communications among these agencies is recognized as a current Achilles heel in the emergency management field. Issues of authority and structure are difficult to resolve, and operations are often performed in an ad hoc fashion; however, improvement in this area is becoming a point of emphasis, and technological advancements are facilitating better communications as well.

The costs of poor coordination and communication can be high. A slow or ineffective initial disaster response disproportionately increases human losses. Also, poorly coordinated and perceived response efforts can damage political careers and the reputations of agencies. After Hurricane Andrew in Florida in 1992, it became apparent early in relief efforts that there were communications and coordination problems between FEMA, the state emergency management system, local agencies, and the governor's office. Many political analysts feel that the poor public perception of the response cost President George H.W. Bush votes in the 1992 presidential election.

Figure 4-5 Hurricane Andrew, Florida, August 24, 1992. Volunteer assistance was received from volunteer organizations, including the American Red Cross and Salvation Army. One million people were evacuated and 54 died in this hurricane. FEMA News Photo.

National Response Plan

The NRP is the major coordination mechanism for the various responding federal agencies during a major disaster. Emergency Support Function (ESF) #5 within the NRP outlines responsibilities for Emergency Management. FEMA has the lead role for this activity, but is supported by most other partner agencies in this respect. Individuals performing this function collect, analyze, process, and disseminate information about a disaster or emergency in order to facilitate the federal government assistance activities. The response is coordinated at the federal, field, regional, and headquarters levels. Daily information updates are provided to the various elements of the operation. The overall purpose of the function is to provide a central collection point where situation information can be compiled, analyzed, and prepared for use by decision makers.

The NRP also includes a communications function (ESF #2), which basically deals with the telecommunications infrastructure and technology. The lead agency for this function is the National Communications System. Its job is to ensure the provision of federal telecommunications support to federal, state, and local response efforts, and to serve as the planning point for use of national telecommunications assets and resources.

FEMA Operations Center

The FEMA Operations Center (FOC) serves as the site of overall coordination and situation assessment operations for major disasters. It maintains a 24-hour capability to monitor all sources of information. Regional Operation Centers (ROCs) are the initial coordinating point for federal response efforts, however. The FEMA Director serves as the Federal Coordinating Officer (FCO) of the FOC, and assumes coordination responsibilities, working with the state coordinating officer and local officials.

Joint Information Center

The Joint Information Center (JIC) is also a valuable tool for getting emergency management partners on the same page. In disasters of catastrophic or nationally significant proportions, a JIC is established to coordinate the dissemination of information about all disaster response and recovery programs. Public Affairs Officers (PAOs) representing all the federal, state, local, and voluntary agencies providing response or recovery services are invited to collocate and be a part of JIC operations. Interagency coordination is one of the central functions of the JIC, and teamwork is a key to implementing successful public information and media affairs programs. JICs involve coordination among the FCO, the lead state PAO, the congressional liaison, community relations and disaster assistance program managers, and other public agency PAOs.

Command and Control versus Coordination

It is generally agreed that some type of mechanism is needed to facilitate coordination and communications among responding partners. What is not agreed upon is the structure of such a mechanism. The argument pits the clear, hierarchical "command and control" model against the more flexible, ad hoc "coordination" model.

The command and control model was adopted from the Incident Command System (ICS) used by fire departments across the United States and has clear lines of authority and responsibility. The coordination model is less rigid and more collaborative. In general, the coordination model is becoming more popular than the traditional command and control structure. For one thing, the new breed of emergency manager is typically more of a recovery coordinator than a field general. Also, command and control structures can sometimes hamper communications. The commanding organization may have a value system and technical language that is distinct from those of partner organizations or the victims. The coordination model takes this variability into account and focuses on providing an open communications forum. The coordination model is also often better for negotiating turf battles among agencies and nongovernmental organizations providing overlapping services.

Technology

There are many examples of technology improving communications among partners. The use of the Internet as such a tool is an obvious trend. The city of Seattle recently integrated its Web site into its emergency communications plan. The site now provides immediate access to information for members of others departments, such as police and transportation. The site also contains a database of press releases and space for current news, which have been coordinated through an information control system. One lesson learned from Seattle's experience is to ensure that staff members who update the site are centrally located with emergency responders within the EOC.

Figure 4-6 West Palm Beach, Florida, October 1999. An inside view of Palm Beach County's state-of-the-art Emergency Operations Center, which enables local, state, and federal emergency management teams to coordinate interagency disaster response. Photo by Ty Harrington/FEMA News Photo.

Many communities are now using wireless systems to improve communications. The City of San Francisco recently developed a wireless voice and data communications system for its public safety agencies. The system overcomes the limited coverage of radio systems and the problem of various departments using incompatible systems. Mobile and portable radios are now in use at the city's fire, police, and emergency agencies. The Departments of Public Health, Public Works, Water, and the mayor's EOC also use the system. Officials indicate that it will go a long way toward helping the city handle the almost 4,000 emergency 911 calls it receives daily.

Wireless communications sometimes have their own limitation, however. After the terrorist attacks of September 11, 2001, cellular phone use overwhelmed wireless networks and prevented some local police and officials from making critical calls. In response, the White House plans to give emergency crews and government officials priority on the nation's cellular telephone system. Already in the United States, about 800 public institutions with emergency communications systems are given priority over regular users during an emergency. A similar effort is being initiated in Japan. After the 1995 Great Hanshin earthquake, incoming calls to Japan increased 50-fold and swamped the network.

CONCLUSION

Responding to disaster events is the most visible activity that any federal, state, or local emergency management agency conducts. The politicians, the media, and the general public rate the success of an emergency management organization by how well it functions in the response phase of a disaster. A successful disaster response at any level of government requires a strong command and control system, clear lines of communication, and coordination of numerous agencies from multiple jurisdictions. Local first responders—fire, police, and emergency medical technicians—are on the scene first. Local and state emergency managers coordinate resources and assess the damage and the capacity of their jurisdictions to respond effectively. For major disaster events, a presidential disaster declaration activates the NRP that delivers the full resources of the federal government in support of local and state authorities.

The key to the success of the emergency management system in the United States is the commitment of this country's elected officials to use the government to come to the aid of its citizens when a crisis occurs. The response process as described in this chapter ensures that government at all levels is capable of fulfilling this commitment.

CASE STUDIES

THE SPACE SHUTTLE COLUMBIA DISASTER

On Feb. 1, 2003, as the Space Shuttle *Columbia* reentered the earth's atmosphere following a successful space mission, it suddenly began to break apart, showering debris over an area of hundreds of square miles in East Texas and western

Louisiana. President Bush issued emergency declarations for Texas and Louisiana, in the absence of requests for assistance from either governor, as the shuttle craft was considered federal property. Within hours federal and state agencies had deployed teams to the disaster area to assist local fire, law enforcement, and emergency management authorities already onsite. More than 60 agencies, including public and private groups, responded with personnel, supplies, and equipment. Disaster Field Offices (DFO) were opened at Barksdale Air Force Base in Los Angeles, CA, and in Lufkin, TX, and a satellite DFO was established in Fort Worth, TX. The Lufkin DFO was the regional center of all search-related operations. This was the first major response performed by the newly created Department of Homeland Security.

As a federally declared disaster, FEMA was in charge of FRP coordination, and also coordinated the response and recovery operations. NASA, with the assistance of the Texas Forest Service (TFS), the U. S. Forest Service (USFS), the Environmental Protection Agency (EPA), and many others, supervised the search for shuttle material. The EPA's role was to assist FEMA and NASA by conducting environmental monitoring and assisting in the cleanup of hazardous materials from the Space Shuttle Columbia. EPA experts from across the country were mobilized to help local, county, and state officials protect public health and the environment, as well as to assist officials in recovering materials from communities and providing for safe transport of these materials to secure locations.

From the onset, the agencies' priorities were threefold: ensure public safety, retrieve evidence-pieces of the shuttle that ultimately could determine the cause of the tragedy, and reimburse expenses of state and local governments and private citizens who may have sustained property damage as a result of the accident and search. NASA quickly identified potential hazardous materials, such as tanks containing toxic substances or unexploded pyrotechnic devices, and once found, the EPA secured the material. The EPA also worked with state and local authorities to clear school campuses and public access areas, and tested air and water samples taken along the flight path for shuttle contaminates. Using the resources of the Emergency Response and Removal Service (ERRS) contractors and the U.S. Coast Guard (USCG), Gulf Strike Team, the EPA found no evidence of hazardous material in the atmosphere or drinking water supplies. Early in the recovery effort, teams from NASA, the FBI, National Guard, Urban Search and Rescue (US&R) organizations, the Department of Public Safety, and others, conducted a successful search in East Texas to recover and bring home the bodies of Columbia's crew.

Three days after the accident, local fire, police, volunteers, Texas Department of Public Safety officers (DPS), Louisiana State Police, and EPA, USFS, TFS and National Guard units from Texas, Louisiana, Oklahoma, and New Mexico began clearing shuttle debris in high-traffic areas. A one-page set of guidelines prepared by the State of Texas, NASA, and EPA enabled the teams to collect, document, tag, and transport nonhazardous debris without prior EPA or NASA clearance. These initial teams ended their search operations on February 17. The TFS, under the direction of NASA, now assumed responsibility for search activities in the field, which involved extensive air and ground searches in a 10-mile by 240-mile corridor along the projected shuttle flight path. The TFS—through the Texas

Interagency Coordination Center—called upon experienced management and firefighting crews from across the nation and Puerto Rico. The air operations, managed by TFS, included up to 36 helicopters and 10 fixed wing aircraft. Also involved in the air search, but not managed by TFS, were motorized para gliders, an ER-2 (similar to the U-2), a specially equipped DC-3, and the Civil Air Patrol (CAP), among others. Volunteers put in more than 800 search-days of flying in the weeks just after the accident and covered the flight corridor area west of Fort Worth to the New Mexico border. The USFS, Bureau of Indian Affairs, Bureau of Land Management, National Park Service, U.S. Fish and Wildlife Service, along with state forestry organizations and contractors, provided the greatest number of crews, drawing from their expertise in wildland firefighting. More than 4,000 people at a time searched 12 hours a day, seven days a week. Camp crews were stationed at sites near Hemphill, Nacogdoches, Palestine, and Corsicana, with a goal of finding as much material as possible before spring vegetation growth made the search more difficult.

The U.S. Navy supervised the water search activities in Lake Nacogdoches and Toledo Bend Reservoir, located at the eastern end of the 2,400 square-mile search area. Beginning on February 22, 60 divers from the Navy, USCG, EPA, DPS, Houston and Galveston police and fire departments, and Jasper County Sheriff's department combed the lakes using sophisticated sonar-equipped boats to help identify shuttle material. As in any operation of this magnitude, the hazards for all the searchers were challenging. Ground crews slogged though mud, dense vegetation, and rocky areas, faced wild hogs, snakes, and other vermin, and dealt with the ever-changing weather. Divers reckoned with the murky waters of the East Texas lakes, along with underwater forests, and various submerged hazards.

Ground and air operations covered over 1.5 million acres, mostly in Texas, with searches also conducted in Louisiana, California, Utah, Nevada and New Mexico. 82,500 shuttle items were recovered and processed by the Kennedy Space Center in Florida, weighing 84,800 pounds, and amounted to almost 40% of the total weight of Columbia. The total cost of the search and recovery operation amounted to $161,945,000. These funds include costs associated with the ground, air, and water search operations, equipment, and personnel. FEMA Public Assistance, working through Texas and Louisiana, reimbursed the two states approximately $4.5 million for their efforts. FEMA turned over control of the recovery operation to NASA on April 30. The same day, NASA opened the Columbia Recovery Operation (CRO) office at the Johnson Space Center in Houston. FEMA closed the Disaster Field Office in Lufkin, TX, on May 10.

Source: www.fema.gov

OKLAHOMA CITY

On April 19, 1995, an explosion rocked the federal plaza in Oklahoma City. The Alfred P. Murrah Federal Building was destroyed after a bomb, which was placed in a rental truck next to the building, was detonated. Upon arriving in the area, first responders witnessed smoke and fire coming from the Water Resource building. Believing that it was a natural gas explosion, it was not until EMS personnel entered this building that they noticed the gaping hole in the Murrah Building. The Fire Chief's first step was to have a single command center, which incorporated all buildings and victims within a one-mile radius. There were 33 fire stations, with at least 1,000 firefighters and 52 pieces of rescue apparatus that responded to the scene.

Within 45 minutes after notification from the Oklahoma Department of Civil Emergency Management, FEMA deployed staff to Oklahoma City. FEMA coordinated the federal response to the Oklahoma City bombing and later worked closely with state and local officials on recovery efforts. The President signed an Emergency Declaration within eight hours of the occurrence. This was the first time section 501(b) of the Stafford Act, granting FEMA the primary federal responsibility for responding to a domestic consequence management incident, was ever used. The President subsequently declared a major disaster on April 26, 1995. Because the disaster site was also a federal crime scene, FEMA appointed a liaison to the FBI to coordinate site access, support requirements, public information, and other issues. The coordinated work among federal agencies in Oklahoma City led to the further clarification of agency and department roles in crisis and consequence management.

Harsh lessons were learned in Oklahoma City. A situation arose when local radio stations requested that all medical personnel should respond to the disaster area. A nurse who answered the call was killed by falling debris while trying to rescue victims in the building. A term constantly used after the bombing was the Oklahoma Standard. Oklahoma had personnel on the scene within 30 minutes. Federal officials were notified within minutes of the disaster. Volunteer services were immediate, and because this was a local disaster, everyone took responsibility to do whatever they could to help. Hospital personnel established an effective and efficient triage system. Phone numbers, Internet sites, and briefings were launched within hours of the disaster. The American Red Cross, as in all disasters, was quick to respond with personnel and supplies to help family members of those who were injured or killed in the bombing. The Salvation Army responded within hours with food and supplies. By the end of the day, the Salvation Army had deployed seven units to provide services to the workers and the victims. Law Enforcement and EMS personnel had up-to-date training. Oklahoma had excellent coordination with the Public Works Department, the National Weather Service, and the National Guard. The Department of Public Safety also had a predetermined disaster plan in place.

Although there were some initial problems with communication, this was resolved within an hour as a result of support from Cellular One and Southwest Bell. They were able to clear lines, reconfigure their systems, and dispatch cell phones to personnel on scene. But most important was that the Oklahoma Highway Patrol could talk directly with personnel from federal agencies that were on the scene. A Department of Safety technician was able to program radios within 45 minutes of the disaster. Like most major cities, Oklahoma is equipped with 800MHz radios that can be linked with systems throughout the region. In any disaster, communication is the first line of defense in a successful response.

It is essential that hospitals, rescue personnel, site commanders, and law enforcement officials have the ability to talk to one another. This was necessary to update the Disaster Field Office about the status of the response as well as obtaining needed personnel and supplies throughout the response. The only glitch was that the police were limited to those with whom they could communicate.

HURRICANE FLOYD

On September 14, 1999, FEMA began mobilizing federal resources in preparation for possible landfall by Hurricane Floyd. Although, in previous years, states had to wait for the disaster to strike before obtaining FEMA assistance, in the case of Hurricane Floyd, FEMA took a proactive stance by activating Emergency Response Teams, allocating funds to local communities for law enforcement, and working with the Tropical Predication Center to monitor Hurricane Floyd's track. The ROC was put into action three days before the actual landfall of Hurricane Floyd.

On September 16, 1999, Hurricane Floyd made landfall near Cape Fear, North Carolina. The Category II hurricane had sustained winds of 110 miles per hour, but unlike Hurricane Andrew, the local first responders in coordination with FEMA were better prepared to handle this disaster. Emergency materials, generators, sheeting, tarp, bottled water, blankets, and clothing were identified and available for immediate delivery. Disaster Medical Assistance Teams (DMATs)

Figure 4-7   Aurora, Illinois, July 1993. Illinois flood victim gets food from the Salvation Army. Photo by Liz Roll/FEMA News Photo.

had been placed on alert to provide medical services. Public works, including engineers, electricians, phone company employees, and public work personnel also were prepared for deployment to the area. Although forecasters thought that Floyd would hit Florida or Georgia, FEMA officials were mobile as the hurricane continued to track farther north. On September 15, 1999, President Clinton signed emergency declarations for North and South Carolina to fund law enforcement officials to help evacuate the areas. More than 2,100 employees were prepared to respond to the disaster. FEMA Urban Search and Rescue Teams from Indiana, Maryland, and Pennsylvania were activated. Upon the hurricane reaching land, FEMA's Mobile Emergency Response System (MERS) provided communication support to the affected communities.

FEMA's proactive response before landfall ensured that those affected by the hurricane would have the needed materials and services to help in the recovery phase. While the rain was still falling, FEMA established their toll-free service line. Within days, people were receiving financial aid to help them through the disaster. Although FEMA took some flack from certain areas of North Carolina and Virginia because of the long-lasting flooding, lives were saved and damage was reduced because of FEMA's and the 27 agencies' response to the hurricane.

HURRICANE ANDREW

On August 24, 1992, Hurricane Andrew, a Category 4 hurricane, made landfall over Dade County, Florida. For everything that went right during the response for Hurricane Floyd, the opposite was true for Hurricane Andrew. "When Hurricane Andrew was approaching Florida and the advance element of the federal emergency response team deployed to the state emergency operations center in Tallahassee, it was evident that the state lacked sufficient space and resources to coordinate an operation to handle a disaster caused by a major hurricane like Andrew" (FEMA, 1993). In a postdisaster audit of FEMA's disaster management performance after Hurricane Andrew, the Inspector General noted that "state officials acknowledged that their initial assessment of requirements for federal assistance were too low, and that at first they were resistant to the idea of a massive flood of federal resources into south Florida" (FEMA, p. 41). Other problems noted by the Inspector General included a failure on the part of the state to request certain federal services because the state was reluctant to incur its 25 percent cost share and the lack of awareness of certain services by both state and local officials.

What became evident in the first weeks after Andrew was that FEMA and the overall federal response as well as the Florida response were uncoordinated, confused, and often inadequate (FEMA, 1993). FEMA requested its Inspector General to conduct a postdisaster audit, and Governor Chiles issued an executive order (92-242) establishing the Governor's Disaster Planning and Response Review Committee "to evaluate current state and local statutes, plans and programs for natural and man-made disasters, and to make recommendations to the Governor and the State Legislature" not later than January 15, 1993 (FEMA, 1993). The national emergency management system was acknowledged as being broken, and both the federal government and the state wanted to know why and what should be done to improve it.

The one key factor was that FEMA had yet to obtain clarification about its authority to supercede all other government agencies during a disaster. The Inspector General's report tasks each area that FEMA failed to perform. From preparation to response and recovery, FEMA and federal officials dropped the ball. If it had not been for DoD intervention, people would have been left to their own device in seeking medical assistance, shelter, food, and water. Because federal agencies had to have a formal declaration declared, they were slow in responding and providing assistance to the people of Florida.

Without electricity, FEMA was unable to disseminate the needed information to the communities. Telephone lines, radio, and TV stations were disrupted for the first few days. People were not aware of the services available to them until days after the hurricane had struck. Although there was a FEMA employee at the Emergency Operation Center, he did not have the resources or the communication capabilities to get the needed response. The Defense Coordinating Officer (DCO) who was assigned to the Emergency Response Team was continually being drawn away from his assignment and also had his role continually expanded or changed during the response (FEMA, 1993).

The Inspector General's 200-page report took every aspect of the response and recovery phase into account and discussed in detail what needed to be done by local, state, and federal agencies for future catastrophic events. The report took into account the duplication of efforts by volunteer organizations and the lack of communication among the multiple federal agencies that had responded to Florida. Most consider President Bush's election loss to be partly attributable to the federal government's inability to manage domestic disasters.

With the Inspector General's report in hand, FEMA director James Witt moved forward on his goal to make FEMA the lead agency in emergency and disaster management. With the Federal Response Plan rewritten and clarification made, FEMA has moved forward successfully in using the FRP as a foundation that can be used during all disasters.

5. The Disciplines of Emergency Management: Recovery

INTRODUCTION

There is often a theoretical debate over when the response function ends and the recovery function begins. For this book, the response function is classified as the immediate actions to save lives, protect property, and meet basic human needs. The recovery function is not so easily classified. This function often begins in the initial hours and days following a disaster event and can continue for months and, in some cases, years, depending on the severity of the event.

Unlike the response function, where all efforts have a singular focus, the recovery function or process is characterized by a complex set of issues and decisions that must be made by individuals and communities. Recovery involves decisions and actions relative to rebuilding homes, replacing property, resuming employment, restoring businesses, and permanently repairing and rebuilding infrastructure. The recovery process requires balancing the more immediate need to return the community to normalcy with the longer-term goal of reducing future vulnerability. The recovery process can provide individuals and communities with opportunities to become more economically secure and improve the overall safety and quality of life.

Because the recovery function has such long-lasting effects and usually high costs, the participants in the process are numerous. They include all levels of government, the business community, political leadership, community activists, and individuals. Each of these groups plays a role in determining how the recovery will progress. Some of these roles are regulatory, such as application of state or local building ordinances, and some, such as the insurance industry, provide financial support. The goal of an effective recovery is to bring all the players together to plan, finance, and implement a recovery strategy that will rebuild the disaster-affected area safer and more secure as quickly as possible.

As noted in the previous chapter, the precipitating event for an area affected by a disaster is the presidential declaration of disaster under the Stafford Act. Recovery activities begin immediately after a presidential declaration as the agencies of the federal government collaborate with the state in the affected area in coordinating the implementation of recovery programs and the delivery of recovery services.

In the period of 1990 to 1999, FEMA spent more than $25.4 billion for declared disasters and emergencies compared to $3.9 billion in current dollars for 1980–1989. For the 1990–1999 period, more than $6.3 billion was provided in grants for temporary housing, home repairs, and other disaster-related needs for individuals and families. An additional $14.8 billion went to states and local governments for cleanup and restoration projects, including more than $1.37 billion for mission-assigned work undertaken by other federal agencies. In the 1990s, a total of 88 declarations were issued for hurricanes and typhoons, for which FEMA obligated more than $7.78 billion for disaster costs. The most costly to FEMA was Hurricane Georges in 1998, followed closely by Hurricane Andrew in 1992.

The most frequently declared disaster type was flooding resulting from severe storms, with more than $7.3 billion committed by FEMA for response and recovery costs. The most costly were the Midwest floods in 1993 and the Red River Valley floods in 1997.

By December 2001, the disaster assistance provided by FEMA, the Small Business Administration (SBA), and the state of New York for the September 11, 2001 World Trade Center event had reached $700 million. Recovery costs for this disaster as of December 5, 2001 included the following:

- More than $344 million in public assistance funds to help New York City repair damaged infrastructure, restore critical services, and remove, transport, and sort debris.
- More than $196 million in individual assistance approved in the form of grants and loans. This assistance includes temporary disaster housing assistance, mortgage and rental assistance, disaster food stamps, individual and family grants, and SBA low-interest loans to homeowners and businesses.
- More than $151 million provided through other agencies, including the U.S. Army Corp of Engineers, Disaster Medical Assistance Teams from the Department of Health and Human Services, and FEMA's Urban Search-and-Rescue Task Force

Without a doubt, the federal government plays the largest role in providing the technical and financial support for recovery. For that reason, this chapter focuses on the federal role in the disaster recovery function. It discusses the structure and the various programs available to assist individuals and communities in the postdisaster environment. The various national voluntary organizations that provide some assistance for recovery are briefly referenced, and several case studies are included to demonstrate the different types of recovery.

As noted earlier, the decisions during recovery are predominantly driven by local government. At the end of the chapter is a listing of potential planning tools for the recovery process. This, along with a more encompassing discussion of the complexities of recovery, and roles and responsibilities of the various players in it, can be found in a book prepared for FEMA by the American Planning Association entitled *Planning for Post-Disaster Recovery and Reconstruction*.

THE NATIONAL RESPONSE PLAN FOR DISASTER RECOVERY OPERATIONS

Issued in 2005, the National Response Plan (NRP) outlines how the federal government implements the Robert T. Stafford Disaster Relief and Emergency Assistance Act, as amended, to assist state and local governments when a major disaster or emergency overwhelms their ability to respond effectively. The NRP describes the policies, planning assumptions, concept of operations, response and recovery actions, and responsibilities of 32 federal departments and agencies, including the American Red Cross, that guide federal operations following a presidential declaration of a major disaster or emergency.

The NRP is built on the template of the National Incident Management System (NIMS), which provides a consistent doctrinal framework for incident management at all jurisdictional levels, regardless of the cause, size, or complexity of the incident. The activation of the NRP and its coordinating structures and protocols—either partially or fully—for specific Incidents of National Significance provides mechanisms for the coordination and implementation of a wide variety of incident management and emergency assistance activities. Included in these activities are federal support to state, local, and tribal authorities; interaction with nongovernmental, private donor, and private-sector organizations; and the coordinated, direct exercise of federal authorities, when appropriate.

A fundamental assumption in the NRP is that recovery is a cooperative effort among federal, state, and local governments, voluntary agencies, and the private sector in partnership. A Principal Federal Official (PFO) is designated by the Secretary of Homeland Security to facilitate federal support to the established ICS Unified Command structure, and to coordinate overall federal incident management and assistance activities.

The Response and Recovery Operations Branch coordinates the request and delivery of federal assistance and support from various special teams. This branch is composed of four groups: Emergency Services, Human Services, Infrastructure Support, and Community Recovery and Mitigation.

When established in coordination with state and local jurisdictions, a Disaster Recovery Center (DRC) is a satellite component of the Joint Field Office (JFO—Includes the Federal Coordinating Officer (FCO), State Coordinating Officer (SCO), and other senior federal officials) and provides a central facility where individuals affected by a disaster can obtain information on disaster recovery assistance programs from various federal, state, local, tribal, private-sector, and voluntary organizations.

The JFO is the central coordination point among federal, state, local, and tribal agencies and voluntary organizations for delivering recovery assistance programs. The JFO Operations Section includes the Human Services Branch, the Infrastructure Support Branch, and the Community Recovery and Mitigation Branch. The Human Services and Infrastructure Support Branches of the JFO Operations Section assess state and local recovery needs at the outset of an incident and develop relevant timeframes for program delivery. These branches ensure federal agencies that have relevant recovery assistance programs are notified of an incident and share relevant applicant and damage information with all involved agencies as appropriate,

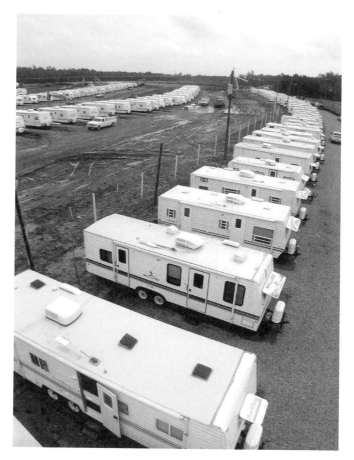

Figure 5-1   Rocky Mount, North Carolina, September 29, 1999. A new life awaits residents whose homes were flooded by the rains from Hurricane Floyd. These manufactured homes, located near Rocky Mount, North Carolina, will house more than 300 families. Photo by Dave Saville/FEMA News Photo.

ensuring that the privacy of individuals is protected. A brief summary of these branches is presented here.

Human Services Branch. Coordinates assistance programs to help individuals, families, and businesses meet basic needs and return to self-sufficiency (see Figure 5-1). This branch also coordinates with volunteer organizations and is involved in donations management, and coordinates the need for and location of DRCs with local and tribal governments. Federal, state, local, tribal, voluntary, and nongovernmental organizations staff the DRCs, as needed, with knowledgeable personnel to provide recovery and mitigation program information, advice, counseling, and related technical assistance.

Infrastructure Support Branch. Coordinates public assistance programs authorized by the Stafford Act to aid state and local governments and eligible private nonprofit organizations with the cost of emergency protective services and the repair or replacement of disaster-damaged public facilities and associated environmental restoration.

Community Recovery and Mitigation Branch. Works with the other operations branches and state and local officials to assess the long-term impacts of an Incident of National Significance, define available resources, and facilitate the development of a course of action to most efficiently apply available resources to restore and revitalize the community as well as reduce the impacts from future disasters.

These branches coordinate with one another to identify appropriate agency assistance programs to meet applicant needs, synchronizing assistance delivery and encouraging incorporation of hazard mitigation measures where possible. Hazard mitigation measures are identified in concert with congressionally mandated, locally developed plans. Hazard mitigation risk analysis; technical assistance to state, local, and tribal governments, citizens, and business; and grant assistance are included within the mitigation framework.

Additionally, these branches work in tandem to track overall progress of the recovery effort, particularly noting potential program deficiencies and problem areas. Long-term environmental recovery may include cleanup and restoration of public facilities, businesses, and residences; re-establishment of habitats and prevention of subsequent damage to natural resources; protection of cultural or archeological sites; and protection of natural, cultural, and historical resources from intentional damage during other recovery operations.

Emergency Support Function #14 (ESF #14). Long-Term Community Recovery and Mitigation provides a framework for federal government support to state, regional, local, and tribal governments, nongovernmental organizations (NGOs), and the private sector designed to enable community recovery from the long-term consequences of an Incident of National Significance. This support consists of available programs and resources of federal departments and agencies to enable community recovery, especially long-term community recovery, and to reduce or eliminate risk from future incidents, where feasible.

Federal disaster assistance available under a major disaster falls into three general categories: Individual Assistance, Public Assistance, and Hazard Mitigation Assistance. Individual Assistance is aid to individuals, families, and business owners. Public Assistance is aid to public and certain private nonprofit entities for emergency services and the repair or replacement of disaster-damaged public facilities. Hazard Mitigation Assistance is funding available for measures designed to reduce future losses to public and private property. A detailed description of the first two types of assistance follows. More information on Hazard Mitigation Assistance can be found in Chapter 3.

FEMA'S INDIVIDUAL ASSISTANCE RECOVERY PROGRAMS

Individual Assistance programs are oriented to individuals, families, and small businesses, and the programs include temporary housing assistance, individual and family grants, disaster unemployment assistance, legal services, and crisis counseling. The disaster victim must first register for assistance and establish eligibility. Three national centers provide centralized disaster application services for disaster victims. FEMA's National Processing Service Centers (NPSCs) are located in Denton, Texas; Berryville, Virginia; and Hyattsville, Maryland.

Since the first national center opened in 1994, more than 2.5 million applications have been processed and 2.8 million calls taken for more than 275 major disasters. These NPSCs house an automated teleregistration service, through which disaster victims apply for Disaster Housing and the Individual and Family Grant program, and through which their applications are processed and their questions answered.

This automated system provides automatic determination of eligibility for about 90 percent of Disaster Housing cases, usually within 10 days of application. The other 10 percent of cases, which may need documentation, take a little longer. Cases are also automatically referred to the state for possible grant assistance if the applicant's needs exceed the Disaster Housing program and the individual cannot qualify for a disaster loan from the Small Business Administration.

Following the September 11 events, FEMA was concerned that many individuals and businesses had not sought help in the aftermath of the attack. Working with the Advertising Council, and a volunteer ad agency, Muezzin Brown & Partners, a public service advertising campaign was developed to let viewers know that assistance was available by calling FEMA's toll-free registration number. The advertisements were distributed to electronic and media outlets in New York, New Jersey, Connecticut, Pennsylvania, and Massachusetts.

Disaster Housing Program

The Disaster Housing Program assures that people whose homes are damaged by disaster have a safe place to live until repairs can be completed. These programs are designed to provide funds for expenses that are not covered by insurance and are

Figure 5-2   South-facing Long Beach on Oak Island, North Carolina, September 17, 1999. Hurricane Floyd brought a devastating 15 feet of storm surge that damaged or destroyed hundreds of houses along this community's oceanfront and flattened its frontal sand dunes. Here, even elevation failed to save this home. Additional strapping, upgraded mitigation work, might have helped. Photo by Dave Gatley/FEMA News Photo.

available to homeowners and renters who are legal residents of the United States and who were displaced by the disaster.

- *Lodging expenses reimbursement* provides a check for reimbursement for the cost of short-term lodging such as hotel rooms, incurred because of damage to a home or an officially imposed prohibition against returning to a home.
- *Emergency minimal repair assistance* provides a check to help repair a home to a habitable condition.
- *Temporary rental assistance* provides a check to rent a place for the predisaster household to live.
- *Mortgage and rental assistance* provides a check to pay the rent or mortgage to prevent evictions or foreclosure. In order to qualify, the applicant must be living in the same house before and after the disaster and have a documented disaster-related financial hardship that can be verified by FEMA.

Individuals and Households Program (IHP)

The Individuals and Households Program (IHP), formerly called the Individual and Family Grant (IFG) Program, provides funds for the necessary expenses and serious needs of disaster victims that cannot be met through insurance or other forms of disaster assistance. The IHP is not designed to cover all of a victim's losses (home, personal property, household goods) that resulted from the disaster, nor is it intended to restore damaged property to its condition before the disaster. Also, the IHP does not cover any business-related losses that resulted from the disaster. By law, the IHP cannot provide any money for losses that are covered by insurance.

The following list illustrates the assistance available through the IHP:

- **Temporary Housing (a place to live for a limited period of time).** Money is available to rent a different place to live, or a government provided housing unit when rental properties are not available.
- **Repair.** Money is available to homeowners to repair damage from the disaster that is not covered by insurance. The goal is to make the damaged home safe, sanitary, and functional.
- **Replacement.** Money is available to homeowners to replace their home destroyed in the disaster that is not covered by insurance. The goal is to help the homeowner with the cost of replacing their destroyed home.
- **Permanent Housing Construction.** Direct assistance or money for the construction of a home. This type of help occurs only in insular areas or remote locations specified by FEMA, where no other type of housing assistance is possible.
- **Other Needs.** Money is available for necessary expenses and serious needs caused by the disaster. This includes medical, dental, funeral, personal property, transportation, moving and storage, and other expenses that are authorized by law.

The IHP covers only repair or replacement of items that are damaged as a direct result of the disaster that are not covered by insurance. Repairs or rebuilding may not improve a victim's home above its predisaster condition unless such improvements are required by current building codes.

- **Housing Needs.** Money to repair a home is limited to making the home "safe and sanitary" so the victim can continue to live there. IHP will not pay to return a home to its condition before the disaster. Grants may be used for housing needs to repair:
  - Structural parts of your home (foundation, outside walls, roof)
  - Windows, doors, floors, walls, ceilings, cabinetry
  - Septic or sewage system
  - Well or other water system
  - Heating, ventilating, and air conditioning system
  - Utilities (electrical, plumbing, and gas systems)
  - Entrance and exit ways from your home, including privately owned access roads
  - Blocking, leveling, and anchoring of a mobile home and reconnecting or resetting its sewer, water, electrical and fuel lines, and tanks
- **Other than Housing Needs.** Money to repair damaged personal property or to pay for disaster-related necessary expenses and serious needs is limited to items or services that help prevent or overcome a disaster-related hardship, injury, or adverse condition. Grants may be used to pay for:
  - Disaster-related medical and dental costs
  - Disaster-related funeral and burial cost
  - Clothing, household items (room furnishings, appliances), tools (specialized or protective clothing and equipment) required for a job, necessary educational materials (computers, school books, supplies)
  - Fuels for primary heat source (heating oil, gas, firewood)
  - Clean-up items (wet/dry vacuum, air purifier, dehumidifier)
  - Disaster damaged vehicle
  - Moving and storage expenses related to the disaster (moving and storing property to avoid additional disaster damage while disaster-related repairs are being made to the home)
  - Other necessary expenses or serious needs as determined by FEMA

Money received from IHP for Housing and Other than Housing Needs must be used for eligible expenses only, as identified by FEMA. If a grantee does not use the money for the reasons defined in the grant application, they may not be eligible for any additional help and may have to return any grant money provided. Grant money:

- Is usually limited to up to 18 months from the date the President declares the disaster.
- Does not have to be repaid.
- Is tax-free.
- Is not counted as income or a resource in determining eligibility for welfare, income assistance, or income-tested benefit programs funded by the federal government.
- Is exempt from garnishment, seizure, encumbrance, levy, execution, pledge, attachment, release, or waiver.
- May not be reassigned or transferred to another person.

FEMA pays 100% of the Housing portion of the grant, and 75% of the Other Needs portion. The state pays the remaining 25% of the Other Needs portion. The states may administer only the Other Needs portion of the grant. The total maximum amount of grant assistant for each family or individual in fiscal year 2005 is $25,000, and this amount is broken down further into the various types of assistance provided. For example, although up to $25,000 may be provided for home repairs, a maximum of $10,000 will be provided for replacement of "owner occupied private residences."

Although some money often is made available through the IHP, most disaster aid from the federal government is provided in the form of loans from the Small Business Administration (SBA), that must be repaid. Applicants to IHP may be required to seek help from SBA first, before being considered for certain types of IHP help.

The SBA can provide three types of disaster loans to qualified homeowners and businesses to repair or replace homes, personal property, or businesses that sustained damages not covered by insurance.

- **Home disaster loans** provide funds to homeowners and renters to repair or replace disaster-related damages to home or personal property.
- **Business physical disaster loans** provide funds to business owners to repair or replace disaster-damaged property, including inventory, and supplies.
- **Economic injury loans** provide capital to small businesses and to small agricultural cooperatives to assist them through the disaster recovery period. If the SBA determines that the individual is ineligible for a loan, or if the loan amount is insufficient to meet the individual's needs, then the applicant is referred to the IFG program.

Source: www.fema.gov

Disaster Unemployment Assistance

The Disaster Unemployment Assistance (DUA) program provides unemployment benefits and reemployment services to individuals who have become unemployed because of major disasters, and who are not eligible for disaster benefits under regular unemployment insurance programs.

Legal Services

The Young Lawyers' Division of the American Bar Association, through an agreement with FEMA, provides free legal assistance to low-income disaster victims. The assistance that the participating lawyers provide is for insurance claims; counseling on landlord/tenant problems; assistance in consumer protection matters, remedies, and procedures; and replacement of wills and other important legal documents destroyed in a major disaster. This assistance is intended for individuals who are unable to secure legal services adequate to meet their needs as a consequence of a major disaster.

Special Tax Considerations

Taxpayers who have sustained a casualty loss from a declared disaster may deduct that loss on the federal income tax return for the year in which the casualty occurred or through an immediate amendment to the previous year's return. Businesses may file claims with the Bureau of Alcohol, Tobacco, and Firearms (ATF) for payment of federal excise taxes paid on alcoholic beverages or tobacco products lost, rendered unmarketable, or condemned by a duly authorized official under various circumstances, including where a major disaster has been declared by the President.

Crisis Counseling

The Crisis Counseling Assistance and Training Program is designed to provide short-term crisis counseling services to people affected by a presidentially declared disaster. The purpose of the crisis counseling is to help relieve any grieving, stress, or mental health problems caused or aggravated by the disaster or its aftermath. These short-term services are provided by FEMA as supplemental funds granted to state and local mental health agencies. The American Red Cross, the Salvation Army, and other voluntary agencies as well as churches and synagogues also offer crisis counseling services.

Cora Brown Fund

Cora C. Brown of Kansas City, Missouri, died in 1977 and left a portion of her estate to the United States to be used as a special fund solely for the relief of human suffering caused by natural disasters. The funds are used to assist victims/survivors of presidentially declared major disasters for disaster-related needs that have not or will not be met by government agencies or other organizations.

FEMA'S PUBLIC ASSISTANCE GRANT PROGRAMS

FEMA, under the authority of the Stafford Act, administers the Public Assistance Program. The Public Assistance Grant Program provides federal assistance to state and local governments and to certain private nonprofit (PNP) organizations. These grants allow them to recover from the impact of disasters and to implement mitigation measures to reduce the impacts from future disasters. The grants are aimed at governments and organizations with the final goal to help a community and its citizens recover from devastating major disasters. The federal share of assistance is not less than 75 percent of the eligible cost for emergency measures and permanent restoration. The state determines how the nonfederal share is split with the applicants.

Eligible applicants include the states, local governments, and any other political subdivision of the state, Native American tribes, Alaska Native Villages, and certain PNP organizations. Eligible PNP facilities include educational, utility, irrigation, emergency, medical, rehabilitation, temporary or permanent custodial care facilities, and other PNP facilities that are open to the public and provide essential services of a governmental nature to the general public. The work must be required as the result of the disaster, be located within the designated disaster area, and be the legal responsibility of the applicant. PNPs that provide critical services such as power, water, sewer, wastewater treatment, communications, or emergency medical care may apply directly to FEMA for a disaster grant. All other PNPs first must apply to the SBA for a disaster loan. If the loan is declined or does not cover all eligible damages, the applicant may reapply for FEMA assistance.

Work that is eligible for supplemental federal disaster grant assistance is classified as either emergency work or permanent work:

- *Emergency work* includes debris removal from public roads and rights-of-way as well as from private property when determined to be in the public interest. This may also include protective measures performed to eliminate or reduce immediate threats to the public.
- *Permanent work* is defined as work that is required to restore an eligible damaged facility to its predisaster design. This effort can range from minor repairs to replacement. Some categories for permanent work include roads, bridges, water control facilities, buildings, utility distribution systems, public parks, and recreational facilities. With extenuating circumstances the deadlines for emergency and permanent work may be extended.

As soon as possible after the disaster declaration, the state, assisted by FEMA, conducts the Applicant Briefings for state, local, and PNP officials to inform them of the assistance that is available and how to apply for it. A Request for Public Assistance must be filed with the state within 30 days after the area is designated eligible for assistance. A combined federal, state, and local team works together to design and deliver the appropriate recovery assistance for the communities. In determining the federal costs for the projects, private or public insurance can play a major role. For insurable buildings within special flood hazard areas (SFHAs) and damaged by floods, the disaster assistance is reduced by the amount of insurance settlement that would have been received if the building and its contents had been fully covered by a standard NFIP policy. For structures located outside of an SFHA, the amount is reduced by the actual or anticipated insurance proceeds.

In 1998, FEMA redesigned the Public Assistance program to provide money to applicants more quickly and to make the application process easier. The redesigned program was approved for implementation on disasters declared after October 1, 1998. This redesigned program placed new emphasis on people, policy, process, and performance. The focus of the program was also modified to provide a higher level of customer service for disaster recovery applicants and to change the role of FEMA from inspection and enforcement to an advisory and supportive role.

OTHER FEDERAL AGENCY DISASTER RECOVERY FUNDING

Other federal agencies have programs that contribute to social and economic recovery. Most of these additional programs are triggered by a presidential declaration of a major disaster or emergency under the Stafford Act; however, the Secretary of

Agriculture and the Administrator of the SBA have specific authority relevant to their constituencies to declare a disaster and provide disaster recovery assistance. All the agencies are part of the structure of the NRP. This section does not provide a complete list of all disaster recovery programs available after a disaster declaration, but provides a summary of many of the federal agencies in addition to FEMA that provide disaster recovery programs. These agencies include the following:

- U.S. Army Corps of Engineers
- Department of Housing and Urban Development
- Small Business Administration
- U.S. Department of Agriculture
- Department of Health and Human Services
- Department of Transportation
- Department of Commerce
- Department of Labor

A more comprehensive list is available in the Catalog of Federal Domestic Assistance (CFDA), available through the Federal Assistance Programs Retrieval System. Each automated edition is revised in June and December.

U.S. Army Corps of Engineers

In a typical year, the Corps of Engineers responds to more than 30 presidential disaster declarations, plus numerous state and local emergencies. Under the NRP, the Army has the lead responsibility for public works and engineering missions. For example, after the events of September 11, 2001, the Army provided technical assistance for the debris removal operation. By December 2001, more than 661,430 tons of debris had been moved to the Staten Island landfill.

Department of Housing and Urban Development

The Department of Housing and Urban Development (HUD) provides flexible grants to help cities, counties, and states to recover from presidentially declared disasters, especially in low-income areas, subject to availability of supplemental appropriations. When disasters occur, Congress may appropriate additional funding for the Community Development Block Grant (CDBG) and HOME programs to rebuild the affected areas and bring crucial seed money to start the recovery process. Because it can fund a broader range of recovery activities than most other programs, CDBG disaster recovery assistance supplements recovery assistance from FEMA and helps communities and neighborhoods that otherwise might not recover because of limited resources.

The CDBG program funds have been especially useful to communities that are interested in incorporating mitigation into their recovery process. These funds have been combined with FEMA assistance to remove or elevate structures from the flood plain and to relocate residents and businesses to safer areas.

The HOME Program helps expand the supply of decent, affordable housing for low- and very low-income families by providing grants to states and local govern-

Figure 5-3   New York, New York, October 30, 2001. FEMA/NY State Disaster Field Office personnel meet to coordinate federal, state, and local disaster assistance programs. Photo by Andrea Booher/FEMA News Photo.

ments. Funds can be used for acquisition, new construction, rehabilitation, and tenant-based rental assistance. HOME disaster recovery grants are an important resource for providing affordable housing to disaster victims.

Small Business Administration

The SBA Disaster Loan Program offers low-interest loans to assist in long-term recovery efforts for those who are trying to rebuild their homes and businesses in the aftermath of a disaster. Disaster loans from SBA help homeowners, renters, businesses of all sizes, and nonprofit organizations fund rebuilding efforts. The SBA Disaster Loan Program reduces federal disaster costs compared to other forms of assistance, such as grants, because the loans are repaid to the U.S. Treasury. The SBA can approve loans only to applicants who have a reasonable ability to repay the loan and other obligations from earnings. The terms of each loan are established in accordance with each borrower's ability to repay. Generally, more than 90 percent of the SBA's disaster loans are made to borrowers without credit available elsewhere and have an interest rate of around 4 percent. The disaster loans require borrowers to maintain appropriate hazard and flood insurance coverage, thereby reducing the need for future disaster assistance.

The SBA is authorized by the Small Business Act to make two types of disaster loans: physical disaster loans and economic injury disaster loans. Physical disaster loans are a primary source of funding for permanent rebuilding and replacement of uninsured disaster damages to privately owned real and/or personal property. Economic injury disaster loans provide necessary working capital until normal operations resume after a physical disaster.

In 2000, the SBA approved 28,218 loans for $1.028 billion. Since the inception of the program in 1953, the SBA has approved more than 1.5 million disaster loans for more than $28.5 billion. In 2001 after the September 11 events, the SBA approved more than $161 million in low-interest loans to more than 2,000 applicants for home repairs, business loans, and loans to assist small businesses suffering economic injury as a result of losses caused by the disaster.

U.S. Department of Agriculture

The U.S. Department of Agriculture (USDA) Farm Service Agency (FSA) provides low-interest loan assistance to eligible farmers and ranchers to help cover production and physical losses in counties declared as disaster areas by the President or designated by the Secretary of Agriculture. The emergency loans can be used to restore or replace essential physical property, pay all or part of production costs associated with the disaster year, pay essential family living expenses, reorganize the farming operation, and refinance debts.

Department of Health and Human Services

The Department of Health and Human Services (DHHS) is the lead federal agency responsible for implementing the health and medical portion of the NRP. Their activities provide support to individuals and communities affected by disasters, state and local mental health administrators, and other groups that respond to those affected by human-caused disasters (such as school violence). The Center for Mental Health Services (CMHS) within the DHHS works with FEMA to implement the Crisis Counseling Assistance and Training Program discussed earlier in this chapter.

The DHHS also provides disaster assistance for older Americans through its Administration on Aging (AoA). Older people often have difficulty obtaining necessary assistance because of progressive physical and mental impairments and other frailties that often accompany aging. Many older people, who live on limited incomes, and are sometimes alone, find it impossible to recover from disasters without special federal assistance service. The AoA's national aging network assists older persons by providing critical support such as meals and transportation, information about temporary housing, and other important services on which older adults often rely.

Department of Transportation

Congress authorized a special program from the Highway Trust Fund for the repair or reconstruction of federal-aid highways and roads on federal lands that have suffered serious damage as a result of natural disasters or catastrophic failures from an external cause. The Department of Transportation (DOT) Federal Highway Administration (FHWA) administers the Emergency Relief Program, which supplements the commitment of resources by states, their political subdivisions, or other federal agencies to help pay for damages resulting from disasters. The applicability of the program to a natural disaster is based on the extent and intensity of the disaster.

Department of Commerce

Within the Department of Commerce, the Economic Development Administration (EDA) administers programs and provides grants for infrastructure development, business incentives, and other forms of assistance designed to help communities alleviate conditions of substantial and persistent unemployment in economically distressed areas and regions. The EDA provides postdisaster economic assistance for communities affected by declared natural disasters. Funding for this program has been a problem over the years.

Department of Labor

The Department of Labor (DOL) Disaster Unemployment Assistance (DUA) Program provides financial assistance to individuals whose employment or self-employment has been lost or interrupted as a direct result of a major disaster and who are not eligible for regular state unemployment insurance. Funding for this program comes from FEMA. The DUA is administered by the state agency responsible for providing state unemployment insurance.

The Workforce Investment Act of 1998 authorizes the U.S. Secretary of Labor to award National Emergency Grants to assist any state that has suffered an emergency or major disaster to provide disaster relief employment. These funds can be used to finance the creation of temporary jobs for workers dislocated by disasters to clean up and recover from the disaster, and to provide employment assistance to dislocated workers. Interestingly, in creating this program, Congress expanded eligibility beyond people affected by the disaster to dislocated workers and certain civilian Department of Defense employees affected by downsizing and certain recently separated members of the armed forces.

NATIONAL VOLUNTARY RELIEF ORGANIZATIONS

National Voluntary Organizations Active in Disaster (NVOAD) coordinates planning efforts by many voluntary organizations responding to disaster in order to provide more effective service to people affected by disaster. Members include 34 national voluntary organizations active in disaster mitigation and response, 52 state and territorial chapters (VOADs), and dozens of local organizations. Once a disaster occurs, NVOAD or an affiliated state VOAD encourages members and other voluntary agencies to convene on site. The member organizations provide a wide variety of disaster relief services, including emergency distribution services, mass feeding, disaster child care, mass or individual shelter, comfort kits, supplementary medical care, cleaning supplies, emergency communications, stress management services, disaster assessment, advocacy for disaster victims, building or repair of homes, debris removal, mitigation, burn services, guidance in managing spontaneous volunteers, and victim and supply transportation. NVOAD maintains a close relationship with FEMA and encourages the state and local affiliates to work closely with the state and local emergency management agencies.

The American Red Cross

Although the American Red Cross is not a government agency, its authority to provide disaster relief was formalized when, in 1905, the Red Cross was chartered by Congress to "carry on a system of national and international relief in time of peace and apply the same in mitigating the sufferings caused by pestilence, famine, fire, floods, and other great national calamities, and to devise and carry on measures for preventing the same." Red Cross disaster relief focuses on meeting people's immediate emergency disaster-caused needs and provides disaster assistance to individuals to enable them to resume their normal daily activities independently. The Red Cross provides shelter, food, and health and mental health services to address basic human needs. The Red Cross also feeds emergency workers, handles inquiries from concerned family members outside the disaster area, provides blood and blood products to disaster victims, and helps those affected by disaster to access other available resources.

The Red Cross is one of the nongovernmental organizations mentioned in the NRP and is the primary agency for ESF #6, Mass Care, Housing, and Human Services. The Red Cross functions as a federal agency in coordinating the use of federal mass care resources in a presidentially declared disaster or emergency, and coordinates federal assistance in support of state and local efforts to meet the mass care needs of victims of a disaster. This federal assistance supports the delivery of mass care services of shelter, feeding, and emergency first aid to disaster victims; the establishment of systems to provide bulk distribution of emergency relief supplies to disaster victims; and the collection of information to operate a Disaster Welfare Information system to report victim status and assist in family reunification.

RECOVERY PLANNING TOOLS

Despite the pressures on politicians and community leaders to return to a period of normalcy as quickly as possible and because of federal incentives, public interest, and insurance retractions, more and more communities are looking at ways to reduce their future vulnerability. As disasters repeat themselves and the public sees the emotional and financial benefits of mitigation, communities are making the long-term investment in mitigation. For example, the devastating 1993 Midwest floods that occurred again in some areas in 1995 had a minimal impact in those towns where buyout and relocation programs were undertaken after the 1993 flood. The following is a partial list of policy areas and tools that should be considered by decision makers as they develop their recovery plan:

- *Land-use planning techniques*, including acquisition, easements, annexation, stormwater management, and environmental reviews

- *Zoning*, including special-use permits, historic preservation, setbacks, density controls, wetlands protection, floodplain, and coastal zone management
- *Building codes*, including design controls, design review, height and type, and special study areas (soil stability ratings)
- *Financial*, including special districts, tax exemptions, special bonds, development rights, property transfer, or use change fees
- *Information and oversight*, including public awareness and education, regional approaches and agreements, global information systems, town hall meetings, and public hearings

CONCLUSION

As this chapter demonstrates, the federal government plays a significant role in initiating and funding the disaster recovery process. But for recovery to be effective, the planning and decision making must be done at the local level. With a disaster comes disruption and tragedy, but in the aftermath comes opportunity. Changes to FEMA's Stafford Act now require communities and states to have mitigation plans approved before the disaster. These plans, developed in the calm before an event happens, can become the blueprint for facilitating recovery and making communities less vulnerable in the postdisaster environment. Communities should strive to integrate pre-event recovery and mitigation planning into their ongoing planning efforts. Such integration will allow for the political process to work, to include citizen participation, and to garner support for changes that will make their communities safer and more secure.

CASE STUDIES

ECONOMIC RECOVERY IN NEW YORK CITY AFTER SEPTEMBER 11, 2001

Prior to September 11, the World Trade Center was the heart of a vibrant downtown business district. The massive complex consisted of seven buildings, including the twin World Trade Center Towers. These 110 story skyscrapers, built in 1970 by the New York Port Authority, contained nearly an acre of space on each floor. Combined, they represented 12 million square feet of office space—14% of the office space in downtown Manhattan—and were the home to 50,000 employees. Together with the other buildings destroyed or damaged on that date, over 25% of the commercial office space in lower Manhattan was immediately uninhabitable.

The economic impact of the attack was immediate and severe. In addition to their physical space, many companies lost all or a large percentage of their work force and operational equipment. The transportation system upon which employees depended was destroyed. The nation's financial system shut down, and air travel was suspended. Shipping ground to a halt, and companies that relied on just-in-time products for production were left without many necessary parts. TV and radio stations lost advertising revenue as reports of the attack went commercial-free for days. Consumer spending and confidence were devastated, and didn't return for weeks. And the insurance industry, heavily invested in the city, realized the costliest single event in its history.

The exact financial impact figures related to the attack are still hard to obtain, as the means to measure them are not standardized. The human casualty figure, in flux for months, was finally set at 2,749. The economic figures are much more amorphous, due to a number of reasons. First, the economy is dynamic, and was affected by several other factors such as the recession that was ongoing, the various scandals (Enron, WorldCom), among other issues. Second, the recovery effort is still underway, and costs related to it will likely remain open until as late as 2015, when all the WTC construction is scheduled to be completed. Insurance payments are still outstanding, and the federal government still has money yet to be allocated.

The economic recovery from the WTC attacks started immediately. To limit the immediate impact on shareholder confidence, the New York financial markets were shut for a period of several days. The Federal Reserve bolstered the system by preparing to inject liquidity into the system to prevent defaults, and interest rates for short-term borrowing were lowered. The Fed also ensured the availability of U.S. dollars overseas, and Congress supported U.S. airlines with $10 billion in guaranteed loans. After electricity and communication were restored, just a few days later, the markets were ready to open and begin returning to normal operations.

These initial actions, however, were superficial, intended to limit the extent of damage that had already been sustained. It was apparent from the start that a much greater amount of recovery actions would be needed in the years to follow. Numerous organizations, governmental agencies, and other groups have participated in this recovery, several of which are profiled next.

FEMA

The response to the attacks on the WTC marked a significant change in the way in which FEMA allocated funds. In a "normal" disaster, FEMA first determines the needs as defined by established eligibility criteria, and then distributes funds from its general disaster relief fund. Congress does not give money for a specific disaster; rather, they allocate money to this pool, from which FEMA operates. There is no predefined upper limit for the disaster; as a result, disaster funding projects can be open for years after the event occurs (events related to the Northridge earthquake, for example, were still being funded nearly 10 years after the event.

In this incident, however, the amount of money that was to be allocated was established early in the process. FEMA received $8.80 billion of the $20 billion in federal funds allocated by Congress, and FEMA was given enhanced flexibility in determining how the money should be used. This strategy allowed FEMA to establish an early close-out process, forcing the city and state to establish priorities early on. It also allowed FEMA to distribute funds in ways that normally would not have been possible under the Stafford Act, such that all of the $8.80 billion would be allocated. This flexibility has also been vital to the economic recovery of the area. It has gone beyond simply getting people back on their feet, to helping Lower Manhattan reestablish itself as core of the New York City economy. FEMA funds have been used to assist owners with the cleaning of WTC dust from their private residences, reimburse the city from losses associated with a reduction in tourism, pay for increased security as a result of the attacks, and fund cost-of-living allowances for the beneficiaries of the pensions of the firemen and police officers killed in the attack.

HUD

The Department of Housing and Urban Development is responsible for the second largest allocation of funds to the WTC site. HUD funds were used to reimburse utility companies for emergency repairs immediately after the attacks. They assisted both individuals and businesses with compensation for disaster-related losses, through mortgage and rental insurance, crisis counseling, grants for disaster-related expenses, and businesses recovery grants and loans. HUD has also been instrumental in both the infrastructure and economic recovery of the WTC site. It has spent $568 million to not only return the utility infrastructure of the site to normal, but to improve it. HUD's Community Development Block Grant (CDBG) has been used to fund several programs, among them the Small Firm Attraction and Retention Grant Program (SFARG), the Job Creation and Retention Program (JCRP), the Employee Training and Assistance Program (ETAP), and the Business Recovery Loan Fund. These funds have been vital to retaining the businesses that make up the economic heart of lower Manhattan.

DOT

The U.S. Department of Transportation has been involved with the effort to rebuild and improve the transportation systems damaged and destroyed at the WTC site. Because of the large number of workers that commute there, having a robust and efficient system is vital to the economic recovery of the site. DOT has been involved in restoring operation to the transportation systems and providing temporary repairs to the roads during the response phase. It is now involved in the permanent replacement of the Port Authority Trans-Hudson (PATH) terminal, and improvements to the Fulton Street Transit Center and South Ferry Subway Station.

IRS

As part of the $20 billion package allocated for New York City, Congress approved the Liberty Zone tax benefit, worth approximately $5 billion. This amount is not money provided by the government; rather, it is a tax break targeted specifically to companies surrounding the WTC site in lower Manhattan, deemed the Liberty Zone. Among its provisions are a business employee credit, special depreciation allowance, tax-exempt private activity bonds (Liberty Bonds), and increased expensing. Some of these breaks have already expired, whereas others will continue on for several more years. The $5 billion figure is an estimate, and the IRS is not tracking the actual usage of these benefits.

State and Local: Empire State Development Corporation

The State of New York's economic development corporation is aiding in the economic recovery of the region through its NY Incentives program, designed to help small business owners realize the benefits of doing business in the area by assisting with the various economic incentive programs.

Lower Manhattan Development Corporation

The Lower Manhattan Development Corporation (LMDC) is a state-city corporation designed to oversee the redevelopment and improvement of the WTC site and the entire lower Manhattan area. It was created shortly after the attack by Mayor Giuliani and Governor Pataki, and consists of eight board members appointed by the state, and eight appointed by the city. It consults with citizen groups on issues such as transportation and infrastructure, residential and commuter concerns, economic development, tourism and the arts, and memorial planning. It approved the plans for the rebuilding of the WTC site, and the included memorial site. Most importantly, it is in charge of channeling the funds received from the federal government.

Port Authority

The Port Authority of New York and New Jersey was founded in 1921 to enhance regional commerce and transportation in the New York City metropolitan area. It is a 12-member board, with six members appointed by the governor of each state. The Port Authority built the World Trade Center in 1970 and owned it until July, 2001, when it leased it to a private party. It owns the land today, and is working closely with the LMDC to rebuild the World Trade Center and its transportation infrastructure.

Other Agencies

Numerous other agencies are involved in the rebuilding of the WTC site and lower Manhattan. Among them are the Metropolitan Transit Authority, NYC Planning Commission, NYC DOT, NYC Dept. of Environmental Protection, and NYC Economic Development Corporation. Local community groups, arts societies, architects, and regional planning associations are also involved.

Insurance

Many of the insurance claims from the WTC attack have yet to be settled. Estimates of actual payout range from $30 to $70 billion, depending upon the estimate source and date. The two World Trade Center towers, each insured for $3.5 billion, were reimbursed for only $3.5 billion total because the two attacks were considered to be part of a single event. For the insurance industry as a whole, this attack was a watershed event. Insurance companies normally operate with thin profit margins and a reliance on actuary tables to determine the likelihood of events, but acts of terrorism are potentially bankrupting and nearly impossible to predict. The answer from the U.S. government has been the passage of the Terrorism Risk Insurance Act of 2003, which provides federal sharing of public and private compensation for insurance of commercial property.

Charitable Contributions

Although charity is present at most disasters, it was especially prevalent in the WTC disaster, especially in regards to funds collected for victims and victim's families. An estimated 600 charities registered with the IRS with the explicit intention of collecting funds related to the disaster. The top 35 of these funds had collected nearly $2.7 billion by October 2002. The largest of these, the American Red Cross Liberty Fund, had collected over $1 billion dollars. (In addition to the funds collected, the American Red Cross served an estimated 11.5 million meals and provided 50,423 disaster workers in the first two months of the disaster.)

Of the money collected by the charities, over 70 percent had been distributed by October 2002. Much of the money went to victims' families, in an effort to recoup lost salaries. The GAO reports that the average nonuniformed victims' families received $90,000 in cash assistance, and uniformed families, because of charities established especially for them, received an average assistance of $715,000 (Port Authority police), $905,000 (NYC police), and $938,000 (NYC firefighters). Other examples of areas where charities donated money to help include mental health counseling, health care provision, employment assistance, and legal and financial help.

Sources:

Cooper, J.C., & Madigan, K. 2002. "Consumers have done their part. Now, businesses will have to pitch in as faster recovery requires more capital spending and hiring," *Business Week*, September 16, **3799**, 19.

Fiscal Policy Institute. 2001. Economic impact of the September 11 World Trade Center attack, preliminary report, September.

Lowenstein, Ronnie. 2002. Federal aid to New York City in the aftermath of September 11th: How much, and for what? February 11. *Testimony of Ronnie Lowenstein, Director, New York City Independent Budget Office before the Joint Hearing of the City Council Finance, Lower Manhattan Redevelopment, and State and Federal Legislation Committees.*

New York Senate Finance Committee. 2002. Financial impact of the World Trade Center attack, January.

Office of the State Comptroller. 2002. The impact of the World Trade Center tragedy on the Metropolitan Transit Authority. Report 9–2002.

Kadlec, D., Baumohl, B., Sieger, M., & Zagorin, A. 2001. "Up from the ashes," *Time*, September 24, V. 158, **13**, 80–82.

Lenain, P., Bonturi, M., & Keon, V. 2002. "The fallout from terrorism: Security and the economy," *Organisation for Economic Cooperation and Development. The OECD Observer*, May, 2002, **231/232**, 9–13.

Civic Alliance to Rebuild Downtown New York. 2002. Listening to the city. February 7.

Loomis, C.J. 2002. "Insurance after 9/11," *Fortune*, June 10. V. 145, **12**, 110–115.

Malik, A. 2002. World Trade Center. *GBER*. V. 2, **1**, 5–10.

GAO. 2002. More effective collaboration could enhance Charitable Organizations' Contributions in disasters. December. GAO-03-529.

Wall Street Journal. 2001. "New York City expected to recoup most losses of Sept. 11," November 16, A2.

GAO. 2003. Overview of federal disaster assistance to the New York City area, October, GAO-04-72.

Ratajczak, D. 2002. "Will insurance be a buffer or a drag to the economy during this recovery?" *Journal of Financial Service Professionals*, May, V. 56, **3**, 18–20.

Wikipedia.Com. N/D. Rescue and recovery effort after the September 11, 2001 attacks. At www.wikipedia.com.

GAO. 2002. Review of studies of the economic impact of the September 11, 2001, terrorist attacks on the world trade center, May, GAO-02-700R.

Starkman, D. & Frangos, A. 2004. "Before Ground Zero rebuilding, $1.3 billion has already been spent," *Wall Street Journal*, February 25, A1.

The Economist. 2003. "United States: two years on; New York City," *The Economist*, September 13, V. 368, **8341**.

NYCP and CC. 2002. Working together to accelerate New York's recovery: Update of the NYC parternship's economic impact analysis of the September 11th attack on New York City, February 11. *New York City Partnership and Chamber of Commerce*.

New York City Independent Budget Office. 2003. "World Trade Center assistance: aid received, aid to come," *New York City Independent Budget Office: Inside the Budget*, July 31, 117.

New York City Independent Budget Office. 2001. "World Trade Center disaster: tracking federal aid for cleanup and rebuilding," *New York City Independent Budget Office: Inside the Budget*, September 28, **89**.

NYCEDC. 2003. "WTC business assistance report, *New York City Economic Development Corporation and New York State Empire State Development Corporation*, February 13.

LMDC. 2003. World Trade Center memorial and cultural program, amended general project plan, September 16. *Lower Manhattan Development Corporation*.

FEDERAL ACTION PLAN FOR THE RED RIVER VALLEY FLOODS

In April 1997, the Red River flooded its banks, displacing more than 60,000 people and affecting communities especially hard in Grand Forks, North Dakota and East Grand Forks, Minnesota. On April 7 and 8, a presidential disaster was declared for severe spring storm conditions in North Dakota, South Dakota, and Minnesota. Following an April 22 visit to these communities, the President announced that all emergency measures and costs of debris removal under the Stafford Act would be covered 100 percent by the federal government so that the state and local governments could concentrate their resources on response and recovery efforts. The President also announced the formation of an interagency task force to develop a long-term recovery plan for the affected states. James Lee Witt, the director of FEMA at that time, chaired the effort.

The Federal Action Plan for Recovery identified three priorities for federal long-term recovery efforts: mitigation of flood hazards, housing, and reestablishing community sustainability. In conjunction with state and local governments, the action plan detailed a wide range of grants, loans, and technical assistance that the federal government would provide to ensure that community recovery needs were addressed. The President also ordered several federal departments to implement efforts to make the communities more disaster resistant. He directed the U.S. Army Corps of Engineers to aggressively pursue the development and implementation of structural and nonstructural flood protection works for the cities of Grand Forks and East Grand Forks.

FEMA and HUD were directed, in partnership with the states, to implement an accelerated program to purchase flood-damaged residences in the most severely devastated areas. FEMA, HUD, the Army Corps, the Economic Development Administration (EDA), and the SBA were directed to use all available authorities to support state and local rebuilding efforts and to incorporate mitigation to make the communities disaster resistant. The President also asked the affected communities to vigorously pursue mitigation and to manage development wisely to avoid future flooding events. He encouraged the residents of these communities to purchase and maintain flood insurance.

To address the issue of immediate and long-term housing availability and to maintain community continuity during the recovery process, the President directed FEMA to continue providing temporary housing on an expedited basis by providing emergency home repair grants, travel trailers to be sited next to unlivable damaged residences under repair, mobile homes for those facing longer-term displacement, and rental assistance. HUD, the Department of Commerce, the EDA, the USDA, and the SBA were directed to establish a recovery office in Grand Forks to help the communities create new housing resources through planning and design assistance, infrastructure funding, and continued low-interest loans to homes and businesses.

The Recovery Action Plan also addressed the challenge of reestablishing the sustainability of the community through preserving historic downtown and residential areas, attracting and retaining a workforce, building and repairing infrastructure and housing, and capitalizing small businesses. To help meet these challenges, the President directed HUD, the EDA, the SBA, the Army Corps, the USDA, FEMA, and the Department of Energy (DOE) to provide short-term and long-term planning and technical assistance to the communities most impacted. The SBA, HUD, the EDA, the USDA, and FEMA were directed to continue to make low-interest loans and targeted grants to support development of new business facilities, assist in relocation of businesses away from highly hazardous areas, stimulate private-sector investment, and address reestablishment and relocation of critical facilities, including water treatment plants. HUD, the SBA, the USDA, and the EDA were also directed to actively seek innovative solutions to the short-term capitalization of businesses, in particular small businesses.

The President directed FEMA to provide temporary classroom and administration facilities for schools and to support the communities' efforts in the design and siting of new construction of schools away from high-risk areas. FEMA was also directed to continue the repair, restoration, and mitigation of damaged infrastructure, including roads, bridges, hospitals, and other public and private nonprofit facilities.

Other agencies also helped address the immediate disaster recovery needs of these three states after the floods. The Department of Labor made nearly $10 million available under the Job Training Partnership Act Title III program to provide temporary jobs for disaster-affected workers in the three states. The Centers for Disease Control and Prevention (CDC) from the DHHS provided emergency assistance to the affected areas on environmental health, disease and injury surveillance, worker safety, and water quality. The Federal Highway Administration (FHA) allocated emergency funds to repair highways. The Environmental Protection Agency (EPA) provided technical assistance to the states on solid waste, pesticides, household hazardous waste, air monitoring, and underground storage tank issues.

In regard to these actions, James Lee Witt stated: "The Long Term Recovery Task Force developed recommendations that transcend our usual disaster programs. This innovative effort demonstrates the federal government's true commitment to the long-term recovery of communities in the three states deluged by the Red River of the North. In addition to helping these communities recover, we are committed to assisting state and local governments with the task of rebuilding safer and smarter to reduce future flood risks."

LONG-TERM RECOVERY ACTION PLAN FOR
HURRICANE GEORGES

On September 21, 1998, Hurricane Georges, sustaining winds as high as 150 miles per hour, struck Puerto Rico and dumped more than two feet of rain on the island. More than 100,000 residences were damaged or destroyed, and 31,500 people were forced to seek refuge in shelters. This was the worst natural disaster to hit Puerto Rico in 70 years, and a major disaster was declared for all 78 of Puerto Rico's municipalities. In response to the severity and scope of the destruction, the President activated the Long-Term Recovery Task Force composed of 15 federal departments, agencies, and offices, and headed by then FEMA director James Lee Witt. The President directed the group to develop an action plan to facilitate the coordination and delivery of federal recovery assistance to Puerto Rico.

The purpose of the Task Force is to coordinate and target the diverse disaster programs of more than a dozen federal agencies to ensure the greatest level of effective federal support. The Task Force worked in collaboration with representatives of the government of Puerto Rico to identify five long-term recovery priorities: mitigation, housing, economic revitalization and sustainability, energy, and transportation.

The government of Puerto Rico identified **mitigation** as one of the core elements of its vision for long-term recovery. Federal mitigation actions emphasized three areas: building codes, planning and coordination, and floodplain management. FEMA provided technical assistance for developing long-term strategies to reduce losses in future disasters and provided funding under the Hazard Mitigation Grant Program. The federal government also worked with Puerto Rico to acquire property and elevate structures in the floodplain. The U.S. Army Corp of Engineers worked with Puerto Rico to identify funding for and expedite construction of flood control projects.

Federal assistance for **housing** focused on repairing existing homes, addressing long-term shelter needs, replacing destroyed homes, restoring public housing, and providing technical assistance and training. FEMA provided funding assistance under the Disaster Housing Assistance program and the Individual and Family Grant program. Additional funding was provided through the SBA Home Disaster Loans and the USDA Rural Housing Service. HUD provided disaster funds through the Community Development Block Grant program. FEMA collaborated with Puerto Rico on improved housing design plans for low-income residents and also provided technical assistance and funding for the development of long-term sheltering options.

The federal government worked with Puerto Rico to put in place improvements to achieve the long-term benefits of **economic revitalization and sustainability**. In the agricultural sector this was accomplished through financial assistance for crop and physical losses, expanding agricultural insurance and coverage, and financial and technical assistance for conservation measures to reduce flooding and erosion. The USDA Risk Management Agency provided funding for crop loss insurance claims. The USDA Natural Resources Conservation Service provided financial and technical assistance to address flooding and soil erosion problems.

In the nonagricultural sector, the federal government provided community development planning assistance, supported small business recovery, encouraged new investment, proposed fiscal assistance, provided unemployment assistance, and promoted flood insurance for homeowners, renters, and businesses. HUD made available technical assistance for economic development strategies and financial packaging. The EDA provided a community planning grant to the University of Puerto Rico's Economic Development University Center and committed funds to Puerto Rico's Economic Development Bank for a revolving loan fund assistance program. The DOL provided funding to create temporary jobs to assist in the immediate and long-term cleanup and recovery efforts. The DOL also provided unemployment assistance.

Hurricane Georges caused 100 percent of the electrical service in Puerto Rico to be disrupted. Its failure crippled other basic services such as water and sewage treatment, telephone service, transportation, and local commerce. Federal assistance for the **energy** sector included providing resources for repairing electrical transmission and distribution lines, and recommendations for design improvements; emergency generators; and assistance for developing a more reliable electrical system. The cost for repairing the island's electrical system was paid by a combination of Puerto Rico's self-insurance coverage and funding through FEMA's public assistance program. Electric utility workers, trucks, and equipment were flown to the island to assist local crews. Emergency generators were provided to keep critical facilities operational, and plans were developed to keep some of the generators in place to provide backup power during future disasters. The Department of Energy, FEMA, and Puerto Rico examined mitigation measures to improve the disaster resistance of the electrical system through enhanced generation/transmission relationships, better power line placement, and placing poles deeper in the ground.

Key **transportation** issues that were addressed included repairing damaged roads and bridges, developing a reliable power source for the Tren Urbano project, and dredging harbors. The Army Corps removed tons of debris from roadways, installed four temporary bridges, and provided financial assistance for critical dredging activities to maintain safe harbor channels. The FHA and FEMA provided financial assistance for rebuilding the island's damaged transportation system. Mitigation measures were incorporated into road and bridge repairs to reduce the risk of such severe damage in the future. The Federal Transit Authority and FEMA worked with the government of Puerto Rico to explore funding options to establish a reliable power source for the Tren Urbano, a San Juan metro-area mass transit system.

The Governor of Puerto Rico, Pedro Rosselló, stated, "From the President on down, the federal government mobilized all of the resources at its disposal—even before the hurricane struck—and has earned the eternal gratitude of Puerto Rico's 3.9 million people for its role in helping us cope with this catastrophe. The scope of the response is illustrated by the fact that the President's Long-Term Recovery Task Force is rarely activated."

UNIVERSITY OF HOUSTON O'QUINN LAW LIBRARY

Tropical Storm Allison formed on Wednesday evening, June 6, 2001, in the Gulf of Mexico southeast of Galveston, Texas, and eventually exited the United States on Sunday night, June 17 after passing through Florida and proceeding up the East Coast. Allison proved to be the most destructive Tropical Storm in U.S. history, costing 43 lives and nearly $5 billion. The storm hit Houston, Texas, especially hard, dumping between 30 to 40 inches of rain and causing an estimated $1 billion in damage. On June 9, 2001, President Bush declared a major disaster for the state of Texas, with 28 counties eligible for public assistance. The University of Houston O'Quinn Law Library was flooded with 8 feet of water after the heavy rains from Tropical Storm Allison.

The lower floor of the library filled nearly to the 12-foot ceilings with a mixture of water, oil, asbestos, and other pollutants. The 35,000 square feet of space in the lower level were equal to nearly two floors of a typical downtown skyscraper. The metal shelves were destroyed, partly by the tremendous weight of waterlogged books and partly by being literally exploded as the wet books began swelling and exerting tremendous sideways pressure. The library lost between 200,000 and 500,000 books, and damages were estimated at $30 million.

Through the Public Assistance program, FEMA approved $21.4 million for the replacement of 174,000 copies of law books and microfiche storage collection. The funding approved by FEMA was for two separate projects: one project in the amount of $1,204,600 was for the microfiche collection, and the other project in the amount of $27,295,196 was for law book replacement. FEMA provided 75 percent of the cost, with the remaining 25 percent coming from local sources. "With the support of all our communities, and major assistance from FEMA, not only have we recovered, but we're putting in place an even stronger and more secure resource for our law center faculty and students as well as the community," said University of Houston President, Arthur K. Smith.

6. The Disciplines of Emergency Management: Preparedness

INTRODUCTION

Preparedness within the field of emergency management can best be defined as a state of readiness to respond to a disaster, crisis, or any other type of emergency situation. Preparedness is not only a state of readiness, but also a theme throughout most aspects of emergency management. If you look back into the history of the United States, you see the predecessors of today's emergency managers focusing on preparedness. The fallout shelters of the 1950s and the air raid wardens were promoting preparedness for a potential nuclear attack from the Soviet Union. An early 1970s study prepared by the National Governor's Association talked about the importance of preparedness as the first step in emergency management.

After the Three Mile Island Nuclear Power Plant incident in 1979, preparedness around commercial nuclear power plants became a major issue for continued licensing of these plants. The increased emphasis on preparing the public for a potential event through planning and education and preparing local responders through required exercises caused an increased focus on preparedness. The Nuclear Regulatory Commission's licensing requirements required local emergency plans, exercise of those plans, and evaluation of the exercises.

This process had a profound impact on the discipline of emergency management. This off-site preparedness planning process became the model for future emergency response plans. The required exercises were some of the first such activities. They brought a legitimacy and level of public and political exposure to the emergency management profession. Most people agree that the radiological emergency preparedness program that was initiated in the aftermath of Three Mile Island and that became part of the newly created federal agency, the Federal Emergency Management Agency (FEMA), was the start of modern emergency management discipline.

Since that time, preparedness has advanced significantly, and its role as a building block of emergency management continues. No emergency management organization can function without a strong preparedness capability. This capability is built through planning, training, and exercising. Preparedness activities have led toward an increased professionalism within the discipline of emergency management. Throughout the 1990s, FEMA was focused on supporting and enhancing these efforts, not just at the federal level but also throughout government and into the private sector.

All organizations in private, public, and government sectors are susceptible to the consequences of a disaster and must consider preparedness. For example, preparedness focuses not only on getting essential government services, such as utilities and emergency services, functioning at predisaster levels, but also on assisting businesses in quickly reopening to the public. Both of these key functions of preparedness help minimize the required time for the affected population to return to predisaster life. Business contingency planning has emerged as a profitable offshoot of government preparedness efforts.

This chapter discusses the preparedness cycle from a systems approach, preparedness programs, hazard preparedness, training programs, and exercise programs. The focus is on federal efforts, predominantly FEMA, and best practices are highlighted through several case studies.

PREPAREDNESS: THE BUILDING BLOCK

With the National Response Plan, there are 15 emergency functions, each of which relies on a level of preparedness. These functions are defined as Transportation, Communications, Public Works and Engineering, Firefighting, Emergency Management, Mass Care, Housing, and Human Services, Resource Support, Public Health and Medical Services, Urban Search and Rescue, Oil and Hazardous Materials Response, Agriculture and Natural Resources, Energy, Public Safety and Security, Long-Term Community Recovery and Mitigation, and External Affairs. Each individual functional area must ensure its own preparedness in order to establish a systemwide posture that is ready to respond and act in an emergency.

All 15 functions are highly dependent on each other. For example, the functions of emergency communications must be prepared to establish emergency telecommunications support in order for the firefighters, who must be prepared with the equipment and training to extinguish the fires, to know where to go and coordinate with the urban search and rescue teams that locate and rescue victims, each of which must be provided timely transportation to reach the disaster scene.

Preparedness is therefore defined more fully by FEMA as

the leadership, training, readiness and exercise support, and technical and financial assistance to strengthen citizens, communities, state, local, and tribal governments, and professional emergency workers as they prepare for disasters, mitigate the effects of disasters, respond to community needs after a disaster, and launch effective recovery efforts. (www.fema.gov)

MITIGATION VERSUS PREPAREDNESS

Preparedness has been defined, and it has been mentioned that preparedness encompasses various aspects of response, but how does mitigation play into the equation? Mitigation is the cornerstone of emergency management. It's the ongoing effort to lessen the impact disasters have on people and property. Mitigation involves keeping homes away from floodplains, engineering bridges to withstand earthquakes, creating and enforcing effective building codes to protect property from hurricanes—and more.

Preparedness deals with the functional aspects of emergency management such as the response to and recovery from a disaster, whereas mitigation attempts to lesson these effects through predisaster actions—as simple as striving to create "disaster-resistant" communities.

A SYSTEMS APPROACH: THE PREPAREDNESS CYCLE

As an academic field, as well as an applied practice in the public and private sector, emergency management just recently has been established. For this reason, it has thus far drawn on the fields of emergency medicine, fire suppression, and law enforcement for many of its foundations. Although these are tried-and-tested specialties, they also are steeped in tradition—relying less on academic or analytic processes. Without a foundation that ties academia and structured analytic methodologies with tradition, the extreme complexity of emergency management, often requiring coordination among tens to hundreds of individual agencies and organizations, will not be managed effectively. Therefore, a systematic approach must be established for emergency management as a whole, and specifically the steps necessary to reach preparedness.

The diagram depicted in Figure 6-1, used in terrorism planning, depicts the planning process, beginning with assessing the threats to a jurisdiction or business, be it natural or manmade, and working in a systematic approach toward a cyclical process to establish preparedness. This systematic and cyclical approach is defined by the continual evolution of the phases on the exterior ring—assessment, planning, preparation, and evaluation.

In this depiction, the interior ring defines each of the steps that organizations must work toward in order to be prepared. The first step is to identify what types of disasters, or threats, a jurisdiction, business, or any entity faces. Next, assessing the current vulnerability, or level of preparedness, will lead toward determining the shortfalls between current preparedness and the requirements to meet an improved

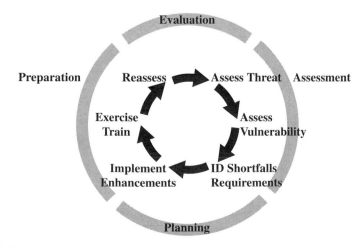

Figure 6-1 Preparedness Planning Cycle.

preparedness posture. This improved posture may be determined through industrial standards set forth by organizations such as the National Fire Protection Association, who sets fire safety standards, or the International Organization for Standardization (ISO), one of the largest developers of standards and certifications. Local, state, and/or federal laws can also statutorily define a required level of preparedness.

Implementing enhancements or revamping complete systems then bridges these identified shortfalls. Exercises and training can then be used to test whether the enhancements or new systems are, in fact, meeting the standards determined in earlier stages. If they are, then the end goal of readiness or preparedness regarding a particular threat such as terrorism or floods is met.

The cyclical nature of this system is the fundamental aspect of the successive steps to be taken after determining whether a jurisdiction, or any type of entity, is prepared. Whether or not those standards are met, the entity must reexamine its threats because both natural and technological threats change constantly. The important realization that preparedness is a dynamic state and can either improve or diminish in a short time must be understood by the emergency management professional. Using the systems approach will ensure that an overall emergency management system is prepared, but more important, each of the individual functional areas are also prepared.

The importance and diversity of this vital aspect of preparedness planning can be demonstrated through the other types of assessment processes available. Another example available to emergency managers is provided by FEMA in its Capability Assessment for Readiness (CAR) program.

FEMA'S CAPABILITY ASSESSMENT FOR READINESS PROGRAM

FEMA and the National Emergency Management Association (NEMA) have joined together in partnership to develop an emergency management readiness and capability assessment system for state and local emergency managers. The result of this effort is the Capability Assessment for Readiness (CAR). This initiative further strengthens the Emergency Management Performance Grant (EMPG) Program that provides federal financial assistance to state and local governments. Also imbedded in the CAR are those important ingredients developed by the National Fire Protection Association (NPFA 1600) termed Emergency Management Standards. The CAR process also provides for the assessment component of the EMPG process that will continue to evolve in coming years.

The states completed the CAR between May and June 2000 and forwarded their data to FEMA for analysis and the subsequent development of a national report. FEMA expects the report to be completed and distributed to the President and the U.S. Congress in the first quarter of 2001.

WHAT: The CAR process is an initiative that is part of the EMPG continuous improvement cycle. This process is designed to assess the operations, readiness, and capabilities of a state to mitigate against, prepare for, respond to, and recover from all disasters and emergencies. The assessment focuses on the following 13 core elements that address major emergency management functions:

- Laws and Authorities
- Hazard Identification and Risk Assessment
- Hazard Mitigation
- Resource Management
- Planning
- Direction, Control, and Coordination
- Communications and Warning
- Operations and Procedures
- Logistics and Facilities
- Training
- Exercises, Evaluations, and Corrective Actions
- Crisis Communications, Public Education, and Information
- Finance and Administration

The CAR provides a common format for a self assessment for state emergency management organizations to communicate strengths and areas needing improvement. The CAR process seeks to answer three basic questions:

- Is the emergency management program comprehensive for the needs of the states?
- Are the goals, objectives, and mission of the organization being achieved?
- Is the state able to direct strategic deployment of resources and help communities and citizens avoid becoming disaster victims?

HOW: Each state and territory will conduct a comprehensive self-assessment, in coordination with their respective FEMA region during Fiscal Year (FY) 2000. The results of the assessment will be used to assist states and FEMA in joint strategic planning and the identification of potential objectives for federal/state partnerships to improve their emergency management, operations, and capabilities. The CAR represents the firm commitment of both NEMA and FEMA to establish a system for assessing the capability and readiness of each state and territory to better fulfill our mission of saving lives and protecting property.

FEMA is also working with NEMA and the International Association of Emergency Management (IAEM) to develop a local CAR that will provide local emergency managers the opportunity to evaluate their emergency management programs. It is designed to complement the state CAR to ensure greater accuracy of the results. A first draft of this document has been completed and is currently under review by NEMA, IAEM, the states, and other organizations.

An additional important initiative is our partnership with the National Congress of American Indians (NCAI) and tribal governments to develop a Tribal Capability Assessment for Readiness (Tribal CAR). The Tribal CAR self-assessment will help tribal governments to determine the strengths and weaknesses of their emergency management programs. It will use the same format

continues

and 13 Emergency Management Functions that are found in the State and Local CAR. When completed, the Tribal CAR will be distributed to all tribal governments that are interested in improving their disaster response within tribal jurisdictions.

The development of the State CAR and ongoing work with local communities and tribal governments is managed by FEMA's Preparedness, Training and Exercises Directorate, and involved all FEMA Directorates, Offices, and Administrators of FEMA, FEMA regions, states, and territories, and NEMA. Again, this initiative has the full support of both NEMA and FEMA and is a key part of the FEMA strategy to expand the emergency management culture to focus on helping communities to avoid becoming disaster victims.

Source: FEMA

PREPAREDNESS PROGRAMS

Preparedness is everyone's job. Not just government agencies but all sectors of society—service providers, businesses, civic and volunteer groups, industry associations and neighborhood associations, as well as every individual citizen—should plan ahead for disaster. As such, preparedness programs are developed to target each of these audiences in order to educate, promote, and test preparedness.

One of these public education programs is the Community and Family Preparedness Program operated by FEMA, which educates the general public about disaster awareness and preparedness. The core message of the Community and Family Preparedness Program is the Family Disaster Plan—four basic steps people can take to prepare for any type of disaster.

1. *Find out what types of disasters are most likely to occur in your community and how to prepare for them.* Contacting your local emergency management office or American Red Cross chapter for information and guidelines is a good way to get started.
2. *Create a Family Disaster Plan.* Hold a family meeting to talk about the steps the family members will take to be ready when disaster happens in the community.
3. *Take action.* Each family member, regardless of age, can be responsible for helping the family be prepared. Activities can include posting emergency telephone numbers, installing smoke detectors, determining escape routes, assembling disaster supply kits, and taking first aid or CPR courses.
4. *Practice and maintain the plan.* The final step emphasizes the need to practice the plan on a regular basis so family members will remember what to do when disaster strikes.

As just one of the many preparedness programs sponsored by FEMA and other public and private disaster response and emergency management organizations, the Community and Family Preparedness Program highlights the foundation of a disaster program that is applicable to a wide range of disasters. Many more programs look specifically at preparedness regarding one type of disaster and can be obtained through agencies such as FEMA, the American Red Cross, and your state and local offices of emergency management.

AMERICAN RED CROSS HURRICANE PREPAREDNESS TIPS

Here's what you can do to prepare for such an emergency.

Know What Hurricane WATCH and WARNING Mean

- WATCH: Hurricane conditions are *possible* in the specified area of the watch, usually within 36 hours.
- WARNING: Hurricane conditions are *expected* in the specified area of the warning, usually within 24 hours.

Prepare a Personal Evacuation Plan

- Identify ahead of time where you could go if you are told to evacuate. Choose several places—a friend's home in another town, a motel, or a shelter.
- Keep the telephone numbers of these places handy as well as a road map of your locality. You may need to take alternative or unfamiliar routes if major roads are closed or clogged.
- Listen to NOAA Weather Radio or local radio or TV stations for evacuation instructions. If advised to evacuate, do so immediately.

Assemble a Disaster Supplies Kit Including the Following Items

- First-aid kit and essential medications
- Canned food and can opener
- At least three gallons of water per person
- Protective clothing, rainwear, and bedding or sleeping bags
- Battery-powered radio, flashlight, and extra batteries
- Special items for infants, elderly, or disabled family members
- Written instructions on how to turn off electricity, gas, and water if authorities advise you to do so (remember, you'll need a professional to turn them back on)

Prepare for High Winds

- Install hurricane shutters or purchase precut 1/2-inch outdoor plywood boards for each window of your home. Install anchors for the plywood and predrill holes in the plywood so that you can put it up quickly.

continues

- Make trees more wind resistant by removing diseased and damaged limbs, then strategically removing branches so that wind can blow through.

Know What to Do When a Hurricane WATCH Is Issued

- Listen to NOAA Weather Radio or local radio or TV stations for up-to-date storm information.
- Prepare to bring inside any lawn furniture, outdoor decorations or ornaments, trash cans, hanging plants, and anything else that can be picked up by the wind.
- Prepare to cover all windows of your home. If shutters have not been installed, use precut plywood as described previously. *Note:* Tape does not prevent windows from breaking, so taping windows is not recommended.
- Fill your car's gas tank.
- Recheck manufactured home tiedowns.
- Check batteries and stock up on canned food, first-aid supplies, drinking water, and medications.

Know What to Do When a Hurricane WARNING Is Issued

- Listen to the advice of local officials, and leave if they tell you to do so.
- Complete preparation activities.
- If you are not advised to evacuate, stay indoors, away from windows.
- Be aware that the calm "eye" is deceptive; the storm is not over. The worst part of the storm will happen once the eye passes over and the winds blow from the opposite direction. Trees, shrubs, buildings, and other objects damaged by the first winds can be broken or destroyed by the second winds.
- Be alert for tornadoes. Tornadoes can happen during a hurricane and after it passes over.
- Remain indoors, in the center of your home, in a closet or bathroom without windows.
- Stay away from floodwaters. If you come upon a flooded road, turn around and go another way.
- If you are caught on a flooded road and waters are rising rapidly around you, get out of the car and climb to higher ground.

Know What to Do After a Hurricane Is Over

- Keep listening to NOAA Weather Radio or local radio or TV stations for instructions.
- If you evacuated, return home when local officials tell you it is safe to do so.
- Inspect your home for damage.

Source: American Red Cross, www.redcross.org

EDUCATION AND TRAINING PROGRAMS

Since its inception in 1979, FEMA has become a leader in developing and teaching courses in emergency management. FEMA manages the Emergency Management Institute (EMI) and the National Fire Academy (NFA), which are colocated on a former college campus in Emmitsburg, Maryland. Thousands of firefighters, fire officers, and emergency managers have been trained by FEMA. Additionally, FEMA has helped establish degree programs in junior colleges, colleges, and universities across the country. Currently, FEMA is expanding its training and education capacities through distance learning programs.

Emergency Management Institute

FEMA defines the mission of the Emergency Management Institute as follows: "EMI provides a nationwide training program of resident courses and nonresident courses to enhance U.S. emergency management practices." According to the EMI catalog:

> Approximately 5,500 participants attend resident courses each year while 100,000 individuals participate in non-resident programs sponsored by EMI and conducted by State emergency management agencies under cooperative agreements with FEMA. Another 150,000 individuals participate in EMI-supported exercises, and approximately 1,000 individuals participate in the Chemical Stockpile Emergency Preparedness Program (CSEPP). Additionally, hundreds of thousands of individuals use EMI distance learning programs such as the Independent Study Program and the Emergency Education NETwork (EENET) in their home communities.

The 2001–2002 EMI Catalog of Activities listed more than 60 resident courses offered at the Emmitsburg Campus and more than 100 nonresident courses in the following subject areas:

- Mitigation
- Readiness and Technology
- Professional Development
- Disaster Operations and Recovery
- Integrated Emergency Management
- Chemical Stockpile Emergency Preparedness Program (CSEPP) Training Courses (nonresident only)

EMI also offered nearly 30 Independent Study courses in the 2001–2002 period. Some of the Independent Study courses offered include the following:

- Emergency Program Manager: An Orientation to the Position
- Radiological Emergency Management
- Hazardous Materials: A Citizen's Orientation
- Managing Floodplain Development through the National Flood Insurance Program (NFIP)
- Animals in Disaster: Module A, "Awareness and Preparedness"
- State Disaster Management
- Multi-Hazard Planning for Schools
- The Professional in Emergency Management

- Introduction to Public Assistance Process
- Role of Voluntary Agencies in Emergency Management
- The Emergency Operations Center's Role in Community Preparedness, Response, and Recovery Operations

Two courses of note offered at EMI are the Integrated Emergency Management Course (IEMC) and the Disaster-Resistant Jobs Train-the-Trainer Courses. The IEMC is a weeklong course for public officials that covers all aspects of a community emergency management function. Community officials from Oklahoma City participated in the IEMC just months before the terrorist bombing in 1995 and credit the lessons they learned at IEMC with helping them respond quickly and effectively to the bombing.

THE INTEGRATED EMERGENCY MANAGEMENT COURSE

Protecting the population is a primary responsibility of government, and fulfilling this responsibility depends on the abilities of emergency personnel to prepare for, respond to, recover from, and mitigate against disaster. It means developing and maintaining a high standard of readiness and an ability to function effectively under crisis conditions. Emergency personnel can attain readiness either through managing emergencies or through participating in exercises. Clearly, exercises are the preferred method of gaining the necessary expertise.

The IEMC, offered by the EMI of FEMA, places public officials and emergency personnel in a realistic crisis situation within a structured learning environment. The course builds the awareness and skills needed to develop and implement policies, plans, and procedures to protect life and property through applications of sound emergency management principles in all phases of emergency management.

Community participants in IEMC include elected, midlevel management, supervisory, and operational personnel from various disciplines, including fire, emergency management, planning, finance, personnel, public health, transportation and public works, and information technology.

Early in the course, an emergency scenario begins to unfold in sequence with classroom-style lectures, discussions, and small-group workshops. As the course progresses, scenario-related events of increasing complexity, threat, and pressure occur. Participants develop emergency policies, plans, and procedures to ensure an effective response. The course culminates in an emergency exercise designed to test participant knowledge, awareness, flexibility, leadership, and interpersonal skills under extreme pressure.

Participants are challenged to use the new ideas, skills, and abilities in addition to their own knowledge and experience. In this way, the IEMC allows individuals to rehearse their real-life roles in a realistic emergency situation, while identifying additional planning needs.

Source: FEMA, www.fema.gov

The Disaster-Resistant Jobs course was developed in cooperation with the Economic Development Administration (EDA) of the U.S. Department of Commerce and is designed to "help small and medium-sized communities protect the economy from the effects of catastrophic events." This course was developed in response to the devastating impact the 1997 floods had on the City of Grand Forks, North Dakota. The EDA and FEMA recognized that more economic development planning could be done to reduce the impacts of future disasters on local economies.

DISASTER-RESISTANT JOBS AND TRAIN-THE-TRAINER COURSES

All too often, communities that have experienced major disasters lose a major portion of their economic base. Studies have shown that after a disaster, 60 percent of small and medium-sized businesses fail within two years. Many never return to business once they are closed for even a few days because of floods, tornadoes, earthquakes, and hurricanes. Not only does the community suffer from the effects of the hazard, but also in the long run the loss of jobs and tax base further reduces the community's ability to return to normal.

The EDA and FEMA have developed the Disaster-Resistant Jobs course that will help small and medium-size communities protect the economy from the effects of catastrophic events. The topics of this three and a half day course are as follows:

Course	Topic
Unit One	The Importance of Disaster-Resistant Jobs
Unit Two	Creating Disaster–Resistant Jobs
Unit Three	Recognizing the Impact
Unit Four	What about Mitigation?
Unit Five	Disaster-Resistant Economic Development Planning Process
Unit Six	Business Recovery

The purpose of the Train-the-Trainer courses are to develop a cadre of trainers who can raise awareness in their own localities. Participants must have the desire and ability to address groups, including local economic development agencies, Chamber of Commerce meetings, service club luncheons, business meetings, and other formats to address the issue of protecting the community's economic base. Participants will be provided with a tool kit of materials that can be used to tailor their presentation before groups.

Source: FEMA, www.fema.gov

FEMA's EMI Higher Education Project works to establish and support emergency management curriculum in junior colleges, colleges, and universities. The project has developed a prototype curriculum for Associate Degrees in Emergency Management. Currently, FEMA lists 82 Emergency Management Higher Education Programs in institutions spread across all 50 states and Puerto Rico.

COMMUNITY EMERGENCY RESPONSE TEAM

Following a major disaster, first responders who provide fire and medical services will not be able to meet the demand for these services. Factors such as number of victims, communication failures, and road blockages will prevent people from accessing emergency services they have come to expect at a moment's notice through 911. People will have to rely on each other for help in order to meet their immediate lifesaving and life-sustaining needs.

If it can predict that emergency services will not meet immediate needs following a major disaster, especially if there is no warning, as in an earthquake, and people will spontaneously volunteer, what can government do to prepare citizens for this eventuality?

1. Present citizens with the facts about what to expect following a major disaster in terms of immediate services.
2. Give the message about their responsibility for mitigation and preparedness.
3. Train them in needed lifesaving skills with emphasis on decision-making skills, rescuer safety, and doing the greatest good for the greatest number.
4. Organize teams so that they are an extension of first-responder services offering immediate help to victims until professional services arrive.

The Community Emergency Response Team (CERT) concept was developed and implemented by the Los Angeles City Fire Department (LAFD) in 1985. The Whittier Narrows earthquake in 1987 underscored the areawide threat of a major disaster in California. Further, it confirmed the need for training civilians to meet their immediate needs. As a result, the LAFD created the Disaster Preparedness Division to train citizens and private and government employees.

The training program that the LAFD initiated makes good sense and furthers the process of citizens understanding their responsibility in preparing for disaster. It also increases their ability to safely help themselves, their family, and their neighbors. FEMA recognizes the importance of preparing citizens. The EMI and the National Fire Academy adopted and expanded the CERT materials, believing them to be applicable to all hazards.

The CERT course will benefit any citizen who takes it. This individual will be better prepared to respond to and cope with the aftermath of a disaster. Additionally, if a community wants to supplement its response capability after a disaster, civilians can be recruited and trained as neighborhood, business, and government teams that, in essence, will be auxiliary responders. These groups can provide immediate assistance to victims in their area, organize spontaneous volunteers who have not had the training, and collect disaster intelligence that will assist professional responders with prioritizing and allocating resources following a disaster. Since 1993, when this training was made available nationally by FEMA, communities in 28 states and Puerto Rico have conducted CERT training.

> The CERT course is delivered in the community by a team of first responders who have the requisite knowledge and skills to instruct the sessions. It is suggested that the instructors complete a CERT Train-the-Trainer course conducted by their State Training Office for Emergency Management or the Emergency Management Institute in order to learn the training techniques that are used successfully by the LAFD.
>
> The CERT training for community groups usually is delivered in two-and-a-half-hour sessions, one evening per week over a seven-week period. The training consists of the following:
>
> * Disaster Preparedness
> * Disaster Fire Suppression
> * Disaster Medical Operations
> * Light Search and Rescue
> * Disaster Psychology and Team Organization
> * Course Review and Disaster Simulation
>
> *Source*: FEMA, www.fema.gov

National Fire Academy

The mission of the National Fire Academy (NFA) is: "Through its courses and programs, the National Fire Academy works to enhance the ability of fire and emergency services and allied professionals to deal more effectively with fire and related emergencies."

Since its inception in 1975 as the delivery mechanism for fire training for the congressionally mandated U.S. Fire Administration (USFA), the NFA estimates it has trained more than 1.4 million students. The NFA delivers courses at its Emmitsburg, Maryland campus, which it shares with the EMI, and across the nation in cooperation with state and local fire training organizations and local colleges and universities.

U.S. FIRE ADMINISTRATION

As an entity of FEMA, the mission of the USFA is to reduce life and economic losses caused by fire and related emergencies, through leadership, advocacy, coordination, and support. It serves the nation independently, in coordination with other federal agencies, and in partnership with fire protection and emergency service communities. With a commitment to excellence, the USFA provides public education, training, technology, and data initiatives.

USFA Programs

USFA programs to prevent and mitigate the consequences of fire are divided into four basic areas:

continues

- *Public Education*. Develops and delivers fire prevention and safety education programs in partnership with other federal agencies, the fire and emergency response community, the media, and safety interest groups.
- *Training*. Promotes the professional development of the fire and the emergency response community and its allied professionals. To supplement and support state and local fire service training programs, the NFA and the EMI develop and deliver educational and training courses with a national focus.
- *Technology*. Works with the public and private groups to promote and improve fire prevention and life safety through research, testing, and evaluation. Generates and distributes research and special studies on fire detection, suppression, and notification systems, and on fire and emergency responder health and safety.
- *Data*. Assists state and local entities in collecting, analyzing, and disseminating data on the occurrence, control, and consequences of all types of fires. The National Fire Data Center describes the nation's fire problem, proposes possible solutions and national priorities, monitors resulting programs, and provides information to the public and fire organizations.

The U.S. Fire Administration has launched a training program focused on the response to terrorist events. The courses, all subtitled "Emergency Response to Terrorism (ERT)," may be taken at the Emergency Management Institute (EMI) in Emmitsburg, MD, at the state level, or online (select courses). Terrorism courses include:

- ERT—Basic Concepts
- ERT Tactical Considerations—Company Officer
- ERT Tactical Considerations—Emergency Medical Services
- ERT Tactical Considerations—Hazardous Materials
- ERT Strategic Considerations for Command Officers
- Emergency Response to Terrorism Self Study

Source: FEMA, www.fema.gov

The NFA's on-campus programs target middle- and top-level fire officers, fire service instructors, technical professionals, and representatives from allied professions. Any person with substantial involvement in fire prevention and control, emergency medical services, or fire-related emergency management activities is eligible to apply for Academy courses. The NFA also delivers courses using CD-ROMs, their simulation laboratory, and the Internet. For those interested in pursuing degrees, the Degrees at a Distance Program extends the NFA's academic outreach through a network of seven colleges and universities. Fire service personnel who cannot attend college because of work hours and locations are able to earn a degree in fire technology and management through independent study.

CURRICULUM OFFERED AT THE NATIONAL FIRE ACADEMY

- Arson
- Emergency Medical Services
- Emergency Response to Terrorism
- Executive Development
- Fire Prevention: Management
- Fire Prevention: Public Education
- Fire Prevention: Technical
- Hazardous Materials
- Incident Management
- Management Science
- Planning & Information Management
- Training Programs

Source: FEMA, www.fema.gov

Other FEMA Education and Training Resources

FEMA provides other education and training resources such as curriculum and activities for teachers to use in the schools, school safety, and fire safety materials and information on how to talk to your kids about terrorism. FEMA has built an award-winning Web site for children called "FEMA for KIDS" that has such features as becoming a disaster action kid, the disaster area, the disaster connection: kids to kids, homework help, games and quizzes, and about FEMA.

EXERCISES

Once a plan is developed and personnel trained to the plan, the next step is exercising the plan. Exercises provide an opportunity to evaluate the efficiency and effectiveness of the plan and its components and to test the systems, facilities, and personnel involved in implementing the plan. Exercises are conducted at all levels of government and in the private sector.

FEMA defines an exercise as "a controlled, scenario-driven, simulated experience designed to demonstrate and evaluate an organization's capability to execute one or more assigned or implicit operational tasks or procedures as outlined in its contingency plan." There are four types of exercises identified by FEMA: full-scale, partial-scale, functional, and tabletop. Full descriptions of these exercise types are provided.

EMERGENCY MANAGEMENT EXERCISE TYPES

Exercises are generally categorized by their scope:

- **Full-scale.** This exercise used is to evaluate the operational capabilities of emergency management systems over an extended period. Usually, most or all of the organization's plan will be tested. The full-scale exercise usually is conducted in conditions as close to an actual event as possible. Field teams and crews will deploy and demonstrate their procedures. The full-scale exercise is designed to stress the organization's ability to accomplish their mission under realistic conditions.

- **Partial-scale.** This is an exercise with limited goals, with a portion of the organization participating; the scope generally is less than that of a full-scale exercise. It may be conducted to evaluate a limited number of objectives or it may be used to evaluate the organization's capability to execute newly developed procedures. Some teams may be deployed to actual field sites, whereas some procedures may be demonstrated under simulated conditions. Partial-scale exercises are generally shorter than full-scale exercises.

- **Functional.** This exercise allows the evaluation of various procedures that are similar to one another, such as medical treatment or communications. It is limited to activities within a specific functional category of the organization. Activities are scenario-driven, as with the full-scale exercise.

- **Tabletop.** This exercise usually involves senior staff, elected or appointed officials in an informal setting. Using a hazard-specific scenario, supporting documentation, and injected messages simulating field-derived information, the participants discuss anticipated actions while in a controlled environment. With a facilitator to keep the discussions focused, the products derived from a tabletop exercise may include emerging policy, plan revisions, and conceptualization of new procedures.

Source: FEMA Comprehensive Exercise Program, July 1995

FEMA manages the Comprehensive Exercise Program (CEP). The goal of the CEP is to develop, implement, and institutionalize a comprehensive, all-hazard, risk-based exercise program. Exercises conducted under the auspices of FEMA's CEP will be used to test and evaluate emergency management plans, policies, procedures, systems, and facilities developed to mitigate against, prepare for, respond to, and recover from the effects of all types of emergencies. CEP exercises include extensive involvement of state and local officials as well as representatives from other federal agencies. The CEP program provides five categories of exercises.

COMPREHENSIVE EXERCISE PROGRAM EXERCISE CATEGORIES

- **State and Local All-Hazard Exercises.** These exercises serve as the focal point for all state and local emergency management exercise activity addressing natural, technological, and manmade disasters as well as national security hazards. They are designed to test and evaluate the operational readiness and capability of emergency management systems, identify systemic deficiencies and efficiencies, and define corrective actions needed to ensure readiness and emergency operations proficiency. Emergency management functions rather than specific scenarios will be examined.

- **FEMA-Sponsored FRP Exercises.** The concept of operations, policies, and procedures set forth in the FRP for providing a federal response to state and local governments under the authorities of the Stafford Act are tested and validated in these exercises. Ideally, detailed headquarters and regional plans and procedures to implement the FRP will also be tested and validated. State and local governments will be encouraged to participate so their EOPs may be similarly tested and validated. The ultimate goal of these exercises is to achieve a seamless federal, state, and local response to and recovery from disasters of all types.

- **Legislatively Mandated Exercises Supported by FEMA.** These exercises focus on plans developed at the state and local level based on guidance and requirements established by the federal government. Federal involvement in the state and local planning process is required to ensure that established standards are met and maintained. This involvement will also ensure that incorporation of hazard-specific material into the jurisdiction's single EOP is accomplished in a manner consistent with the plans of federal departments and agencies responsible for incident response.

- **FEMA-Supported National and International Security Exercises.** National and international security exercises are designed to improve the capability of organizations and individuals to execute emergency management responsibilities and familiarize members of the federal government with the issues that might be encountered during a major emergency, including national security emergencies requiring the invocation of emergency authorities. These exercises also provide opportunities to validate/identify for subsequent correction, national security emergency management plans, policies, procedures, and systems. Sponsorship of these exercises is usually by the DoD or the North Atlantic Treaty Organization (NATO). For these types of exercises, FEMA coordinates federal civil government counterpart exercise activities.

- **Special/Extraordinary Event Exercises Sponsored or Supported by FEMA.** These exercises focus on events for which overall planning rests primarily at the federal level with other government jurisdictional elements brought in as necessary. These exercises provide opportunities to evaluate

continues

system interoperability for communications, automated data processing (ADP), and other electronic media. Exercises in this category are also designed to deal with a wide range of contingencies ranging from nuclear, chemical, and biological terrorism; continuity of government; satellite re-entry; VIP visits; presidential inaugurations; Olympic Games support; and, regional events such as large-scale civil disturbances. These exercises provide the opportunity to evaluate the effectiveness of memorandums of understanding (MOUs) between various federal departments and agencies as well as other plans, policies, and procedures designed to guide the interaction between them. They can also provide opportunities to explore issues and requirements for the management of emergencies for which there are no plans, policies, procedures, or MOUs in existence.

Source: FEMA, www.fema.gov

Office for Domestic Preparedness

The Office for Domestic Preparedness (ODP) is the principal component of the Department of Homeland Security responsible for preparing the United States for acts of terrorism. In carrying out its mission, ODP is the primary office responsible for providing training, funds for the purchase of equipment, support for the planning and execution of exercises, technical assistance, and other support to assist states and local jurisdictions to prevent, plan for, and respond to acts of terrorism.

Training

The ODP provides tailored training to enhance the capacity of states and local jurisdictions to prevent, deter, and respond safely and effectively to incidents of terrorism involving weapons of mass destruction (WMD). This includes reaching multiple disciplines, through training at the awareness, performance, and planning/management levels, and employing the most appropriate mediums and vehicles for the particular audience:

- Direct delivery
- Train-the-trainer
- Computer-based training
- Web-based training
- Video teleconferencing

Training and Data Exchange (TRADE) Group

A significant component of the validation process for ODP courses is the Training and Data Exchange (TRADE) Group—a federal interagency group that reviews member courses for consistency, avoidance of unnecessary duplication, and use of the most up-to-date information and protocols available. The TRADE Group is composed of the following agencies:

- United States Fire Administration's National Fire Academy
- Federal Bureau of Investigation
- Department of Justice
- Federal Emergency Management Agency
- Environmental Protection Agency
- Department of Energy
- Department of Health and Human Services
- Centers for Disease Control and Prevention
- Emergency Management Institute
- Federal Law Enforcement Training Center
- Department of Homeland Security

Equipment

The terrorist attacks of September 11, 2001, demonstrated that response to an incident of terrorism can rapidly deplete local supplies and equipment. In order to further enhance the capability of state and local units of government to prevent, deter, respond to, and recover from incidents of terrorism involving the use of chemical, biological, radiological, nuclear, and explosive (CBRNE) weapons and cyber attacks, ODP offers a suite of equipment programs that support ODP grant funding.

The ODP equipment programs provide a means of direct support to first responders to enable them to acquire additional, specialized equipment, as well as training and technical assistance on those necessary equipment items. ODP equipment programs allow state and local units of governments to meet the challenges presented by terrorist organizations and to strengthen the capabilities of first responders to safely and effectively prepare for and respond to a terrorist incident.

The various equipment programs of ODP include:

- Information Technology and Evaluation Program (ITEP)
- Interoperable Communications Technical Assistance Program (ICTAP)
- The Responder Knowledge Base (RKB)
- Rapid Assistance Team (RAT)
- Equipment Purchase Assistance Program
- Homeland Defense Equipment Reuse (HDER) Program
- Domestic Preparedness Equipment Technical Assistance Program (DPETAP)
- Prepositioned Equipment Program (PEP)
- Interoperable Communications User's Handbook

Technical Assistance

ODP's Homeland Security Preparedness Technical Assistance (TA) Program provides direct assistance to state and local jurisdictions to improve their ability to prevent, respond to, and recover from threats or acts of terrorism involving chemical, biological, radiological, nuclear, or explosive (CBRNE) weapons. TA provides a process to help resolve a problem and/or create innovative approaches. All TA services are available to eligible recipients at no charge.

TA programs in place or currently under development within ODP include:

- **Homeland Security Assessment and Strategy Technical Assistance.** Assists states and local jurisdictions with assessment process, ability to conduct assessments, and development of a comprehensive homeland security strategy.

- **Initial Strategy Implementation Plan (ISIP) Technical Assistance.** Assists states with completing the ISIP template. Workshops address 1) developing a list of projects based on the State and/or Urban Area Homeland Security Strategy and 2) enhancing understanding of how to complete the ISIP template and the process for ISIP submission.

- **Domestic Preparedness Equipment Technical Assistance Program (DPETAP).** Provides equipment-specific training on CBRNE detection, decontamination, and personal protection equipment.

- **Terrorism Early Warning Group Replication.** Replicates program that enhances capabilities for analyzing strategic and operational information needed to respond to terrorism and protect critical infrastructure.

- **Interoperable Communication Technical Assistance Program (ICTAP).** Enhances public safety communications interoperability with regard to CBRNE terrorism threats.

- **Port and Mass Transit Planning Technical Assistance.** Assesses the needs of port/mass transit agencies to prepare for and counter post-9/11 terrorist threats.

- **Rapid Assistance Team Technical Assistance.** Deploys teams on short notice to support targeted projects, such as identifying equipment needs or equipment procurement plans.

- **General Technical Assistance.** Provides specialized assistance to enhance state and local strategies to prevent, respond to, and recover from CBRNE terrorism.

- **Prevention Technical Assistance.** New initiative to facilitate terrorism prevention efforts, such as collaboration, information sharing, risk management, threat recognition, and intervention.

- **Plans and Planning Synchronization Technical Assistance.** Provides planning support for multijurisdictional terrorism response using innovative software tool.

Exercises

The ODP's goal is to help states, cities, towns and villages gain an objective assessment of their capacity to prevent or respond to and recover from a disaster so that modifications or improvements can be made before a real incident occurs. This is conducted primarily through three mechanisms—the Homeland Security Exercise and Evaluation Program (HSEEP), the National Exercise Program (NEP), and the Models, Simulations, and Games (MS&G) Review Program.

- **Homeland Security Exercise and Evaluation Program (HSEEP).** The Homeland Security Exercise and Evaluation Program (HSEEP) is both doctrine and policy for designing, developing, conducting, and evaluating exercises. HSEEP is a threat- and performance-based exercise program that includes a cycle, mix, and range of exercise activities of varying degrees of complexity and interaction. HSEEP includes a series of four reference manuals to help states and local jurisdictions establish exercise programs and design, develop, conduct, and evaluate exercises (each of which can be downloaded from the ODP Web site):

- Volume I: Overview and Doctrine
- Volume II: Exercise Evaluation and Improvement
- Volume III: Exercise Program Management and Exercise Planning Process
- Volume IV: Sample Exercise Documents and Formats

- **National Exercise Program (NEP).** The National Strategy for Homeland Security directed the establishment of a National Exercise Strategy. Homeland Security Presidential Directive #8 (HSPD 8) directed Secretary Tom Ridge to establish a National Exercise Program (NEP). Secretary Ridge charged ODP to develop a program that identifies and integrates national level exercise activities to ensure those activities serve the broadest community of learning.

    In addition to full-scale, integrated national level exercises, the NEP provides for tailored exercise activities that serve as DHS's primary vehicle for training national leaders and staff. The NEP enhances the collaboration among partners at all levels of government for assigned homeland security missions. National-level exercises provide the means to conduct "full-scale, full system tests" of collective preparedness, interoperability, and collaboration across all levels of government and the private sector.

    The cornerstone of national performance-based exercises is the Top Officials (TOPOFF), biennial exercise series. TOPOFF included a functional exercise in 2000 (TOPOFF I) and a full-scale exercise in 2003 (TOPOFF II). The next TOPOFF exercise will occur in 2005.

- **Models, Simulations, and Games (MS&G).** One hundred Models, Simulations, & Games (MS&G) have been reviewed for their ability to support domestic preparedness training and exercising (T&E). For each product, the review considered the product's functionality from a T&E perspective, its hardware and software requirements, and cost. Product functionality was compared to key T&E attributes that were summarized from over 1100 T&E requirements.

Grant Programs

ODP grant programs provide funding to enhance the capacity of state and local jurisdictions to prevent, respond to, and recover from incidents of terrorism involving chemical, biological, radiological, nuclear, or explosive (CBRNE) weapons and cyber attacks. ODP's grant programs were initiated in 1998, and currently provide funds to all 50 states, the District of Columbia, the Commonwealth of Puerto Rico, American Samoa, the Commonwealth of Northern Mariana Islands, Guam, and the U.S. Virgin Islands.

The FY 2004/2005 ODP Grant Programs include:

- 2004 Grants
  - Information Technology and Evaluation Program (ITEP)
  - Competitive Training Grants Program
  - Assistance to Firefighters Grant Program
  - Homeland Security Grant Program
  - Urban Areas Security Initiative
- 2005 Grants
  - Homeland Security Grant Program
  - Operation Safe Commerce

The FY05 HSGP provides a single application kit and program guidance for six programs, including:

- The State Homeland Security Program (SHSP)
- The Urban Areas Security Initiative (UASI)
- The Law Enforcement Terrorism Prevention Program (LETPP)
- The Citizen Corps Program (CCP)
- The Emergency Management Performance Grants (EMPG)
- The Metropolitan Medical Response System (MMRS) Program Grants

Through this program, state and local emergency prevention, preparedness, and response personnel will be provided with over $2.5 billion in grant funding to enhance and improve homeland security efforts.

Source: www.ojp.usdoj.gov/odp/

BUSINESS CONTINUITY PLANNING AND EMERGENCY MANAGEMENT

Business continuity planning (BCP) provides focus-driven preparedness for businesses. At its simplest, BCP is the act of setting up a plan to ensure the survival of an organization. Since the early concern with the restoration of computer data, the concept of continuity has evolved in response to a changing environment. Major events have demanded that BCP encompass a growing number of concerns. The severe consequences of September 11 have raised many implications about how BCP will evolve in response to the disaster. How BCP evolves will directly influence business as a whole.

The implications of BCP are:

1. Terrorism must be considered as a real threat to the survival of business.
2. BCP will expand to include concern for the physical safety of employees.
3. BCP may involve the decentralization of business operations.
4. BCP may have to expand its sphere of concern to include the regional impacts of a disaster (including economic) to the area where a business is located.
5. The human relationships that a business depends on for its survival should be a major concern.
6. A recovery time of zero.

7. The renewed importance of critical data backup systems.
8. The inclusion of physical security concerns.
9. The increased importance of and pressure on business continuity planners.

The events of September 11 raised awareness of the fact that the survival of business depends on many external factors. External factors such as infrastructure and public safety authorities play a key role in whether BCP is ultimately successful. After September 11, infrastructure vital to business has even come under the control of public safety authorities. In this case, BCP is doubly dependent on public safety authorities. This awareness has led to attempts at greater communication between business and government since the attacks. In early March 2002, the newly created Office of Homeland Security unveiled its Homeland Security Advisory System.

Business immediately responded with its own proposal, the Critical Emergency Operations Link, which is intended to be a direct, two-way communication link to government at all levels. Business is demanding interaction with government so that it can anticipate how to react in the event of not only terrorist attacks, but also any catastrophe that threatens its survival. The attempt at greater communication and interaction by business is a proactive effort to turn its reliance on public safety authorities into an opportunity to ensure the success of BCP.

This approach suggests that business will demand a more extensive role for emergency management in BCP. The connection between emergency management and BCP is natural because it is the authority that has the responsibility of public safety planning. By demanding that emergency management play an extensive role in BCP, businesses can interact with government to ensure their survival. Emergency management should meet this demand with an outstretched arm because it represents a great opportunity for the field. If emergency management sincerely cooperates, then business may demand that government at all levels allocate more resources to emergency management in order to ensure that it can provide effective assistance. Ultimately, with business as its advocate, emergency management may gain the influence it needs to assume a greater role in leading the local and national public safety agenda.

CONCLUSION

Preparedness consists of three basic elements: preparing a plan, training to the plan, and exercising the plan. Preparedness planning at the community level is critical to reducing the effects of disaster events. FEMA sponsors numerous planning, training, and education activities designed to assist communities and states in developing effective preparedness plans and training personnel to implement these plans. Through its Comprehensive Exercise Program, FEMA helps local and state governments to exercise these plans. After-action evaluation of these exercises refines the plans.

Business Continuity Planning is a significant growth area for the emergency management community. The devastating impacts of September 11 have resulted in increased coordination and cooperation between business and emergency managers. It is hoped that the emergency management community will exploit this opportunity and get businesses more active in supporting the other phases of emergency management, particularly mitigation.

CASE STUDIES

HOW WERE BUSINESSES AFFECTED BY THE SEPTEMBER 11 ATTACKS?

Six months later, how has BCP been affected by the attacks? The severe destruction at the World Trade Center (WTC) has led to many significant implications that are redefining BCP. In order to look at these implications, this case study first lists the latest damage estimates for businesses in the WTC and the Lower Manhattan area.

- *Death Toll.* According to a February 16, 2002 *Washington Post* article "A Towering Task Lags in New York," the attacks killed more then 2,800 people (Powell and Haughney, 2002).
- *Estimated Dollar Amount of Damage.* As of February 1, 2002, Chris Hawley writes in his article, "Globalization and Sept. 11 are pushing Wall Street off Wall Street," that the attacks caused an estimated $83 billion in damage, and only about $50 billion will be covered by insurance. Taxpayers may have to cover some of the rest (Hawley, 2002).
- *Displaced Tenants of the WTC.* According to Gary Stock of the Unblinking Web site, the final tally of WTC tenants has not been completed because many sources of information contained outdated tenant lists. On the day of the attacks, the number of tenants ranged from 435 to 500. By October 19, the number increased to at least 700 (Stock, 2002).
- *Estimated Job Losses.* As of February 1, 2002, analysts predicted Manhattan would lose about 125,000 jobs after the attacks. Nearly 53,000 financial services jobs were expected to move out of Lower Manhattan—the Wall Street district—and 19,000 jobs had already left the city completely (Hawley, 2002). By February 16, 2002, one in four jobs in downtown Manhattan had disappeared—a job loss total that is thousands more than analysts had predicted immediately after September 11 (Powell and Haughney, 2002).
- *Estimated Loss of Office Space.* As of March 11, 2002, according to the article "Return to Downtown," the destruction of office space caused by the attacks equaled about 12 million square feet at the WTC and damage to another 20 million square feet in the surrounding area (Wax and Diop, 2002).
- *Communication Infrastructure Damage.* On October 29, 2001, in the article "Despite its losses, Verizon went right back to work restoring communication services," John Rendleman writes that on the day of the attacks a Verizon switching center was destroyed by the collapse of the WTC. This caused telecom service failure to 14,000 businesses and thousands of residential customers in lower Manhattan (Rendleman, 2001). According to the article "Out of the Ashes," Verizon shared its infrastructure with some 40 competitive local

exchange carriers whose services were similarly affected (Gilbert, 2002). By October 29, 2001, 90 percent of the service was restored.

- *Cleanup Concerns.* As of March 11, 2002, the cleanup of the Ground Zero site was expected to be complete by the end of May. Plans to reopen the No. 1 and No. 9 subway line stops were expected to be completed later in 2002. The reopening of the first downtown retailer was completed two weeks earlier (Wax and Diop, 2002).

A significant issue that has been raised by the devastation to office space concerns the relocation of employees. Since the attacks, 55 percent of businesses displaced by September 11 have indicated that they will return (Wax and Diop, 2002). Wax and Diop add that, "Businesses that aren't returning have largely relocated to midtown, New Jersey, and elsewhere" (Wax and Diop, 2002). The issue of relocation is important given the number of employees that have moved out of the affected area. In the article "Consultants push Wall St. to leave," Stephen Gandel writes that, "In all, 39,610 financial services jobs have been relocated from downtown in the last six months (Wax and Diop, 2002). More than half, 24,376 of those employees, have been moved to midtown (Gandel, 2002). In the article "Seeking Safety, Downtown Firms are Scattering," Charles V. Balgi adds, "that another 144,000 jobs are in jeopardy in a second wave of departures" (Balgi, 2002).

CHIMACUM HIGH SCHOOL EARTHQUAKE PREPAREDNESS PROGRAM

Program Description: This program involves high school students teaching elementary school students about earthquake preparedness. Each class designs its own project for communicating this information. School staff see the value of such peer education. For example, the class of 1997 designed a community service project. One element of the project was to participate in the school district's earthquake preparedness committee and provide input from the students. The students also researched the needs of classroom teachers, purchased supplies, and stocked each classroom with a "teacher's kit." They also researched and prepared personal "kid kits," which are sold for $7. The kid kits are a voluntary purchase. In addition, the students prepared an earthquake preparedness course script based on information from FEMA "Earthquake Dudes" and FEMA literature, a videotape, and an earthquake simulation with sound effects, which is available on request. Each class restocks the teacher's kit. High school students have taken American Red Cross courses, so shelters could be opened in high schools if needed.

Evaluation Information: Formal evaluation forms are completed after every class session by the regular classroom teacher and class students. All forms are on file. There are increased signs of school and community concern and awareness as elementary students discuss what they have learned with their parents and siblings.

Annual Budget: The school district budgeted $800 to $1,000 to purchase supplies for the teacher's kits.

Sources of Funding: The Chimacum school district and Chimacum class of 1997 fundraising.

Program Type: Teaching earthquake preparedness.

Target Population: Chimacum elementary school students.

Setting: Rural Western Washington Olympic Peninsula, in a community located near a newly documented, active earthquake fault line.

Project Startup Date: 1993.

From FEMA's *Partnerships in Preparedness*, A Compendium of Exemplary Practices in Emergency Management, Volume II, May 1997.

NEIGHBORS FOR DEFENSIBLE SPACE

Program Description: A grassroots volunteer program, Neighbors for Defensible Space developed out of a need to reduce the risk of uncontrolled wildfire in and around the fire-dependent district of Lake Tahoe, which has prevented catastrophic wildfires for more than 90 years.

There are three basic components in such a wildfire situation: weather, topography, and fuels. Fuels are the one element Neighbors for Defensible Space can control, and the program relies on its ability to either reduce, remove, or modify fuels. The North Lake Tahoe District program has been a model in public education and cooperative efforts in this area, and has been able to demonstrate that both fire protection and environmental concerns can be addressed when dealing with wildfires. Neighbors for Defensible Space is in its second year of a five-year plan of "prescribed burning," a program that returns low-intensity fire to the forest system. In addition, the community is in the process of adopting a joint long-range master plan with its Incline Village General Improvement District, which provides water, sewage, water treatment, recreational facilities, and sanitation.

The U.S. Forest Service owns more than 650 parcels of land in the community, which has obtained approximately $900,000 in congressional funds to manage the land. In 1991 the community's taxes paid to selectively harvest 750 acres of dead and dying timber at a cost of approximately $1 million. Forty-eight percent of property owners have involved their private lands in the effort (approximately 3,500 parcels).

Evaluation Information: Defensible Space was recognized by the National Commission on Wildfire Disasters (a congressional committee) as a model of public education and cooperative efforts that produce results in reducing wildfire risk to urban interface communities. Their publications are used by other fire and forestry agencies.

Annual Budget: $5,584 in 1995 from donations.

Sources of Funding: Primarily donations and outside agencies' earmarked funds. Local taxes, congressional funds, state forest stewardship funds, community donations, and property owners provide additional monies.

Program Type: Wildfire mitigation for the Reno/Lake Tahoe/Carson City region.

Target Population: 10,000 district residents.

Setting: Within and surrounding the Reno/Lake Tahoe/Carson City, Nevada region.

Project Startup Date: 1986.

From FEMA's *Partnerships in Preparedness*, A Compendium of Exemplary Practices in Emergency Management, Volume II, May 1997.

SPECIAL NEEDS AWARENESS PROGRAM (SNAP)

Program Description: After flooding occurred in areas of southeast Texas in October 1994, students in the Community Problem Solving class of Austin Middle School, Beaumont, Texas, responded to stories they had heard about people having difficulty during emergency evacuations. The students originated the idea for SNAP and established a pilot program in their community.

The goal of SNAP is to identify those persons, such as the elderly, mentally and physically challenged, or homebound, who would have difficulty in an emergency evacuation. These residents are given special SNAP signs for display only during an emergency. SNAP also notifies police, fire, and emergency management personnel that they should look for the SNAP signs to determine where assistance is needed in an evacuation.

SNAP distributes information on the program to civic organizations, churches, and government agencies in the area through letters, speaker's bureaus, and videotapes. The program has spread throughout the United States and internationally via the Internet and magazine articles.

Evaluation Information: Information on the program has been requested by agencies in 31 states, the Dominican Republic, and Australia. Three magazines—*Natural Hazard Observer*, *Wanted Magazine*, and *D.E.M. Digest*—have featured articles on the program. The 41 SNAP students from Beaumont Middle School who originated the program won first place in the intermediate division in the 1995 International Future Problem Solving (Community Problem Solving) Competition in Providence, Rhode Island.

Annual Budget: $1,200.

Sources of Funding: Beaumont Public Schools Foundation, Inc., FAD (Falcons Against Drugs), funds raised by SNAP team members, and personal donations.

Source for Additional Information: Mrs. Lynne Buchwald, Austin Middle School, Beaumont, Texas (409-866-8143).

Program Type: Emergency evacuation assistance.

Target Population: Elderly, physically and mentally challenged, and homebound residents who would require special assistance during an emergency.

Setting: Any residential area in any state; the SNAP program originated in Beaumont, Texas.

Project Startup Date: 1994.

From FEMA's *Partnerships in Preparedness*, A Compendium of Exemplary Practices in Emergency Management, Volume II, May 1997.

ARCADIA CHAMBER OF COMMERCE EMERGENCY PREPAREDNESS COMMITTEE FOR BUSINESS OWNERS

Program Description: The Arcadia Chamber of Commerce Emergency Preparedness Committee for Business Owners provides local business owners with a disaster identification packet. The informational packet contains instructions for self-assessment of damage by the owner, along with color-coded placards that correspond to the level of need (e.g., major, moderate, or minor/no damage). Immediately following a disaster, a business owner, using the guidelines provided in the packet, would determine the extent of help needed and display the appropriate color placard. Emergency service units surveying the city instantly would be able to identify areas that required immediate assistance and thus focus available resources on those areas with the greatest need. Instructions also are provided on what supplies are needed and what activities to perform after an earthquake.

Evaluation Information: Other cities and counties have requested information about the disaster identification packet and indicated an interest in replicating the program. Following a presentation to the Arcadia Coordinating Committee, the PTA expressed an interest in adapting the program for use in schools.

Annual Budget: None. Projects are funded individually.

Sources of Funding: Funds come from the Chamber of Commerce and the fire department; printing companies and manufacturers have donated printing and materials.

Program Type: Emergency preparedness information to help businesses identify their extent of need following a disaster.

Target Population: Arcadia business owners.

Setting: Arcadia, California.

Project Startup Date: 1992.

From FEMA's *Partnerships in Preparedness*, A Compendium of Exemplary Practices in Emergency Management, Volume II, May 1997.

PACIFIC GROVE, A MODEL FOR SMALL CITY DISASTER PREPAREDNESS

Program Description: In 1990, Pacific Grove, California (60 miles from the epicenter of the 1989 Loma Prieta earthquake) decided to prepare a comprehensive earthquake and disaster plan, following a study that showed the likelihood of a complete loss of utilities, sewer systems, and telephone services, as well as an overload of cellular systems and damage to streets and highway overpasses during an earthquake. City employees were sent to earthquake preparedness training courses given at the Governor's Office of Emergency Services' California Specialized Training Institute in San Luis Obispo. A disaster coordinator was hired to update the city's disaster plan. A Volunteers in Preparedness program was formed to train neighborhood emergency response teams, which include amateur radio operators and Boy Scouts, in earthquake preparedness, disaster medicine, how and when to turn off the gas, how to rescue victims trapped under earthquake debris, and firefighting. Lacking funding, the disaster coordinator enlisted retirement homes, volunteer organizations, public utilities, and emergency service agencies to join in the state's "Duck, Cover, and Hold" earthquake drill.

Evaluation Information: In 1994 Pacific Grove was cited as the only city (of 12) in Monterey County having an emergency planner and the only city to hold earthquake drills regularly. Pacific Grove received the Institute of Local Self Government's California Cities Helen Putnam Award for Excellence (honorable mention, public safety) in 1995. The city's preparedness programs have received innumerable media mentions.

Annual Budget: $28,000 (FEMA: $11,000 toward the disaster coordinator's salary; $14,000 from the city's fire department budget; and $3,000 from the city budget).

Sources of Funding: FEMA and city budgets.

Program Type: Disaster preparedness.

Target Population: Residents of Pacific Grove (17,000).

Setting: Pacific Grove, California.

Project Startup Date: 1990.

From FEMA's *Partnerships in Preparedness*, A Compendium of Exemplary Practices in Emergency Management, Volume II, May 1997.

DELAWARE CITY, COMMUNITY AWARENESS AND EMERGENCY RESPONSE COMMITTEE (DC-CAER)

Program Description: The DC-CAER, comprising representatives of the chemical industry; volunteer organizations; and public, state, and local governments, addresses mutual concerns involving a chemical plant complex near Delaware City. Formed voluntarily in 1985, the DC-CAER strives to meet three goals: to enhance emergency response capabilities, to test and evaluate these capabilities, and to foster knowledge about chemical-related hazards and protective measures. The DC-CAER maintains a comprehensive emergency response plan to deal with chemical emergencies at the plant; conducts training programs for emergency responders; coordinates annual field emergency response exercises and tabletop drills; conducts community outreach programs to disseminate emergency information; makes presentations about its programs to community, government, and professional organizations throughout Delaware and in other states; and has produced a video that is distributed to Delaware's Extremely Hazardous Substance facilities.

Evaluation Information: The county has received awards from the Chemical Manufacturers Association, National Coordinating Council on Emergency Management, and U.S. Environmental Protection Agency. There have been actual emergencies without injuries.

Annual Budget: None, but special projects have received more than $12,000 since 1985.

Sources of Funding: Shared among 11 chemical plants.

Program Type: Chemical emergency preparedness planning.

Target Population: 6,000 residents, emergency responders, and employees and visitors of 11 chemical plants.

Setting: Suburban environment with one small town.

Project Startup Date: 1985.

From FEMA's *Partnerships in Preparedness*, A Compendium of Exemplary Practices in Emergency Management, Volume II, May 1997.

ARLINGTON COUNTY EMERGENCY MANAGEMENT SYSTEM

Program Description: Arlington County's (VA) Emergency Management System was designed to provide the ability to respond to natural and/or technological disasters in a rapid and efficient manner. The system has three basic components: the Emergency Management Team (EMT), the Emergency Planning Team (EPT), and six functional task group teams. The EMT is composed of the directors of police, fire, public works, public affairs, and the County Manager's office. It is the core of the system and the decision-making body. The EPT is the think tank that anticipates future issues and makes recommendations to the EMT. The EPT and task groups brief the EMT hourly in the early stages of an incident (less frequently as the incident diminishes). During normal business, the EPT reviews the emergency operations plan to ensure that it is current. The EPT includes personnel from departments throughout the county, such as the police, sheriff, fire department, public works, public affairs, County Manager's office, parks and recreation, schools, technology and information services, and Department of Human Services. The six functional task group teams each have a different area of responsibility: shelters, communications, resources, routing and traffic control, employee support, and recovery. Members also include personnel from outside county government who have special expertise. Any of the EMT members can convene the entire team. Through the chain of command, fire and police chiefs would invoke the system. The emergency communications center would call system members,who would assemble in the emergency operations center (EOC). Each team is in a separate area of the EOC. They can communicate in person or by 800 MHz radio. As an incident unfolds, the task groups monitor it on primary radio channels to anticipate resource needs and so on.

Evaluation Information: The program has undergone independent evaluation and has received feedback from participants in the program. Two Air Force Reserve officers, both Individual Mobilization Augmentees, have reviewed the program and participated in annual disaster exercises in which the program is evaluated. Both commented that Arlington's emergency management system was extraordinarily well developed and considerably ahead of most jurisdictions in emergency management. After each exercise, participants fill out a critique to assess their knowledge of the exercise. Results indicate a high knowledge/comfort range.

Annual Budget: No funds were specifically allocated for this program. The Staff Assistant to the Fire Chief was responsible for maintaining the program, so that the only outlay was a portion of his annual salary. Currently, there are only ancillary costs: printing of manuals and documents and a portion of personnel expenses.

Sources of Funding: Arlington County Fire Department budget.

Program Type: Disaster preparedness and emergency management.

Target Population: All workers and residents of the county.

Setting: Countywide.

Project Startup Date: 1992.

THE TsunamiReady PROGRAM

TsunamiReady is an initiative that promotes tsunami hazard preparedness as an active collaboration among federal, state, and local emergency management agencies, the public, and the NWS tsunami warning system. This collaboration functions for the purpose of supporting better and more consistent tsunami awareness and mitigation efforts among communities at risk. Through the TsunamiReady program, NOAA's National Weather Service gives communities the skills and education needed to survive a tsunami before, during, and after the event. TsunamiReady was designed to help community leaders and emergency managers strengthen their local tsunami operations (NOAA, N/D).

The TsunamiReady program is based on the NWS StormReady model (which can be viewed by accessing http://www.stormready.noaa.gov/). The primary goal of TsunamiReady is the improvement of public safety during tsunami emergencies. As just stated, TsunamiReady is designed for those coastal communities that are at known risk of the tsunami hazard (tsunami hazard risk maps can be seen by accessing http://www.pmel.noaa.gov/tsunami/time/).

Traditionally, tsunami hazard planning along the U.S. West Coast and Alaska has been widely neglected because of the statistically low incidence of tsunamis. As result of that perceived "rarity," many individuals and communities have not worked to become as "tsunami-aware" as they could and should be. Among those communities that are considered to be prepared, that level of exhibited preparedness varies significantly (NWS, N/D).

However, as is true with the earthquakes and other rare events that generate tsunamis, avoidable casualties and property damage will only continue to rise unless these at-risk communities become better prepared for tsunamis. As previously mentioned, readiness involves two key components: *awareness* and *mitigation*. Awareness involves educating key decision makers, emergency managers, and the public about the nature (physical processes) and threat (frequency of occurrence, impact) of the tsunami hazard; mitigation involves taking steps before the tsunami occurs to lessen the impact (loss of life and property) of that event when it does occur. As is true with earthquakes, there is no question tsunamis will strike again.

The National Weather Service (NWS) TsunamiReady program was designed to meet both of the recognized elements of a useful readiness effort: it is designed to educate local emergency management officials and their public, and to promote a well-designed tsunami emergency response plan for each community.

Program Objectives

TsunamiReady promotes tsunami hazard readiness as an active collaboration among federal, state, and local emergency management agencies, the public, and the NWS tsunami warning system. This collaboration supports better and more consistent tsunami awareness and mitigation efforts among communities at risk. The main goal is improvement of public safety during tsunami emergencies. To meet this goal, the following objectives need to be met by the community:

- Create minimum standard guidelines for a community to follow for adequate tsunami readiness
- Encourage consistency in educational materials and response among communities and states
- Recognize communities that have adopted TsunamiReady guidelines
- Increase public awareness and understanding of the tsunami hazard
- Improve community preplanning for tsunami disasters

Program Methodology

The processes and guidelines used in the TsunamiReady program were modeled to resemble those of the National Weather Service "StormReady" program. TsunamiReady established minimum guidelines for a community to be awarded the TsunamiReady recognition, thus promoting minimum standards based upon expert knowledge rather than subjective considerations. Communities that accept the challenge to become TsunamiReady, and are deemed to have met these requirements set by the NWS TsunamiReady program, are designated as "TsunamiReady Communities." Guidelines to achieve TsunamiReady recognition are given in the following table, and discussed in detail in the pages immediately following. Four community categories (based upon the population of the community, and provided in the table's heading) are used to measure tsunami readiness.

Note that Guideline 3 has been skipped as it refers exclusively to the StormReady program, which shares these guidelines with the TsunamiReady program. This is a key factor to consider, as it ensures by default that all communities that are StormReady will also be TsunamiReady (as of 2002). As such, all communities being certified for TsunamiReady also must pass all StormReady criteria. StormReady requires access to local weather monitoring equipment (Guideline 3) and some further administrative requirements (Guideline 6). Other than that, the requirements are identical.

Guidelines	Population			
	<2,500	2,500–14,999	15,000–40,000	>40,000
1: Communications and Coordination				
24-hr warning point (WP)	X	X	X	X
Emergency Operations Center		X	X	X
2: Tsunami Warning Reception				
Number of ways for EOC/WP to receive NWS tsunami messages (if in range, one must be NWR with tone-alert; NWR-SAME is preferred)	3	4	4	4
3: Warning Dissemination				
Number of ways for EOC/WP to disseminate warnings to public	1	2	3	4
NWR tone-alert receivers in public facilities (where available)	X	X	X	X
For county/borough warning points, county/borough communication network ensuring information flow between communities	X	X	X	X
4: Community Preparedness				
Number of annual tsunami awareness programs	1	2	3	4
Designate/establish tsunami shelter/area in safe zone	X	X	X	X
Designate tsunami evacuation areas and evacuation routes, and install evacuation route signs	X	X	X	X
Provide written, locality-specific, tsunami hazard response material to public	X	X	X	X
In schools, encourage tsunami hazard curriculum, practice evacuations, and provide safety material to staff and students	X	X	X	X
5: Administrative				
Develop formal tsunami hazard operations plan	X	X	X	X
Yearly meeting/discussion by emergency manager with NWS	X	X	X	X
Visits by NWS official to community at least every other year	X	X	X	X

Guideline 1: Communications and Coordination Center

It is well known that key to any effective hazards management program is effective communication. This could not be truer when considering tsunami-related emergencies, since the arrival of the giant waves can occur within minutes of the initial precipitating event. These so-called "short-fused" events, therefore, require an immediate, but careful, systematic, and appropriate response. To ensure such a proper response, TsunamiReady requires that communities establish the following:

1. **24-Hour Warning Point.** It is the NWS, not the community, that determines a tsunami threat exists. Therefore, in order to receive recognition under the TsunamiReady Program, an applying agency needs to establish a 24-hour warning point (WP) that can receive NWS tsunami information in addition to providing local reports and advice to constituents. Typically, the functions of this type of facility merely are incorporated into the existing daily operation of a law enforcement or fire department dispatching (Emergency Communications Center (ECC)) point.

   For cities or towns without a local dispatching point, a county agency could act in that capacity for them. In Alaska, where there may be communities that have populations of less than 2,500 residents and no county agency to act as a 24-hour warning point, the community is required to designate responsible members of the community who are able to receive warnings 24 hours per day, and who have the authority to activate local warning systems. Specifically, the warning point is required to have:
   - 24-hour operations
   - Warning reception capability
   - Warning dissemination capability
   - Ability and authority to activate local warning system(s)

2. **Emergency Operations Center.** Agencies serving jurisdictions larger than 2,500 people are required to have the ability to activate an emergency operations center (EOC). It must be staffed during tsunami events to execute the warning point's tsunami warning functions. The following list summarizes the tsunami-related roles required of the EOC:
   - Activate, based on predetermined guidelines related to NWS tsunami information and/or tsunami events.
   - Staff with emergency management director or designee.
   - Establish warning reception/dissemination capabilities equal to or better than the warning point.
   - Maintain the ability to communicate with adjacent EOCs/warning points.
   - Maintain the ability to communicate with local NWS office or Tsunami Warning Center.

Guideline 2: Tsunami Warning Reception

Warning points and EOCs each need multiple ways to receive NWS tsunami warnings. TsunamiReady guidelines to receive NWS warnings in an EOC/WP require a combination of the following, based on population:

- NOAA Weather Radio (NWR) receiver with tone alert. Specific Area Message Encoding (SAME) is preferred. Required for recognition only if within range of transmitter.

- NOAA Weather Wire drop: Satellite downlink data feed from NWS.
- Emergency Managers Weather Information Network (EMWIN) receiver: Satellite feed and/or VHF radio transmission of NWS products.
- Statewide Telecommunications System: Automatic relay of NWS products on statewide emergency management or law enforcement system.
- Statewide warning fan-out system: State authorized system of passing message throughout warning area.
- NOAA Weather Wire via Internet NOAAport Lite: Provides alarmed warning messages through a dedicated Internet connection.
- Direct link to NWS office; e.g., amateur or VHF radio.
- E-mail from Tsunami Warning Center: Direct e-mail from Warning Center to emergency manager.
- Pager message from Tsunami Warning Center: Page issued from Warning Center directly to EOC/WP.
- Radio/TV via Emergency Alert System: Local radio/TV or cable TV.
- US Coast Guard broadcasts: WP/EOC monitoring of USCG marine channels.
- National Warning System (NAWAS) drop: FEMA-controlled civil defense hotline.

Guideline 4: Warning Dissemination

1. Upon receipt of NWS warnings or other reliable information suggesting a tsunami is imminent, local emergency officials must be able to communicate this threat information with as much of the population as possible. This is fundamental to making the preparedness program effective. As such, receiving TsunamiReady recognition requires that communities have one or more of the following means of ensuring timely warning dissemination to their citizens (based upon population, as described in the preceding table):
   - A community program that subsidizes the purchase of NWR. (NWR receiver with tone alert. SAME is preferred. Required for recognition only if within range of transmitter.)
   - Outdoor warning sirens.
   - Television audio/video overrides.
   - Other locally-controlled methods; e.g., local broadcast system or emergency vehicles.
   - Phone messaging (dial-down) systems.
2. It is required that at least one NWR, equipped with a tone alert receiver, be located in each critical public access and government-owned building, and must include 24-hour warning point, EOC, School Superintendent office, or equivalent. Critical public access buildings are defined by each community's tsunami warning plan. Locations that are recommended for inclusion by the NWS include all schools, public libraries, hospitals, fairgrounds, parks and recreational areas, public utilities, sports arenas, Departments of Transportation, and designated shelter areas. (SAME is preferred. This is required for recognition only if the community exists within range of a transmitter.)

3. Counties/Boroughs only: A county-/borough-wide communications network ensuring the flow of information among all cities and towns within those administrative borders. This would include provision of a warning point for the smaller towns, and fanning out of the message as required by state policy.

Guideline 5: Community Preparedness

Public education is vital in preparing citizens to respond properly to tsunami threats. An educated public is more likely to take the steps required to receive tsunami warnings, recognize potentially threatening tsunami events when they exist, and respond appropriately to those events. Therefore, communities that are seeking recognition in the TsunamiReady Program must be able to:

- Conduct or sponsor tsunami awareness programs in schools, hospitals, fairs, workshops, and community meetings (the actual number of talks that must be given each year is based upon the community's population).
- Define tsunami evacuation areas and evacuation routes, and install evacuation route signs.
- Designate a tsunami shelter/area outside the hazard zone.
- Provide written tsunami hazard information to the populace, including:
  - Hazard zone maps
  - Evacuation routes
  - Basic tsunami information

  These instructions can be distributed through mailings (utility bills, for example), within phone books, and posted at common meeting points located throughout the community, such as libraries, supermarkets, and public buildings.
- Local schools must meet the following guidelines:
  - Encourage the inclusion of tsunami information in primary and secondary school curriculums. NWS will help identify curriculum support material.
  - Provide an opportunity biennially for a tsunami awareness presentation.
  - Schools within the defined hazard zone must have tsunami evacuation drills at least biannually.
  - Provide written safety material to all staff and students.
  - Have an earthquake plan.

Guideline 6: Administrative

No program can be successful without formal planning and a proactive administration. The following administrative requirements are necessary for a community to be recognized in the TsunamiReady Program:

1. A tsunami warning plan must be in place and approved by the local governing body. This plan must address the following:
   - Warning point procedures
   - EOC activation guidelines and procedures
   - Warning point and EOC personnel specification
   - Hazard zone map with evacuation routes

- Procedures for canceling an emergency for those less-than-destructive tsunamis
- Guidelines and procedures for activation of sirens, cable TV override, and/or local system activation in accordance with state Emergency Alert System (EAS) plans, and warning fan-out procedures, if necessary
- Annual exercises

2. Yearly visits or discussions with local NWS Forecast Office Warning Coordination Meteorologist or Tsunami Warning Center personnel must be conducted. This can include a visit to the NWS office, a phone discussion, or e-mail communication.
3. NWS officials will commit to visit accredited communities, at least every other year, to tour EOCs/Warning Points and meet with key officials.

Administration of the TsunamiReady Program

Oversight of the TsunamiReady program is accomplished within the NWS by the National StormReady Board (the Board). The Board is responsible for changes in community recognition guidelines. Proposed guideline changes shall be directed to the Board for action. The Board consists of the NWS Regional Warning Coordination Meteorologist (WCM) Program Leaders, the National WCM Program Manager, a Federal Emergency Management Agency (FEMA) representative, a National Emergency Management Association (NEMA) representative, and an International Association of Emergency Managers (IAEM) representative.

Oversight of the TsunamiReady program at the local level is provided by the appropriate local StormReady board. The local StormReady board has the authority to enhance TsunamiReady to fit regional situations. At a minimum, this board consists of:

- NWS Weather Forecast Office's Meteorologist-in-Charge
- NWS Weather Forecast Office's Warning Coordination Meteorologist
- State emergency service director or designee
- Local emergency management association president or designee
- Tsunami Warning Center's Geophysicist-in-Charge
- Tsunami Hazard Mitigation Program representative

The Local StormReady Board is responsible for all steps leading to the recognition of the TsunamiReady community. This includes implementing procedures for site verification visits and application review.

Benefits of the TsunamiReady Program

The following benefits of participation in the TsunamiReady Community program include:

- The community is more prepared for the tsunami hazard
- Regularly scheduled education forums increase public awareness of existing dangers

- Contact with experts (emergency managers, researchers, NWS personnel) is increased and likewise, enhanced
- Community readiness resource needs are identified
- Positioning to receive state and federal funds is improved
- Core infrastructure to support other community concerns is enhanced
- The public is allowed the opportunity to see firsthand how their tax money is being spent in hazard programs

Conclusion

Through the TsunamiReady program, NOAA's National Weather Service gives communities the skills and education needed to survive a tsunami before, during and after the event. TsunamiReady helps community leaders and emergency managers strengthen their local tsunami operations. TsunamiReady communities are better prepared to save lives from the onslaught of a tsunami through better planning, education, and awareness. Communities have fewer fatalities and property damage if they plan before a tsunami arrives. No community is tsunami proof, but TsunamiReady can help communities save lives.

Sources:

FEMA. 2004. Fact Sheet: Tsunamis. http://www.govexec.com/dailyfed/1104/111804h1. htm.

Folger, Tim. 1994. "Waves of Destruction," *Discover Magazine*, May, 69–70.

NOAA (National Oceanic & Atmospheric Administration). N/D. The National Tsunami Hazard Mitigation Program Brochure. http://wcatwc.arh.noaa.gov/tsunamiready/ trbrochure.pdf.

NTHMP (National Tsunami Hazard Mitigation Program). 2003. Frequently Asked Questions. http://www.pmel.noaa.gov/tsunami-hazard/tsunami_faqs.htm.

NWS. N/D. TsunamiReady; The Readiness Challenge. http://www.prh.noaa.gov/ptwc/ tsunamiready/tsunami_ready_full_document.pdf.

7. The Disciplines of Emergency Management: Communications

INTRODUCTION

Communications has become an increasingly critical function in emergency management. The dissemination of timely and accurate information to the general public, elected and community officials, and the media plays a major role in the effective management of disaster response and recovery activities. Communicating preparedness, prevention, and mitigation information promotes actions that reduce the risk of future disasters. Communicating policies, goals, and priorities to staff, partners, and participants enhances support and promotes a more efficient disaster management operation. In communicating with the public, establishing a partnership with the media is key to implementing a successful strategy.

This chapter defines the mission of an effective disaster communications strategy, outlines four critical assumptions that serve as the foundation for such a strategy, and identifies the various audiences or customers for disaster communications. The requirements for establishing a disaster communications infrastructure are defined, the difficulties in communicating risk are explored, and a strategy for communicating disaster mitigation and preparedness messages is discussed. Essential to any communications strategy is a practical guide to working with the media, which is also provided. Throughout the chapter, FEMA and the FEMA Public Affairs experiences are used as the principal example. In defining the elements of a crisis communications infrastructure used during the disaster response and recovery, the public affairs operations of FEMA are used as a model.

MISSION

The mission of an effective disaster communications strategy is to provide timely and accurate information to the public in all four phases of emergency management:

- *Mitigation*—to promote implementation of strategies, technologies, and actions that will reduce the loss of lives and property in future disasters
- *Preparedness*—to communicate preparedness messages that encourage and educate the public in anticipation of disaster events
- *Response*—to provide to the pubic notification, warning, evacuation, and situation reports on an ongoing disaster

- *Recovery*—to provide individuals and communities affected by a disaster with information on how to register for and receive disaster relief

ASSUMPTIONS

The foundation of an effective disaster communications strategy is built on the following four critical assumptions:

- Customer Focus
- Leadership Commitment
- Inclusion of Communications in Planning and Operations
- Media Partnership

Customer Focus

An essential element of any effective emergency management system is a focus on customers and customer service. This philosophy should guide communications with the public and with all partners in emergency management. A customer service approach includes placing the needs and interests of individuals and communities first, being responsive and informative, and managing expectations. The FEMA emergency information field guide illustrates the agency's focus on customer service and its strategy of getting messages out to the public as directly as possible. The introduction to the guide states the following:

> As members of the Emergency Information and Media Affairs team, you are part of the frontline for the agency in times of disaster. We count on you to be ready and able to respond and perform effectively on short notice. Disaster victims need to know their government is working. They need to know where and how to get help. They need to know what to expect and what not to expect. Getting these messages out quickly is your responsibility as members of the Emergency Information and Media Affairs team. (FEMA, 1998)

The guide's Mission Statement reinforces this point further:

> To contribute to the well-being of the community following a disaster by ensuring the dissemination of information that:

> - Is timely, accurate, consistent, and easy to understand
> - Explains what people can expect from their government

> Demonstrates clearly that FEMA and other federal, state, local and voluntary agencies are working together to provide the services needed to rebuild communities and restore lives (FEMA, 1998)

The customers for emergency management are diverse. They include internal customers, such as staff, other federal agencies, states, and other disaster partners. External customers include the general public, elected officials at all levels of government, community and business leaders, and the media. Each of these customers has special needs, and a good communications strategy considers and reflects their requirements.

Leadership Commitment

Good communications starts with a commitment by the leadership of the emergency management organization to sharing and disseminating information both internally and externally. The director of any emergency management organization must openly endorse and promote open lines of communications among the organization's staff, partners, and publics in order to effectively communicate. This leader must model this behavior in order to clearly illustrate that communications is a valued function of the organization.

In the 1990s, FEMA Director James Lee Witt embodied FEMA's commitment to communicating with the FEMA staff and partners, the public, and the media. Director Witt was a strong advocate for keeping FEMA staff informed of agency plans, priorities, and operations. Director Witt characterized a proactive approach in communicating with FEMA's constituents. His accessibility to the media was a significant departure from previous FEMA leadership. Director Witt exhibited his commitment to effective communications in many ways:

- He held weekly staff meetings with FEMA's senior managers and required that his senior managers hold regular staff meetings with their employees.
- He published an internal newsletter to employees entitled "Director's Weekly Update" that was distributed to all FEMA employees in hard copy and on the agency electronic bulletin board that updated employees on agency activities.
- He made himself and his senior staff available to the media on a regular basis, especially during a disaster response, to answer questions and to provide information.
- During a disaster response, he held media briefings daily and sometimes two to three times a day.
- He would hold special meetings with victims and their families.
- He led the daily briefings among FEMA partners during a disaster response.
- He devoted considerable time to communicating with members of Congress, governors, mayors, and other elected officials during both disaster and nondisaster times.
- He met four to five times per year with the State Emergency Management Directors, FEMA's principal emergency management partners.
- He gave speeches all over this country and around the world to promote better understanding of emergency management and disaster mitigation.

Through his leadership and commitment to communications, FEMA became an agency with a positive image and reputation. Communications led to increased success in molding public opinion and garnering support for the agency's initiatives in disaster mitigation.

Inclusion of Communications in Planning and Operations

The most important part of leadership's commitment to communications is inclusion of communications in all planning and operations. This means that a communications specialist is included in the senior management team of the emergency management organization. It means that communications issues are considered in the decision-making processes and that a communications element is included in all organizational activities, plans, and operations.

In the past, communicating with external audiences, or customers, and in many cases internal customers, was not valued or considered critical to a successful emergency management operation. Technology has changed that equation. In today's world of 24-hour television and radio news and the Internet, the demand for information is never-ending, especially in an emergency response situation. Emergency managers must be able to communicate critical information in a timely manner to their staff, partners, the public, and the media.

To do so, the information needs of the various customers and how best to communicate with these customers must be considered at the same time that planning and operational decisions are being made. For example, a decision process on how to remove debris from a disaster area must include discussion of how to communicate information on the debris removal operation to community officials, the public, and the media.

During the many major disasters that occurred in the 1990s, FEMA Director Witt assembled a small group of his senior managers who traveled with him to the sites of disasters and worked closely with him in managing FEMA's efforts. This group always included FEMA's Director of Public Affairs. Similarly, when planning FEMA's preparedness and mitigation initiatives, Director Witt always included staff from Public Affairs in the planning and implementation phases. Every FEMA policy, initiative, or operation undertaken during this time included consideration of the information needs of the identified customers and a communications strategy to address these needs.

Media Partnership

The media plays a primary role in communicating with the public. No government emergency management organization could ever hope to develop a communications network comparable to those networks already established and maintained by television, radio, and newspaper outlets across the country. To effectively provide timely disaster information to the public, emergency managers must establish a partnership with their local media outlets.

The goal of a media partnership is to provide accurate and timely information to the public in both disaster and nondisaster situations. The partnership requires a commitment by both the emergency manager and the media to work together, and it requires a level of trust between both parties.

Traditionally, the relationship between emergency managers and the media has been tenuous. There often has been a conflict between the need of the emergency manager to respond quickly and the need of the media to obtain information on the response so it can report it just as quickly. This conflict sometimes resulted in inaccurate reporting and tension between the emergency manager and the media. The loser in this conflict is always the public, which relies on the media for its information.

It is important for emergency managers to understand the needs of the media and the value they bring to facilitating response operations. An effective media partnership provides the emergency manager with a communications network to reach the public with vital information. Such a partnership provides the media with access to the disaster site, access to emergency managers and their staff, and access to critical information for the public that informs and ensures the accuracy of their reporting.

An effective media partnership helps define the roles of the emergency management organizations, to manage public expectations and to boost the morale of the relief workers and the disaster victims. All these factors can speed the recovery of a community from a disaster event and promote preparedness and mitigation efforts designed to reduce the loss of life and property from the next disaster event.

AUDIENCES/CUSTOMERS

In order to effectively communicate disaster information, emergency managers must clearly identify their various audiences and customers. Included in many of these audiences are both partners and stakeholders. Basic emergency management audiences include the following:

- *General public.* The largest audience of which there are many subgroups, such as the elderly, the disabled, minority, low income, youth, and so on, and all are potential customers.
- *Disaster victims.* Those individuals affected by a specific disaster event.
- *Business community.* Often ignored by emergency managers but critical to disaster recovery, preparedness, and mitigation activities.
- *Media.* An audience and a partner critical to effectively communicating with the public.
- *Elected officials.* Governors, mayors, county executives, state legislators, and members of Congress.
- *Community officials.* City/county managers, public works, department heads.
- *First responders.* Police, fire, and emergency medical services.
- *Volunteer groups.* American Red Cross, Salvation Army, the NVOADs, and so on who are critical to first response to an event.

Communications with some of these customers such as the first responders is accomplished principally through radio and phone communications, as described in Chapter 4. Communicating with most of these other audiences is accomplished through briefings, meetings, provision of background materials, and, in some instances, one-on-one interviews. Communications strategies, plans, and operations should be developed to meet the information needs of each of these customers and staffed and funded accordingly.

CRISIS COMMUNICATIONS: RESPONSE AND RECOVERY

Communicating with the public in the midst of a disaster response and recovery effort can be difficult. There are often conflicting reports on casualties and damages and usually some level of confusion among responders. Add to this situation the expectation of the public to get information almost instantaneously and the demands made by the new 24-hour news culture.

The provision of timely and accurate information directly to the public and to the media is critical to the success of any response and recovery effort. An effective communications strategy allows emergency managers and community officials at all levels of government to provide information and comfort to disaster victims and, at the same time, to manage expectations. Regular communications with the public and the media helps ensure that accurate information is being disseminated and reduces the chances for misinformation and rumors. Monitoring direct communications with victims and media reports helps identify potential problems with misinformation and rumors and allows emergency officials to address these issues before they become too widespread and damaging.

In the 1990s, FEMA built a communications infrastructure designed to disseminate critical information to the public and the media and to monitor and correct misinformation during FEMA's disaster response and recovery operations. The two key elements of FEMA's crisis communications infrastructure are staff support and technology.

Staff Support

FEMA's Office of Public Affairs (which for a time was called the Office of Emergency Information and Media Affairs) was responsible for managing day-to-day communications activities for the agency and, during a disaster, for managing a cadre of public affairs Disaster Assistance Employees (DAEs). Public Affairs staff was responsible for establishing and managing Joint Information Centers both at FEMA headquarters and in the field and for working cooperatively with FEMA's Community Relations staff.

Public Affairs Officers

The individuals primarily responsible for carrying out this mission are the FEMA public affairs officers (PAOs). PAOs develop and implement strategies to instill confidence in the community that all levels of government are working in partnership to restore essential services and help individuals begin to put their lives back together. They manage expectations so that disaster victims have a clear understanding of all disaster response, recovery, and mitigation services available to them. An overarching goal is to provide authoritative information to the public to combat misinformation.

Joint Information Center

The structure FEMA uses to implement public affairs activities after a disaster is the Joint Information Center (JIC). FEMA determines the need for a JIC, and if one is established it becomes the central point for coordination of emergency public information, public affairs activities, and media access to information about the latest developments. The JIC is a physical location where PAOs from involved agencies come together to coordinate the release of accurate and consistent information to the media and the public. For a major disaster, a JIC may be established at both FEMA headquarters and on the disaster site. The on-site JIC is preferably co-located with the disaster field office. The chief spokesperson for headquarters JIC is the FEMA director of public affairs, and the chief spokesperson at the on-site JIC is the lead FEMA PAO.

Community Relations

A partner in FEMA's public affairs operation is the Community Relations staff. The community relations function typically is performed jointly by federal and state personnel, but may also include locally hired people who know the community well. Field officers are organized into teams and deployed into affected communities to gather and disseminate information about the response and recovery operation that becomes part of the communications process. They work closely with affected states to identify community leaders and neighborhood advocacy groups to assist in disseminating information and identifying unmet needs.

Technology

A valuable means of communications in postdisaster scenarios is the toll-free number, which has become a core element of FEMA recovery initiatives. The toll-free number is used to inform victims about the type of assistance they may be available to receive and allows them to apply for such assistance. The toll-free number is included in all forms of information and communication generated by the disaster event. An example of its usage is that during the first month after the terrorist events of September 11, 2001, more than 20,000 people called the toll-free FEMA number.

The Internet has become an increasingly popular and effective method of disseminating information to the public, and this trend will continue. FEMA's Web site traffic has grown from an average of 20,000 people per week to more than 3 million. This includes users from more than 50 countries. During major disasters, the Office of Public Affairs immediately posts a special section and keeps it updated. Real-time situation reports, maps, graphics, and links to other Internet sites are posted. In addition, nearly 6,000 clients receive FEMA updates via e-mail. The interactive nature of the Internet has not yet been completely harnessed by the emergency management community and provides an opportunity to expand relationships with the public in the future.

During the 1990s, the FEMA Office of Public Affairs developed several innovative ways of disseminating information to the public. These methods have now been used in more than 200 disasters, including the Midwest floods, the Northridge earthquake, the Oklahoma City bombing, and record hurricane seasons. FEMA credits the new methods with improving its ability to get vital information out to the public and helping rebuild the agency's credibility and the nation's comfort level with its emergency management system. Some of the information dissemination methods are described as follows:

- *The Recovery Channel* provides television coverage of briefings and interviews with experts in multiple languages. Using portable satellite dishes, the signal is beamed into shelters. Network and local television news use this material. Cable television has cooperated, and a network of cable systems is committed to live Recovery Channel coverage. After the Northridge earthquake, Recovery Channel programming reached 680,000 victims on 125 cable systems in Los Angeles, with an additional potential audience of 4 million.

- *The Recovery Times* combines the latest desktop publishing technology with electronic transmission of stories and images to one printing contractor for all disasters. Prepackaging information has enabled quick publication and distribution of emergency information in an extraordinary community outreach effort. During the Midwest floods, FEMA published and distributed *Recovery Times* newspapers in nine states.
- *FEMAFAX/Spectrafax* uses the latest computerized facsimile system. Technology, comprehensive databases, and 48 telephone lines allow rapid, targeted information distribution. The system also has a fax-on-demand service. Clients select from more than 2,000 documents and material is transmitted automatically.
- *The FEMA Radio Network (FRN)* is a digitized audio production and distribution system. Radio stations can record soundbites and public service announcements with disaster officials and scientific experts. The state-of-the-art studio supports news conferences and interviews. Stations reach all this through a toll-free number.
- *The Recovery Radio Network* distributes live broadcasts of emergency public information. It uses the Emergency Alert System (EAS) network to provide a pool feed to local radio stations that are still operating.
- *The FEMA Automatic Internet Emergency News and Situation Report Distribution Service* sends subscribers news releases and disaster situation reports via e-mail.

More information on these programs can be obtained on the FEMA Web site: www.fema.gov/about/eima.htm.

COMMUNICATING PREPAREDNESS AND MITIGATION MESSAGES

The objective of communicating preparedness and mitigation messages to the public is to educate, inform, raise awareness, and promote support for taking action before a disaster strikes.

Risk communication and public awareness programs can be undertaken in the wake of disasters or during times of normalcy. Communication of risk is an area of growing interest in the field and is discussed in more length later in this section. Public awareness is needed to gain approval for any type of emergency management measure. To implement programs, the public has to agree that a hazard exists, that it should be reduced, and that the proposed program is an appropriate measure. To achieve this consensus, the public must be involved as a partner in the process. In today's political climate, new programs usually are negotiated with the public, not decreed from officials. The case study on FEMA's Project Impact illustrates this type of approach well.

CASE STUDY

PROJECT IMPACT

FEMA's promotion of Project Impact provides an excellent example of how to sell disaster mitigation programs to the public. The FEMA public affairs team engaged and involved the public and explained the program in terms they could understand and value, partnered with the media to get its message out, and made effective use of policy windows.

Project Impact is a community-based mitigation initiative, facilitated and partially funded by FEMA. It includes getting local businesses to partner with the local government and community organizations to prepare for and reduce the effects of future disasters. Preliminary surveys had indicated that communities were interested in reducing risk, so Project Impact was born.

The communications team's first challenge was to frame the program in terms that the public could understand. Although the program is a mitigation initiative, the team wanted to move away from emergency management jargon and describe the program in a manner with which the public would be more familiar. The slogan "put FEMA out of business" was developed. The term *mitigation* was replaced with *disaster-resistant*, and then *prevention*, and finally *risk reduction*. The slogans "prevention pays" and "prevention power" were used to reinforce the message.

A public affairs campaign was launched, both at the grassroots levels within target communities and through the print and television media when possible. The communications model employed was based on the following guidelines:

- *Keep the message simple and understandable*. Literature was developed at the fourth-grade level. A "three little pig's analogy" was used to help explain the difference between preparedness and prevention.
- *Stick to the message or point*. Spokespeople used a "remember three things" tactic, whereby three main points are repeatedly mentioned in straight, clear language. Also, the Project Impact pamphlet was reduced to one page, containing five simple prevention tips.
- *Explain what's in it for the public*. The selling point to the public was that Project Impact would result in fewer losses from future disasters.
- *Educate the media on mitigation*. A media partner guide was developed to help Project Impact proponents explain to the media why mitigation is a story, why it's important, and how the media could help spread the message.
- *Involve partners*. The Salvation Army and Red Cross were solicited as partners in promoting Project Impact.
- *You are the message*. Project Impact hats and T-shirts were provided to team members.

From a media standpoint, articles were placed in the *USA Today* Op/Ed section and *Parade* magazine, and Al Roker of the *Today Show* did a spot on Project Impact. The team also took advantage of policy windows by sneaking prevention messages into interviews during major disaster operations. Spokespeople such as FEMA's Kim Fuller promoted Project Impact in interviews during Hurricanes Irene and Floyd. An animated video on mitigation steps was provided to the networks and displayed during the interviews. Also, pre-prepared press releases on how people could rebuild better for the future were provided to the media.

Source: Interview with Kim Fuller, October 2001

Communicating Risk

Most emergency management professionals believe that a more concerted effort to define and communicate risk to the public needs to be made. The value of warning and evacuation systems have been proven time and again but are still often underused. Knowledge of risk does not help if the public is not informed of the danger and the actions they can take to reduce it. Bridging this knowledge gap between the scientific community and the public at large is a major area of emergency management study today.

Risk Communication Theory

The book *Disasters by Design* by Dennis Mileti provides some valuable information on risk communication. Mileti breaks information sources for hazard awareness programs into three categories: authorities, news media, and peers. Obviously, official sources provide the most credibility. Research has shown that hazard awareness campaigns are most effective when they rely on a mix of techniques and information sources. Typically, radio and television are best for initiating or maintaining awareness, and printed materials may be best at providing detailed information.

SEVERE WEATHER WATCHES AND WARNINGS DEFINITIONS

Flood Watch: High flow or overflow of water from a river is possible in the given time period. It can also apply to heavy runoff or drainage of water into low-lying areas. These watches generally are issued for flooding that is expected to occur at least six hours after heavy rains have ended.

Flood Warning: Flooding conditions are actually occurring or are imminent in the warning area.

Flash Flood Watch: Flash flooding is possible in or close to the watch area. Flash Flood Watches generally are issued for flooding that is expected to occur within six hours after heavy rains have ended.

Flash Flood Warning: Flash flooding is actually occurring or is imminent in the warning area. It can be issued as a result of torrential rains, a dam failure, or an ice jam.

Tornado Watch: Conditions are conducive to the development of tornadoes in and close to the watch area.

Tornado Warning: A tornado has actually been sighted by spotters or indicated on radar and is occurring or imminent in the warning area.

Severe Thunderstorm Watch: Conditions are conducive to the development of severe thunderstorms in and close to the watch area.

Severe Thunderstorm Warning: A severe thunderstorm has actually been observed by spotters or indicated on radar and is occurring or imminent in the warning area.

Tropical Storm Watch: Tropical storm conditions with sustained winds from 39 to 73 mph are possible in the watch area within the next 36 hours.

Tropical Storm Warning: Tropical storm conditions are expected in the warning area within the next 24 hours.

Hurricane Watch: Hurricane conditions (sustained winds greater than 73 mph) are possible in the watch area within 36 hours.

Hurricane Warning: Hurricane conditions are expected in the warning area in 24 hours or less.

Source: FEMA, www.fema.gov

Different message characteristics include the amount of material, speed of presentation, number of arguments, repetition, style, clarity, ordering, forcefulness, specificity, consistency, accuracy, and extremity of position advocated. Information characteristics should be tailored for the communications goal (i.e., awareness or adoption) and for the target audience. For example, the Red Cross publishes awareness guides and manuals specific to targeted groups, such as schools, hospitals, corporations, city managers, emergency managers, and the media.

Message types vary as well. Some programs focus on content, such as scientific data or technical information about a hazard, but such information generally is processed and obtained by a small number of people. Conversely, practical instructions focus on the protective response, not the hazard itself. The simplest form of practical instruction is the "prompt," a sign that defines a single contingency and action, such as "pull lever in case of fire." Prompts are more likely to attract attention, be readily comprehended, and retained for future use. Other message styles, such as "attribute portrayal strategy," emphasize the advantages of a proposed hazard adjustment, and "fear appeals" describe the potential negative consequences of not taking the desired risk-reduction action.

Risk communication theory is based on the assumption that people leave themselves vulnerable because they are uninformed or unconvinced about the consequences of their actions. Providing accurate, helpful information would then change people's beliefs about a hazard and lead to an adoption of appropriate mitigation strategies. This is a bit of an oversimplification because many other factors and obstacles are involved, but it illustrates the general principle. The major obstacles to communicating risk and changing people's behavior include competing demands for attention, complacency, denial, and conflicts with existing beliefs.

Mileti breaks the risk communication and new behavior adoption process into the following eight steps:

1. Hearing the warning
2. Believing that it is credible
3. Confirming that the threat exists
4. Personalizing the warning and confirming that others are heeding it
5. Determining whether protective action is needed
6. Determining whether protective action is feasible
7. Determining what protective action to take
8. Taking the protective action

The field is still evolving to determine how best to influence people at each stage of the process. Most public awareness campaigns have been designed to improve disaster preparedness for near-term, high-probability threats. Less is known about what it takes to motivate people to prepare for longer-term, lower-probability events during times of normalcy. This will be an important area of study in the future.

Risk Communication Concerns

One risk communications dilemma is how to get accurate risk information to the public when there are so many other competing, and possibly conflicting, information sources. The government has no control over what unofficial sources say because it can't regulate talking heads, so-called experts, and Web sites. Partnering with the media to provide a steady stream of consistent and accurate information from responsible authorities is the best way to overcome this obstacle.

Other major issues affecting risk communication programs are when to warn the public and how much information to provide. The hurricane scenario provides the ideal model: forecasters identify the storm, watches and warnings are issued, time-frames and probabilities are provided, and the public is given clear instruction on when and how to take protective action. Communicating the risk of other hazards is not always so clear-cut, however. In the wake of the September 11, 2001 terrorist attacks, several general and unspecified terrorism threats were issued by the federal government. Weighty issues to be considered by public officials were (1) with hundreds of tips pouring in, at what point is the risk considered legitimate enough to pass on to the public, and (2) how much information on the threats should be shared.

With the first issue, officials must balance the duty to warn citizens of impending danger with concerns about unnecessarily panicking people and disrupting society. There are political and economic concerns as well. Too many false warnings could lead to a loss of credibility and public inattention to future warnings. With the second issue, officials must balance concerns about frightening the public with unthinkable rumors, and perhaps compromising important information sources, against the need to provide practical, helpful information. General, unspecified warnings may protect intelligence channels, but they do not do much to help the public prepare for the event. These are delicate issues, and a consensus on how best to responsibly educate the public about risk without unnecessarily alarming them has yet to be reached. The case study on earthquake risk in Parkfield, California, explores these issues as well.

CASE STUDY

RISK COMMUNICATION—PARKFIELD, CALIFORNIA

One of the issues facing emergency managers is when to notify the public of a disaster risk. A desire to protect citizens must be weighed against concerns about unnecessarily alarming people, disrupting the economy, and upsetting public officials. The tension between the sheriff and the beach town mayor in the movie *Jaws* exemplifies this issue well. Even though the sheriff warned the mayor of the continuing risk of shark attacks, the mayor would have none of such talk during the busiest tourist weekend of the season and kept the information from the public.

A real-life scenario with a different end result, to date anyway, involved a U.S. effort at earthquake prediction in Parkfield, California, a town adjacent to the San Andreas fault. In 1985, a U.S. Geological Survey (USGS) analysis of previous earthquakes on a particular fault section indicated a strong likelihood of a repeat event by the end of the decade. The director of the USGS issued a formal public broadcast of the quake warning in April 1985 stating there was a 90 percent probability of a magnitude 5.5 to magnitude 6.0 earthquake some time between 1985 and 1993 in the Parkfield area. It also stated that a 10 percent probability existed for a magnitude 7.0 quake. By November 1988, the National Earthquake Prediction Evaluation Council (NEPEC) and the California Earthquake Prediction Evaluation Council had endorsed the prediction.

The release of the information became a national media event and precipitated a media campaign in central California involving newspapers, radio, and television that lasted years. In 1988, the California Governor's Office of Emergency Services published a detailed brochure and mailed it to 120,000 households considered at risk. It covered information about the earthquake hazard, the prediction, a possible short-term warning, and how to take action.

But the expected earthquake has not occurred. Further analysis showed that, though the successive repeat of similar but not identical quakes might be expected on individual fault sections, the amount of time between them may be highly variable. Also, confidence in predictors based on estimates of recurrence intervals has decreased in the scientific community. This case raises the issue of what to do with risk information. The duty to warn and protect the public must be balanced with fears about disrupting society with potentially unreliable risk information. It remains to be seen whether the correct decision was made for Parkfield, California.

Source: *Disasters by Design*, by Dennis Mileti

WORKING WITH THE MEDIA

General

The media has always been naturally drawn to disasters and emergencies because they are compelling human interest stories and provide dramatic footage. With the advent of 24-hour news stations and near real-time coverage via the Internet, the role of the media in disaster response has been magnified. In the response phase, the media often provides the most effective and efficient means for providing timely and accurate information to disaster victims and the general public. In addition, the media can play a critical role in communicating recovery information and in building support for preparedness and mitigation activities.

The biggest development in the media world over the last decade is the 24-hour news cycle. Between CNN, the major networks' all-news stations, their respective Web sites, and the emergence of other independent reporting mechanisms, there is simply more air time and copy to be filled. This translates into increased coverage of disasters and emergencies and creates a demand for timely information. These pressures are only likely to grow in the future. As television becomes increasingly specialized and the number of cable channels expands, it would not be surprising within the foreseeable future to see the advent of a 24-hour *Disaster News Network*, replete with "hurricane-cams" and "on-the-fault" reporting.

The media can make a strong contribution to emergency management. Effective warnings broadcast through the media are widely credited with reducing casualties from hurricanes, tornadoes, and floods. There is often no better or quicker way to get warning messages out. The media can also facilitate assistance to disaster-stricken areas and provide reassurance to the public about the welfare of victims. Also, good science reporting can inform the public about hazards and educate them on hazard-reduction behaviors.

Media as a Partner

Working with the media provides both a challenge and an opportunity. As discussed, the media can be a valuable element of emergency operations, disseminating important information and calling attention to urgent issues, or can be a thorn in the emergency management official's side, distributing misleading information and misguided criticism. The key to a beneficial and productive relationship is to view the media as an important partner and treat them as such.

A great example of this approach is FEMA of the 1990s. In the early 1990s, FEMA was an agency under fire, with legislators pondering its abolishment and the media producing a steady stream of criticism after a series of poorly perceived disaster response efforts. When James Lee Witt was appointed FEMA Director in 1992, he recognized communications as a key area for improvement and took appropriate measures to establish a more open and productive relationship with the media.

As a precursor to this step, the communications staff was provided with the tools and equipment needed to get the job done. Press veteran Morrie Goodman was brought on board, the office began identifying actions it could take to better partner with the media, and a host of new practices were implemented. FEMA provided the media with flyover pictures and videos from closed sites. It posted transcripts and audio of news conferences on the Internet. It created an on-site press and studio room. It provided press conferences via satellite link. It partnered with *USA Today* to include FEMA informational inserts in certain editions of the paper. The press was even provided with an area in the emergency operations center at major crises.

FEMA in turn used the press to promote key information to the public, such as toll-free numbers for victims to call to apply for assistance. Director Witt made it a point to constantly thank the media for their role in helping to get important messages out.

As a result of this open, collaborative approach, the public was better informed, FEMA received better press, and this translated into more support from Capitol Hill, the administration, and the public at large. Much of FEMA's success during the Clinton years can be attributed to the agency's improved ability to deal well with the press.

MAKING INFORMATION PUBLIC AND WORKING WITH THE MEDIA

Established credibility and productive working relationships with representatives of the media is critical. In most instances, the media will be cooperative in publishing important disaster recovery information. In an ideal world, the media would simply use all news releases as issued; however, sometimes media outlets, especially in major media markets, do not use disaster recovery information as important news after the initial stories about the event. It is important to try to make the news media understand the important public service role they play in the recovery effort. Use the following guidelines concerning media relationships:

- Be aware of and sensitive to media deadlines.
- Respond promptly to all media inquiries. Always answer requests for information, even if only to report that the information is not available or will not be available until a given time in the future.
- Reply to questions thoroughly and accurately. Do not provide more information than is requested.
- Be honest and open. If you don't know, say so and get back to the reporter as quickly as possible with the correct answer. Ask about deadlines.
- Do not go into in-depth discussions with reporters about the programs of other agencies.
- Always be diplomatic. Especially if a request seems unreasonable, deal with it tactfully.

Source: FEMA Emergency Information Field Guide (condensed), October 1998

This practice extends to nongovernmental organizations (NGOs) as well. The action of the American Red Cross in the immediate aftermath of the World Trade Center terrorist attacks in 2001 provides a good example. Within a half-hour of the first plane crash, the Red Cross deployed a 35-member rapid-response team to the World Trade Center with a mission to work with media to inform the public of what was happening and what they could do to help. The Red Cross then called in a 65-member volunteer force to their offices in New York, Washington, DC, and Pennsylvania to assist with media calls. Although their persistent solicitation of aid and their subsequent plans for aid distribution eventually came under criticism, the Red Cross's immediate actions illustrate how NGOs are able to partner with the press to get important messages out.

Managing Information

Beyond the general philosophy of treating the media as a partner, basic communications protocols must be followed. Information management is the most basic competency that must be developed. Managing information means developing a coordinated, consistent message in order to prevent confusion and maintain credibility. The release of information should be coordinated with responding partners, such as emergency management officials from other levels of government, law enforcement officials, or public health officials.

As noted earlier, FEMA achieves this through its Joint Information Center (JIC). A variation of this approach is now used by most emergency management organizations in all disaster events.

Telling Your Own Story

Although the careful management of information flows is a critical element of any communications strategy, the desire to distribute perfect, accurate, and coordinated information must be balanced with the need to get information out quickly. The object is to tell your story before someone else tells it for you. This goal goes hand in hand with partnering with the media because the better your relationship with the media is, the more likely you will be to have this opportunity.

This was another focus of FEMA under Witt. In prior years, during major crises such as Hurricanes Iniki and Andrew, and the Loma Prieta earthquake, FEMA generally attempted to shield itself from the press while it coordinated and undertook its response and recovery activities. The resulting vacuum of information left an opening for the media to portray the FEMA response as they perceived it, and coverage of these events was generally negative toward FEMA. Conversely, during major incidents of the Witt years, such as the Midwest floods, the Northridge earthquake, and the Oklahoma City bombing, FEMA made itself as accessible to the media as possible and distributed a constant stream of information on what activities were underway and what victims could do to receive assistance. Rather than reporting on perceived deficiencies, the press shared the information with the public and FEMA's public image improved.

Another excellent example of this strategy in action is New York City Mayor Rudy Giuliani after the World Trade Center attacks. Giuliani generally is perceived as the hero of the tragedy, largely because of his effective communications via the media. He made himself constantly accessible to press, provided continual updates on the status of response and recovery efforts, and reassured citizens that the city would rebound. By putting himself in front of the camera and articulating the story, he built public confidence and goodwill and was able to rally people together toward recovery. Even though he didn't always have all the answers, he was open, honest, and forthcoming, which fostered trust as well as good press.

The point is that if not provided with good information from good sources, the press will continue to look elsewhere. The information they find may not necessarily be accurate or fair, so it is critical to seize the communications agenda and get your story in front of the public.

Message Objectives

The objectives of the message will obviously vary depending on the situation, but in general a media partnership can help educate, inform, reassure, and rally the public. The media can help to garner support and lay the groundwork for future emergency management measures. In times of normalcy, the media partnership can educate the public on disaster mitigation issues, although exposure may be difficult to obtain. Unfortunately, media interest in disasters is usually short-lived and does not last long into the recovery phase. Nevertheless, the media is one means of promoting mitigation with the public.

In times of a crisis or emergency, the media partnership can communicate situation reports regarding the nature and scope of the incident, the estimated human and economic damages, and what recovery measures are underway. This provides the public with a perspective of the incident and lets them know what to expect. Public officials can go on the airwaves to reassure citizens that the government is taking action and soothe the public psyche with recovery updates. Most important, the media partnership can mobilize the public toward action—whether the instruction is to call a toll-free number, evacuate homes, or open mail with gloves, there is no better way to rally the public than through the media.

COMMUNICATIONS MEANS/PRODUCTS

Media Lists and Contacts

FEMA's core media list consists of the following: newspapers, city and regional magazines, local trade and business publications, state bureaus of National Wire Services, local radio and television stations, local cable stations, public broadcasting stations, and public information officers at military bases. The specific contacts that an emergency management agency typically will deal with are metro desk/city reporters, public affairs reporters, business reporters, news assignment editors, and public service announcement directors.

Press Releases

The press release is perhaps the most fundamental communications product. A press release can take the form of news releases, daily summaries, media advisories, feature articles, fact sheets, public service announcements, or other written materials. FEMA describes the objectives of its press releases as to demonstrate that FEMA and its partners are working to provide critical disaster response, recovery, and mitigation programs, and also to provide victims with accurate and timely information about the availability, details, and limits of these programs. FEMA press releases are routed through an established approval process.

The FEMA emergency information field guide offers some basic tips on preparing press releases. One point of emphasis for standard press releases is to never assume that information in previous disasters is appropriate for the current disaster—

always review generic releases for accuracy, timeliness, and appropriateness for each specific disaster. Also, releases and advisories should be kept brief and to the point, in order to increase the likelihood that it will be used in its entirety. An example of a FEMA press release follows. It is notable for its brevity, as it concisely lists essential information such as the who, what, when, and how of victim assistance.

FEMA PRESS RELEASE: FEDERAL DISASTER AID ORDERED FOR MISSISSIPPI STORMS

Washington, DC, December 7, 2001. The head of the Federal Emergency Management Agency (FEMA) announced today that federal disaster aid has been made available for Mississippi families and businesses victimized by tornadoes and other extreme weather that struck the state late last month.

FEMA Director Joe M. Allbaugh said the assistance was authorized under a major disaster declaration issued for the state by President Bush. The declaration covers damage to private property from the severe storms, tornadoes and flooding that began November 24.

Immediately after the President's action, Allbaugh designated the following 10 counties eligible for federal funding to help meet the recovery needs of affected residents and business owners: Bolivar, DeSoto, Hinds, Humphreys, Madison, Panola, Quitman, Sunflower, Tate, and Washington.

The assistance, to be coordinated by FEMA, can include grants to help pay for temporary housing, minor home repairs and other serious disaster-related expenses. Low-interest loans from the U.S. Small Business Administration also will be available to cover residential and business losses not fully compensated by insurance.

Allbaugh said federal funds also will be available to the state on a cost-shared basis for approved projects that reduce future disaster risks. He indicated that additional designations may be made later if requested by the state and warranted by the results of further damage assessments.

Gracia Szczech of FEMA was named by Allbaugh to coordinate federal relief operations. Szczech said residents and business owners who sustained losses in the designated counties can begin the disaster application process by calling 1-800-621-FEMA, or 1-800-462-7585 (TTY) for the hearing and speech impaired. The toll-free telephone numbers will be available starting Saturday, December 8 from 8 a.m. to 6 p.m. seven days a week until further notice.

Updated: December 7, 2001

Source: FEMA, www.fema.gov

Press Conferences

Press conferences allow information to be directly relayed to the media and the public. They provide officials with an opportunity to inform the public, reassure them, and mobilize them toward action. It is expected that in the aftermath of major

Figure 7-1 FEMA Director James Lee Witt addresses the media's questions at the site of the Laguna Canyon mudflows that led to at least one death and caused a great deal of damage (February 26, 1998). Photo by Dave Gatley/FEMA.

crises and emergencies, elected or appointed officials will come out and show the flag via a press conference and help calm public fears. This is an important step toward recovery and a return to normalcy.

Press Inquiries

In contrast to press releases and press conferences, press inquiries involve the media taking the communications initiative. For this reason, a dose of caution should be used when responding. The FEMA emergency information field guide provides the following general tips for interviews with the press:

- Listen to the entire question before responding.
- Avoid answering questions that call for speculation on your part.
- Be aware of false assumptions and erroneous conclusions.
- Avoid answering hypothetical conclusions.
- Be alert to multiple questions.

FEMA also has standard operating procedures to be used in receiving, responding to, and monitoring inquiries in the field. Key points of emphasis include the following:

- Never discuss program specifics or policy issues. Questions about FEMA policies or programs must always be referred to the Public Affairs Officer to be answered by the appropriate designated spokesperson.
- Ask the media to help FEMA help the disaster victims.

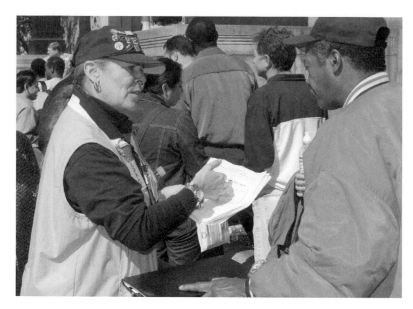

Figure 7-2 New York, New York, October 2, 2001. FEMA Community Relations worker answers questions from victims of the World Trade Center incident. Photo by Andrea Booher/FEMA News Photo.

- Be sure to tell the media about the JIC—the single source of accurate, up-to-date, official information about the disaster.

Web Sites

Web sites related to emergency management have become ubiquitous. From a media communications perspective, Web sites provide easy access to a repository of press releases, situation reports, general news, fact sheets, and general organizational and programmatic information. Diligence must be made to keep the site current, accurate, and easily navigable, or it loses its value as a resource. The FEMA policy for its Web site (www.fema.gov) is to keep news items on the site for 30 days. The same coordination and information management practices used for press releases apply to information posted on the Internet.

Situation Reports

Situation reports are used to provide basic information and statistics regarding emergency response efforts. The reports provide the press with facts that can be used in articles and stories and inform partner response agencies of the status of operations. FEMA produces a steady stream of situation reports in the aftermath of major events. This is consistent with the objectives of telling your story before the press does it for you and partnering with the press by being sensitive to their needs for hard data. Situation reports typically are posted on the Internet or distributed by e-mail.

An example of a typical situation report issued by FEMA during disaster response and recovery efforts is provided. Reliefweb (www.reliefweb.int/w/rwb.nsf) is an excellent source for situation reports on international crises and emergencies posted by the United Nations and other international organizations. The U.S. Office of Foreign Disaster Assistance (www.usaid.gov/hum_response/ofda/) also does a great job of providing situation reports on its assistance programs around the globe.

FEMA SITUATION REPORT

From the Federal Emergency Management Agency (FEMA) "National Situation Update" for Tuesday, October 09, 2001 (www.fema.gov/emanagers/natsitup .htm):

World Trade Center Update*

The City reported that as of yesterday, 393 bodies have been recovered from the World Trade Center (WTC). Of those, 335 have been identified. The number of injured is 8,786 (415 remain hospitalized) and 4,979 persons are registered as missing.

As of yesterday, 206,831 tons of debris had been removed from the WTC site (not including steel) to a landfill on Staten Island. The official estimate for total debris at the WTC is 1.4 million tons.

4,776 New Yorkers have registered for housing assistance. $9.8 million in housing assistance payments have been approved for disbursement.

3,426 New Yorkers have registered for Individual and Family Grants. $32,624 has been approved for disbursement to eligible registrants.

The Small Business Administration has approved $16,984,300 in low-interest loans to businesses and individuals.

$126,325,305 has been obligated as the Federal share for Public Assistance (as of October 8). (Manhattan DFO)

*As of October 7, 2001
Source: FEMA, www.fema.gov

Spokespeople

Spokespeople can lend credibility to a message, but their words must be coordinated with the rest of the communications strategy in order to avoid multiple or contradictory messages. For this reason, it is often wise to select a single spokesperson to deliver information to the press. The lead local official is often the best person to assume this role because he or she will be best informed on the local response and the community's needs.

The FEMA press information guide for its Project Impact initiative provides some valuable pointers for spokespeople:

- Repeat information to reinforce key message points.
- Correct inaccuracies, otherwise they will be accepted as fact.
- Pair use of statistics with stories or case studies that bring them to life.
- Stay out of other people's business. Let other emergency agencies answer their own questions.
- Always be honest. If you don't know an answer to a question, say so and offer to find the answer or refer the reporter to someone who can.

CASE STUDY

FEDERAL GOVERNMENT COMMUNICATIONS DURING ANTHRAX CRISIS

The anthrax outbreak in October 2001 provides some important communications lessons, both from the perspective of media relations and communicating risk to the public. It highlights the importance of providing a consistent, coordinated message through a single spokesperson and also highlights the need to balance a desire to reassure the public with the need to be accurate and credible.

There were two main challenges involved with the crisis. First, medical and public health officials had more questions than answers. Anthrax is a very rare disease in humans, and anthrax spores spread via the mail was basically an unknown commodity altogether. Second, there were multiple responding agencies from various levels of government involved and no established protocol for distributing information.

As a result, the public was given conflicting messages about the nature of the anthrax and misinformation about the true risk. Media criticism of the public response ensued, but it should be pointed out that in November 2001 a *USA Today* survey found that 77 percent of U.S. citizens were confident that the government could handle a major anthrax outbreak, and a Harris interactive poll showed the Center for Disease Control's (CDC) approval rating at 79 percent. Apparently, the public was in a forgiving mood, or perhaps they were just still confused.

The first problem with the anthrax communications was that there was no clear spokesperson. A sole authority was needed to provide uniformity and consistency to the message and reduce fears. After the early conflicting messages, Tom Ridge was appointed the quasi-spokesperson for anthrax and terrorism threats, as part of his duties with the newly created Office of Homeland Security.

Beyond the issue of who should have been providing the message, there were questions about what information should have been provided. The case illustrates a classic communications conundrum. Officials were under pressure to provide current information to the public, which was seeking reassurances, while there was still much uncertainty about the true nature of the threat. Marc Shannon, director of Ketchum's Washington D.C. healthcare practice summed up the dilemma well: "If you don't get out enough information you're accused of being secretive. And if you give too much information you are criticized for stirring up anxiety." As Shannon points out, a key to communications in these instances in not to be afraid to say I don't know.

Tommy Thompson of the Department of Health and Human Services might be accused of erring in this respect. During an interview on *60 Minutes* early in the crisis he said, "We've got to make sure that people understand that they're safe, and that we're prepared to take care of any contingency, any consequence that develops from any kind of bioterrorism attack." After new cases of anthrax continued to be reported, and two D.C. postal workers and a Connecticut woman later died of inhalation anthrax, it became apparent that this was a case of an official going too far in trying to assuage public fears. These remarks were in contrast to those of New York City Mayor Rudy Giuliani, who after the death of the Connecticut woman said words to the effect that the government can't guarantee that every single person will be completely safe from anthrax, and that individuals need to exercise a certain amount of due diligence. Although these remarks may not have been completely comforting, they were accurate, practical, and fostered public trust.

Source: *PR Week*

CONCLUSION

Whether dealing with the media, the public, or partners, effective communication is a critical element of emergency management. Media relations should be open and cooperative, the information stream must be managed to provide a consistent, accurate message, and officials need to be proactive about telling their own story before it is done for them. A customer service approach is essential to communicating with the public, a collaborative approach should be taken to promoting programs, and great care should be given as to how and when risk is communicated to citizens. Multiple agencies and unclear lines of responsibility make communications among partners a challenge; political skill and acumen are needed to overcome such hurdles, and efforts are under way to improve communications in this area.

8. International Disaster Management

INTRODUCTION

People of all nations face risks associated with the natural and technological hazards described throughout this book, and almost all nations eventually become victim to disaster. Throughout history, civilizations have adapted to their surroundings in the hopes of increasing the likelihood of survival. As societies became more organized, complex systems of response to these hazards were developed on local, national, and regional levels. The capacity to respond achieved by individual nations can been linked to several factors, including propensity for disaster, local and regional economic resources, organization of government, and availability of technological, academic, and human resources; however, it is becoming increasingly common that the response ability of individual nations is insufficient in the face of large-scale disaster, and outside assistance must be called upon. Disasters that affect whole regions are not uncommon and require these same international response mechanisms.

This chapter introduces the conglomeration of agencies, including the U.S. government, international organizations, nongovernmental organizations (NGOs), and financial institutions, that prepare for and respond to the natural, technological, and complex humanitarian emergencies (CHEs) that overwhelm the capacity of any one sovereign nation. The mission and goals of each of these entities and groups are described (although their performance is not detailed). In conclusion, a comprehensive case study is presented on the international response to the Gujarat, India, earthquake of January 26, 2001.

DISASTERS IN DEVELOPING NATIONS

Disasters of all kinds strike literally every nation of the world, although these events do not occur with uniformity of distribution. The developing nations suffer the greatest impact of nature's fury, and these same nations are also most often subject to the internal civil conflict that leads to CHEs. Furthermore, the greatest incidence of natural disasters occurs within developing countries, with 90 percent of disaster-related injuries and deaths sustained in countries with per-capita income levels that are below $760 per year (UNICEF).

Although disaster preparedness and mitigation are widely accepted by international development agencies to be integral components in the overall development process, it comes as no surprise that countries ranking lower on development indices have placed disaster management very low in budgetary priority. These nations' resources tend to be focused on more socially demanded interests such as education and base infrastructure, or on their military, instead of on projects that serve a preparatory or mitigative need, such as retrofitting structures with hazard-resistant construction. Because disasters are chance events, and thus not guaranteed to happen, disaster management programs in poor countries tend to be viewed as superfluous. Delegating disaster management responsibilities to the military is also commonly seen even in countries with a moderate level of development, although these agencies rarely are trained to carry out the necessary response tasks required. To compound the situation further, poverty and uncontrolled urbanization often force large populations to concentrate in perilous, high-risk urban areas that contain little or no defense against disasters.

INTERNATIONAL INVOLVEMENT

A disaster requires the involvement of the international community of responders when a nation's capability to respond has become overwhelmed. This threshold is determined by many factors, including the availability of economic resources, the level of local responder training, the resilience of the infrastructure, the public opinion of the government's ability to manage the crisis, and the availability of specialized assets, among many others. Of course, this threshold is crossed much earlier in the poorer countries. It must be recognized, however, that even the wealthiest nations regularly find themselves in need of help from the international community, whether for supplies, manpower, money, or a specific skill or asset that cannot be found locally. Appeals for assistance are made in many ways and are often simultaneously met with unsolicited offers of aid and support. With the global interconnectivity brought about through television and the Internet (the so-called CNN effect), news of a disaster can circle the globe within minutes, stirring the machine of response into action.

There are three types of emergencies that normally involve an international humanitarian response: natural disasters, technological disasters, and complex humanitarian emergencies (CHEs). The first two are clearly defined; however, the CHEs have been subject to diverse interpretations and changing standards, and thus, for the purposes of this book, are characterized by the definition established by the United Nations (UN). They classify a CHE to be a "humanitarian crisis in a country or region where there is total or considerable breakdown of authority resulting from the internal and/or external conflict and which requires an international response that goes beyond the mandate or capacity of any single agency." (DODCCRP) Andrew Natsios, Director of the U.S. Agency for International Development (USAID), identifies five characteristics most commonly seen in CHEs in varying degrees of intensity:

- Civil conflict, rooted in traditional ethnic, tribal, and religious animosities (usually accompanied by widespread atrocities)
- Deteriorated authority of the national government such that public services disappear and political control dissolves

- Mass movements of population to escape conflict or search for food, resulting in refugees and internally displaced people (IDPs)
- Massive dislocation of the economic system, resulting in hyperinflation and the devaluation of the currency, major declines in gross national product, skyrocketing unemployment, and market collapse
- A general decline in food security, often leading to severe malnutrition and occasional widespread starvation (Natsios, 1997)

Although these emergencies are fundamentally different from natural and technological disasters in regards to their generally political and intentional sources, they share many characteristics in terms of their requirements for response and recovery. In accordance, many of the organizations and entities described in this chapter respond to all three types of disasters indiscriminately.

IMPORTANT ISSUES INFLUENCING THE RESPONSE PROCESS

Several issues must be addressed when responding to international disasters. The first, *coordination*, is a vital and immediate component because of the sheer numbers of responding agencies that almost always appear. It is not uncommon in larger disasters to see several hundred local and international NGOs, each with a particular skill or service to offer. Successful coordination and cooperation can lead to great success and many lives saved, but infighting, turf battles, and nonparticipation can lead to confusion and even cause a second disaster (PAHO).

The UN has become widely recognized as the central coordinating body, with specialized UN agencies handling the more specific needs associated with particular disaster consequences. Most often, the UN capitalizes on longstanding relationships with the host country to form a partnership on which they establish joint control. In addition to the UN, several organizations and associations have come up with standards of conduct, such as the Red Cross Code of Conduct (www.ifrc.org/publicat/conduct/index.asp), the Sphere Project Humanitarian Charter and Minimum Standards in Disaster Response (www.sphereproject.org/handbook_index.htm), and the Oxfam Code of Conduct for NGOs (www.oxfam.org).

The second issue is that of *sovereignty of the state*. State sovereignty is based on the recognition of political authority characterized by territory and autonomy. Accordingly, a foreign nation or organization cannot intercede in domestic matters without the prior consent of the ruling government. This can be a major hurdle in CHEs that have resulted from civil war, such as the peacekeeping mission in Somalia where there existed no official government in place with which to work. Although not as commonly seen, sovereignty has also been an issue in matters of natural and technological disasters, particularly when a nation does not want to be viewed as weak or unable to take care of its people. Examples of such behavior include Japan's refusal to allow access to international agencies for several days after the earthquake in Kobe and the actions of the Former Soviet Union following the nuclear power plant accident in Chernobyl.

The third issue is *equality in relief distribution*, and it applies to any type of disaster. Situations often arise where, for any number of cultural or political reasons, certain groups in need of aid are favored over others. The first example of this discrimination is the result of gender bias, which is most commonly found in societies where gender roles are strictly defined and women are traditionally tasked with duties related to the home and children (which tend to be increased in times of crisis). In these cultures, the men are more likely to have opportunities to wait in relief lines for supplies, and the women (as well as children and the elderly) become even more dependent on them for survival. This situation is exacerbated if a woman is a widow or single parent and has no ability to compete for distributed aid.

The second form of inequality in relief is that of class bias. Although most obvious in social systems explicitly based on caste identity, underlying ethnic and racial divides often present similar problems. Avoiding these forms of bias is difficult because the agencies involved must be aware of the discrimination in order to counteract its influence. Often, host-country nationals are "hired" by humanitarian agencies to assist in relief distribution, and inadvertent hiring of specific ethnic or social groups can lead to unfair distribution along those same ethnic/social lines. At the same time, humanitarian agencies are quick to focus on those groups most visibly affected by a CHE, such as IDP populations, causing an inordinate percentage of aid to be directed to them, while other needy groups go unnoticed.

Many of the international response agencies are continuously developing systems of relief and distribution that work to counteract the complex problems associated with these biases; however, the difficult nature of this issue is highlighted in the fact that specifically targeting groups, such as women or children, can lead to reverse discrimination. Any of these biases can lead to a decline in perceived legitimacy or impartiality of the assisting agency and/or result in exacerbation of the needs being addressed (Maynard).

A fourth issue is the importance of *capacity building* and *linking relief with development*. Responding agencies have an obligation to avoid using a bandage approach in assisting the affected country. Disasters almost always present a window of opportunity to rebuild old, ineffective structures and develop policy and practice in a way that leaves behind a more empowered, resilient community. Because these goals mirror those of most traditional development agencies, linking relief and development should not be a major deviation from either type of agencies' missions. These opportunities are greatest in situations that require the complete restoration of infrastructure and basic social services, and are found equally in disaster and CHE scenarios. In the reconstruction phase, it is vital that training and information exchanges occur and that local risk is fully incorporated to mitigate for repeat disasters. These repeat disasters often contribute greatly to a nation's lag in development, and therefore fully addressing them is vital to increasing the nation's likelihood of being developed sustainably.

THE UNITED NATIONS SYSTEM

The UN began in 1945, when representatives from 51 countries met in San Francisco to establish the United Nations Charter as a commitment to preserve peace in the aftermath of World War II. Later that year, the Charter was ratified by the five permanent members: China, France, the Soviet Union, the United Kingdom, and the United States, as well as several other countries. Today, 189 countries are members of the UN, and the Charter (which is similar to a sovereign state's constitution and establishes the rights and responsibilities of Member States) is amended as is necessary to reflect the changing needs of current world politics.

The UN itself is not a government body, nor does it write laws; however, the autonomous Member States do have the ability through the UN to resolve conflict and create international policy. No decision or action can be forced on a sovereign state, but as global ideals are naturally reflected through these collaborative policies, they usually are given due consideration.

Through the major UN bodies and their associated programs, the UN has established a presence in most countries throughout the world and fostered partnerships with Member State governments. Although more than 70 percent of UN work is devoted to development activities, several other issues are central in their mission, including disaster mitigation, preparedness, response, and recovery. In the event of a disaster, the UN is quite possibly the best equipped to coordinate disaster relief and to work with the governments to rehabilitate and reconstruct. This is especially true in the case of the developing countries, where regular projects are ongoing and must be adjusted to accommodate for damages to infrastructure and economy caused by recurrent disasters, and where disasters quickly exhaust the response capabilities.

Upon onset of a disaster, the UN responds immediately and on an ongoing basis by supplying aid in the form of food, shelter, medical assistance, and logistical support. The UN Emergency Relief Coordinator heads the international UN response to crises through a committee of several humanitarian bodies, including the UN Children's Fund (UNICEF), the UN Development Programme (UNDP), the World Food Programme (WFP), the UN High Commissioner for Refugees (UNHCR), and other associates as deemed necessary in accordance with the problems specific to the event. Each of these agencies, as shown in this section, fulfills a specific need presented by most humanitarian emergencies, be they natural or manmade.

The UN also promotes prevention and mitigation activities through its regular development projects. By encouraging the building of early warning systems and the conducting of monitoring and forecasting routines, they are working to increase local capacity to adequately boost local and regional preparedness. In conclusion of the International Decade for Natural Disaster Reduction of the 1990s (which strove to focus on a shift from disaster response-oriented projects to disaster mitigation), the UN adopted its International Strategy for Disaster Reduction (ISDR) to promote the necessity of disaster reduction and risk mitigation as part of its central mission. This initiative seeks to enable global resilience to the effects of natural hazards in order to reduce human, economic, and social losses, through the following mechanisms:

- Increasing public awareness
- Obtaining commitment from public authorities
- Stimulating interdisciplinary and intersectoral partnership and expanding risk-reduction networking at all levels

- Enhancing scientific research of the causes of natural disasters and the effects of natural hazards and related technological and environmental disasters on societies

These strategies are carried out through the country offices and local governments, in the most vulnerable communities. Mitigation and preparedness strategies are implemented at all levels of society via public awareness campaigns, secured commitment from public authorities, intersectoral cooperation and communication, and technical knowledge transfer.

The United Nations Development Programme

The United Nations Development Programme (UNDP) was established in 1965, during the UN Decade of Development, to conduct investigations into private investment in developing countries, to explore the natural resources of those countries, and to train the local population in development activities such as mining and manufacturing. Over the years, as the concept and practice of development expanded greatly, the UNDP took on much greater responsibilities within host countries and within the United Nations as a whole.

Historically, the UNDP was not considered an agency on the forefront of the crisis and disaster management scene because, although they worked on development issues, they did not focus specifically on emergency response systems, which were considered to be the focal point of crisis and disaster management for many years. As mitigation and preparedness received greater emphasis in the field, however, the vital role that the UNDP has played all along is being increasingly recognized. Capacity building has always been central to the mission of the UNDP, in terms of empowering host countries to be better able to address issues of national importance, eventually without foreign assistance.

In the execution of UNDP projects, there was a natural, although unintended, move toward activities that indirectly filled mitigation and preparedness roles. Projects that worked to strengthen government institutions also improved the capacity of such institutions to respond with appropriate and effective policy, power, and leadership in the wake of a disaster. By its very nature, therefore, capacity building could clearly be considered a mitigation activity (although early on, the mitigation of disasters was not as widely understood or practiced as was the response to them).

Attention to disaster management increased through time as natural and manmade disasters were affecting greater populations and causing greater financial impacts, and the developing nations felt the greatest inability to prepare and/or respond to them. It was widely recognized that the unguided development trends typified by these lesser-developed nations led to their greater vulnerabilities. For example, developing countries generally have a severe deficiency in physical infrastructure from which response could be based, they participate in environmental misuse and destruction that exacerbates certain natural hazards, and they often contain migrant populations that settle in concentrated groups within disaster-prone regions.

Considering that 90 percent of natural disasters occur in developing countries, and likewise that 90 percent of victims of disasters live in developing countries, it becomes apparent that the issue of their management cannot be kept separate from the mission of the UNDP (which works primarily in these countries).

Today, the UNDP believes that vulnerability to disasters is strongly linked to a lack of or weak infrastructure, poor environmental policy, misuse of land, and rising populations in areas that are prone to repeat disasters. In many cases, these disasters can literally set a country back years, if not decades, in terms of development achievement. For instance, the president of Honduras has declared that the country has gone back to early 1950s levels of development because of the devastating effects of Hurricane Mitch. It is also recognized that small- to medium-sized disasters in the least developed countries, can "have a cumulative impact on already fragile household economies and can be as significant in total losses as the major and internationally recognized disasters" (UNDP). It is their modern objective, then, to "achieve a sustainable reduction in disaster risks and the protection of development gains, reduce the loss of life and livelihoods due to disasters, and ensure that disaster recovery serves to consolidate sustainable human development" (UNDP).

In 1995, as part of the UN's changing approach to better assisting the relief community as a whole, the Emergency Response Division (ERD) was created within the UNDP. This move drastically augmented the organization's role in responding to disasters. Additionally, 5 percent of UNDP budgeted resources were allocated for quick-response actions in special development situations by ERD teams, thus drastically reducing delays in bureaucratic decision making. Specifically, the ERD helps in creating a collaborative framework among the national government, UN agencies, donors, and NGOs that immediately respond to disasters, provides communication and travel to disaster management staff, and distributes relief supplies and equipment. Following the deployment of ERD teams (generally 30 days), a detailed project plan is submitted, and a full UNDP project can be applied to a disaster. ERDs work in strengthening coordinating mechanisms, and their central strong organization role has shaped future UNDP involvement in disasters.

In 1997, under the UN Programme for Reform, the responsibilities and operational activities of the Emergency Relief Coordinator, regarded as being part of national capability or capacity, were formally transferred to the UNDP. In response, the UNDP created the Disaster Reduction and Recovery Programme (DRRP) within the ERD. The broad-ranging duties of this program pertaining to disaster mitigation, prevention, and preparedness were defined as follows:

- Mainstream disaster reduction into development policy, strategies, plans, and programs
- Strengthen capacity of institutions at all levels for enhanced disaster management
- Develop innovative approaches to accelerate sustainable postdisaster recovery, promoting the inclusion of disaster reduction measures into rehabilitation and reconstruction
- Build partnerships, promote networks, and facilitate cooperation at international, regional, and national levels
- Facilitate the development and delivery of high-quality training and human resource development activities

- Promote and develop disaster-reduction policies and strategies
- Represent UNDP at interorganizational fora on the topic
- Provide direct substantive support to multisectoral integrated country programs

In terms of response, recovery, and reconstruction, the UNDP pledged to support activities in eight subject groups that handle emergencies of natural, technological, and CHE disasters. The groupings are such that any combination can be applied to meet the needs of virtually any type of emergency situation that arises. They are listed as follows, with their summarized roles included:

- *Emergency Interventions.* Establish the nature and scope of the emergency, collect and distribute timely information to all parties involved, and track and coordinate donations from domestic and foreign sources. As the UNDP resident representative of the country leads the effort (through the Disaster Management Team), there is long-term institutional knowledge to manage the disaster response.
- *Programming for Peace and Recovery.* Determine the major difficulties to be addressed and the priority needs in terms of external support, assist the current or new government in addressing these issues, and provide the planning and financial coordination that is required. What is most unique about the UNDP activities is that they seek to break the dependence that has been created by outpourings of international relief that can hamper a return to normalcy after natural or manmade disasters.
- *Area Rehabilitation to Resettle Uprooted Populations.* Create or expand on the capacity of the communities where IDPs or refugees are to be resettled. The UNDP utilizes many of its standard project schemes, such as creation of income-generating activities, building stronger infrastructure, and promoting local participation in the process.
- *Reintegrating Demobilized Soldiers.* Initiate reintegration projects and coordinate the funding from the international community. The UNDP has been able to provide much of the administrative duties of this task and the follow-up once the agencies more specifically concerned with demobilization have left. In the 1990s alone, the UNDP channeled more than $150 million to reintegration and demobilization programs.
- *Demining.* Provide general management input for the conduct of operations and coordinate financial contributions from international donors. Because it is necessary to clear mines before development can continue with any chance of success, the UNDP sees this task as integral in their goal of linking relief to development.
- *Rebuilding Institutions and Improving Government.* Make overall assessments of the state of governance, identify problems that need to be addressed, and assist the government in the coordination of reform/restructure/repair. The long partnerships and the assumed neutrality make the UNDP an ideal body for this role.
- *Organizing National Elections.* Provide local coordination and technical assistance. In many cases, UNDP involvement can give a sense of legitimacy to

an election at a time when stability is fragile, especially after a new government has come to power after a civil conflict. This stability is vital if the country is to emerge from its crisis.

- *Managing Delivery of Program Aid.* Manage UN Office for Project Services in the delivery of program aid, assist in the procurement of services, and assist in the administration of loans. "The donors themselves, bilateral and multilateral, are not often coordinated at the central level where major decisions on allocation of funds are usually made."

In addition to the aforementioned roles and responsibilities, the UNDP leads several interagency working groups. One such group, which consists of representatives from the World Food Programme (WFP), the World Health Organization (WHO), the Food and Agriculture Organization (FAO), the UN Populations Fund (UNFPA), and the UN Children's Fund (UNICEF), works to develop principles and guidelines in order to incorporate disaster risk into the Common Country Assessment and the UN Development Assistance Framework. They have included in their goals capacity building for the central governments consulted, assessment of vulnerability, creation of early warning systems, development and maintenance of a framework for contingency planning, greater efforts toward mine removal, strengthening of country disaster management programs and teams, and national development programs that include the all-hazard spectrum.

The ISDR Working Group on Risk, Vulnerability, and Disaster Impact Assessment works on the setting of guidelines for social impact assessments. The UNDP also coordinates a Disaster Management Training Programme (DMTP) in Central America, which runs a conference on "The use of microfinance and microcredit for the poor in recovery and disaster reduction," and has created a program to elaborate financial instruments to enable the poor to manage disaster risks.

The UNDP currently dedicates more than 40 percent of its resources to emergency relief operations. It is clear that they are not a self-contained disaster relief organization, but that is not how the UN system was developed to function. The many agencies and offices that are involved would not act efficiently without a central coordinating body, and the UNDP recently has been deemed the most able to handle that duty. As should be clearly presented, these duties do not strain the established role as lead development agency within the UN system.

The United Nations Office for the Coordination of Humanitarian Affairs

The United Nations Office for the Coordination of Humanitarian Affairs (OCHA) was created under the UN Secretary-General's Program for Reform in 1998, to accommodate the needs of victims of disasters and emergencies. Their specific role in the broad range of disaster management tasks is to coordinate assistance provided by the UN system in emergencies that exceed the capacity and mandate of any individual agency. The OCHA response to disasters can be categorized under three main groupings, including coordinating the international humanitarian response, providing support and policy development to the humanitarian community, and advocating humanitarian issues to ensure that that the overall direction of relief reflects the general needs of recovery and peace-building.

The head of OCHA is the UN Emergency Relief Coordinator (ERC) and is responsible for the coordination of the response efforts of the UN through the Inter-Agency Standing Committee (IASC). The IASC is a group consisting of both UN and outside humanitarian organization leaders, which analyzes crisis scenarios and formulates a joint response to ensure maximum effectiveness and minimal overlap of relief. The ERC works to deploy appropriate personnel from throughout the UN to assist the resident coordinators and lead agencies in the response, thus increasing the likelihood that on-site coordination will be strong.

The Disaster Response System, established by OCHA, constantly monitors the onset of natural and technological disasters. This system includes training the assessment teams before disasters strike, as well as evaluations conducted postdisaster. When a disaster is identified, the OCHA response is activated, and a situation report is generated to provide the international response community with detailed disaster-specific information (which includes damage caused, actions taken, needs assessed, and current assistance being provided). OCHA may then, if deemed necessary, deploy a UN Disaster Assessment and Coordination (UNDAC) team to assist in the coordination of relief activities and help assess damages and needs (these duties are not as overreaching as in complex emergencies; see the previous UNDP section).

An Operations Coordination Center may be set up in the field in order to assist local first-response teams in their coordination of the often overwhelming international representation of relief agencies that respond. Finally, OCHA can set up communications capabilities if they have been damaged or do not exist to the capacity required by the UN responding agencies. OCHA responsibilities generally are concluded when the operation moves from response to recovery.

The United Nations Children's Fund

Like most other major UN agencies, the UN Children's Fund (UNICEF, formerly known as the United Nations International Children's Emergency Fund) was established in the aftermath of World War II. Its original mandate was to aid the children suffering in postwar Europe, but its mission has been expanded to address the problems that affect poor children throughout the world. UNICEF is mandated by the General Assembly to serve as an advocate for children's rights, to ensure that each child receives at least the minimum requirements for survival, and to increase their opportunities for a successful future. Under the Convention on the Rights of the Child (CRC), a treaty adopted by 191 countries, the UNHCR holds wide-reaching legal authority to carry out its mission.

Before the onset of disasters, it is not uncommon for UNICEF to have established itself as a permanent in-country presence, with regular budgetary resources. In the situations of disaster or armed conflict where this is the case, UNICEF is well poised to serve an immediate role as aid provider to its specific target groups. This rapid response is important because young mothers and children are often the most marginalized groups in terms of aid received. UNICEF works on a regular basis to ensure that children have access to education, healthcare, safety, and protected child rights.

In the response and recovery periods of humanitarian emergencies, these roles are merely expanded to suit the rapidly extended requirements of victims. In countries where UNICEF has not yet established a permanent presence, the form of aid is virtually the same; however, the timing and delivery are affected, and reconstruction is not nearly as comprehensive.

UNICEF maintains that humanitarian assistance should include programs aimed specifically for child victims. Relief projects generally work to provide a rapidly needed response in the form of immunizations, water and sanitation, nutrition, education, and health. Women are recipients of this aid as well because UNICEF considers them to be vital in the care of children. UNICEF also works through recovery and reconstruction projects, providing for the basic rights of children. UNICEF is currently working in 161 countries.

The World Food Programme

The World Food Programme (WFP) is the arm of the UN tasked with reacting to hunger-related emergencies throughout the developing world. The WFP was created late in 1961 by a resolution adopted by the UN General Assembly and the UN Food and Agriculture Organization (FAO). Chance enabled the program to prove the necessity of their existence when the WFP provided relief to more than 5 million people several months before they were deemed officially operational in 1963. In the year 2000 alone, the WFP fed 83 million people through its relief programs. Over the course of its existence, the WFP has provided more than 43 million metric tons of food to countries worldwide.

Because food is a necessity for human survival, it is a vital component of development. The WFP works throughout the world to assist the poor who do not have sufficient food to survive "to break the cycle of hunger and poverty." Hunger alone can be seen as a crisis because more than 800 million people across the globe receive less than the minimum standard requirement of food for healthy survival. Hunger is often associated with other crises, including drought, famine, and human displacement, among others.

In rapid-onset events such as natural disasters, the WFP is activated as a major player in the response to the immediate nutritional needs of the victims. Food is transported to the affected location and delivered to storage and distribution centers. The distribution is carried out according to preestablished needs assessments performed by OCHA and the UNDP. The WFP distributes food through contracted NGOs who have vast experience and technical skills required to plan and implement such projects of transportation, storage, and distribution. The principal partners in their planning and implementation are the host governments (who must request the aid of the WFP to begin with, unless the situation is a CHE where there is no established government, and the UN Secretary General makes the request). The WFP works closely with all responding UN agencies to coordinate an effective and broad-reaching response because food requirements are so closely linked to every other vital need of disaster victims.

In the aftermath of disasters, during the reconstruction phase, it is often necessary for the WFP to remain an active player through continued food distribution.

Rehabilitation projects are implemented in a way that fosters increased local development, and include providing food aid to families, who as a result will have extra money to use in rebuilding their lives, and food for work programs, which break the chains of reliance on aid as well as provide an incentive to rebuild communities.

The World Health Organization

The idea for the World Health Organization (WHO) was proposed during the original meetings to establish the UN system in San Francisco in 1945. In 1946, at the United Health Conference in New York, the WHO constitution was approved, and on April 7 (World Health Day), it was signed and made official. Like the WFP, WHO proved its value by responding to an emergency (a cholera epidemic in Egypt) months before it was an officially recognized organization.

WHO was established to serve as the central authority on sanitation and health issues throughout the world. They work with national governments to develop medical capabilities and healthcare and assist them in the suppression of epidemics. WHO supports research for the eradication of disease and provides expertise on these subjects when requested. They provide training and technical support and develop standards for medical care.

In the event of a disaster, WHO responds in several ways that address the health of victims. Most important, it provides ongoing monitoring of diseases traditionally observed within the unsanitary conditions of disaster aftermath. WHO also provides technical assistance to the responding agencies and host governments who are establishing disaster medical capabilities and serves as a constant source of expertise as needs arise.

Since its inception, regional offices have been established. These offices, which comprise the 191 separate member states of WHO, focus on the health issues most directly related to each regional area's needs and concerns. These regions include the following:

- African Regional Office (AFRO)
- Pan American Health Organization (PAHO)
- South-East Asia Regional Office (SEARO)
- Regional Office for Europe (EURO)
- Eastern-Mediterranean Regional Office (EMRO)

NONGOVERNMENTAL ORGANIZATIONS

The number of nongovernmental organizations (NGOs) focusing on international humanitarian relief has grown exponentially in the past few decades. These organizations have come to play a vital role in the response and recovery to disasters, filling gaps left by national and multilateral organizations. They have significantly improved the ability of international relief efforts to address the needs of victims with a diverse range of skills and supplies. Some of the larger NGOs, like the International Committee of the Red Cross (ICRC), have established an international presence similar to that of the UN and have developed strong local institutional partnerships and a capacity to respond almost immediately with great effectiveness. These grassroots-level organizations are so successful in their activities that the major funding organizations such as USAID, OFDA, and the UN regularly arrange for relief projects to be implemented by them rather than their own staff.

There are several classifications of humanitarian organizations, and for the purpose of clarification, they are described as follows. The following broad categorical definitions are widely accepted among the agencies of the international relief community. These are not definitive categories into which each organization will neatly fit, but they have become part of standardized nomenclature in disaster response:

- *Nongovernmental organization (NGO).* The general term for an organization made up of private citizens, with no affiliation with a government of any nation other than the support from government sources in the form of financial or in-kind contributions. These groups are motivated by greatly varying factors, ranging from religious belief to humanitarian values. NGOs are considered national if they work in one country, international if they are based out of one country but work in more than four countries, and multinational if they have partner organizations in several countries. Oxfam and the ICRC are examples of multinational NGOs. NGOs can be further defined according to their functionality. Examples of these would be the religious groups, such as the Catholic Church; interest groups, such as Rotary International; residents' organizations; occupational organizations; educational organizations, and so on.
- *Private voluntary organization (PVO).* An organization that is nonprofit, tax-exempt, and receives at least a part of its funding from private donor sources. PVOs also receive some degree of voluntary contributions in the form of cash, work, or in-kind gifts. This classification is steadily being grouped together under the more general NGO classification. It should be mentioned that although all PVOs are NGOs, the opposite is not true.
- *International organization (IO).* An organization with global presence and influence. Although both the UN and ICRC are IOs, only the ICRC could be considered an NGO. There exists international law providing a legal framework under which these organizations can function.
- *Donor agencies.* Private, national, or regional organizations whose mission is to provide the financial and material resources for humanitarian relief and subsequent rehabilitation. These donated resources may go to other NGOs, other national governments, or to private citizens. Examples of donor agencies are USAID, the European Community Humanitarian Organization (ECHO), and the World Bank.
- *Coordinating organizations.* Associations of NGOs that coordinate the activities of hundreds of preregistered member organizations to ensure response with maximized impact. They can decrease the amount of overlap and help distribute need to the greatest range of victims. Also, they have the ability to analyze immediate needs assessments and recommend which member organizations would be most effective in response. Examples of coordinating orga-

nizations include InterAction and the International Council for Voluntary Agencies (ICVA).

NGOs bring to the field several resources. First, they are well regarded as information-gathering bodies, and thus are vital in establishing accuracy in the development of damage and needs assessments. They tend to provide a single skill or group of specific technical skills, such as the medical abilities of Medicin sans Frontiers (MSF, Doctors without Borders) or Oxfam's ability to address nutritional needs. The sheer number of helping bodies that are provided by the involvement of NGOs allows for a greater capability to reach a larger population in less time. Finally, the amount of financial support provided as result of the fundraising abilities of NGOs brings about much greater cash resources to address the needs of victims.

These organizations can be characterized by several commonly seen characteristics:

1. *They value their independence and neutrality.* In situations of civil conflict, being perceived as independent is vital to safety and success because they could become targets if associated with an enemy group, or denied access to victims located in territory under the control of a certain warring faction. For this reason, there is often great reluctance on the part of NGOs to share all information to involved governments, to be seen as assisting one group over another, or to report observed war crimes to international tribunals. This independence is advantageous in situations where one national government does not want to be seen as needing the assistance of another national government but is willing to accept the help of autonomous bodies.

2. *They tend to be decentralized in their organizational structure.* For instance, they tend to work without definitive hierarchy and succeed through greater field-level management.

3. *They are committed.* NGOs often are involved not only in the disaster relief, but also in the long-term recovery efforts that follow for months or years. NGO employees are often so dedicated as to repeatedly put themselves in harm's way to deliver aid to victims.

4. *They are highly practice-oriented.* They tend to improvise in the field as necessary and provide on-site training as part of their regular procedures. They rarely use field guides to direct their work, relying rather on the individual experience of employees and volunteers. (CDMHA)

Perhaps the most well-known and most widely established NGO, the Red Cross, will be discussed as an example.

The International Red Cross

The International Red Cross/Red Crescent Movement consists of the International Federation of Red Cross and Red Crescent Societies (IFRC) and the International Committee of the Red Cross (ICRC). The concept of the Red Cross was initiated by Henry Dunant in 1859, following a particularly brutal battle in Italy that he witnessed. Dunant gathered a local group to provide care for the battle-wounded through medical assistance, food, and ongoing relief. Upon returning to Switzerland, he began the campaign that led to the International Committee for Relief of the Wounded in 1863, and eventually the ICRC. The Committee, and their symbol of a red cross on a white background, has become the standard of neutral wartime medical care of wounded combatants and civilians.

The IFRC was founded in 1919 and has grown to be the world's largest humanitarian organization. After World War I, American Red Cross War Committee president Henry Davison proposed a creation of a League of Red Cross Societies, so that the expertise of the millions of volunteers from the wartime efforts of the ICRC could be used in a broader scope of peacetime activities. Today, the IFRC includes 195 member societies, a Secretariat in Geneva, and more than 60 additional delegations dispersed throughout the world.

The IFRC conducts complex relief and recovery operations in the aftermath of disasters throughout the world. Their four areas of focus include promoting humanitarian values, disaster response, disaster preparedness, and health and community care. Through their work, they seek to "improve the lives of vulnerable people by mobilizing the power of humanity," as stated in their mission. These people include those who are victims of natural and manmade disasters and postconflict scenarios.

Like the UN, the IFRC is well established in most countries throughout the world and is well poised to assist in the event that disaster strikes. Volunteers are continuously trained and utilized at the most local levels, providing a solid knowledge base before a major need presents itself. Cooperation among groups, through the federation, provides an enormous pool of people and funds from which to draw when local resources are exhausted.

When a disaster strikes and the local capacity is exceeded, an appeal by that country's national chapter is made for support to the Federation's Secretariat. As coordinating body, the Secretariat initiates an international appeal for support to the IFRD and many other outside sources and provides personnel and humanitarian aid supplies from its own stocks. These supplies, which can be shipped in if not locally available, pertain to needs in the areas of health, logistics and water specialists, aid personnel, and relief management.

The appeal for international assistance is made an average of 30 times per year, and these assistance projects can continue for years. Long-term rehabilitation and reconstruction projects, coupled with the goal of sustainable development and increased capacity to handle future disasters, have become the norm in regards to major disasters in the poorer countries. The following is how the IFRC responds to international disasters.

Depending on the complexity of the required response, a Field Assessment and Coordination Team (FACT) may be deployed to assist the local chapter in determining the support needs for the event. The teams, which are deployable to any location with only 24 hours' notice, consist of Red Cross/Red Crescent disaster managers from throughout the IFRC, bringing with them skills in relief, logistics, health, nutrition, public health, epidemiology, water and sanitation, finance, administration, and psychological support. The team works in conjunction with local counterparts and host-government representatives to assess the situation and determine what the IFRC response will consist of. An international appeal is drafted, and then launched, by the Secretariat in Geneva. The teams stay in-country to coordinate the initiation of relief activities. Once the effort has stabilized and has become locally manageable, the FACT concedes its control to the local Red Cross headquarters.

In 1994, following a spate of notably severe disasters (i.e., the Armenian earthquake, the Gulf War Kurdish refugee problem, and the African Great Lakes Region crisis), the IFRC began to develop an Emergency Response Unit (ERU) program to increase disaster response efficiency and efficacy. These ERUs are made up of preestablished supplies, equipment, and personnel, who respond as a quick-response unit on a moment's notice and are trained and prepared to handle a much wider range of scenarios than before. This concept, similar to the UNDP Emergency Response Division (ERD), has already proven effective in making IFRC response faster and better, through several deployments, including Hurricane Mitch in Honduras. The teams, upon completion of their response mission, remained in-country to train the locals in water and sanitation issues, thus further ensuring the sustainability of their efforts. ERU teams are most effective in large-scale, sudden-onset, and remote disasters.

Finally, the IFRC is heavily engaged in disaster preparedness and has identified several strategies toward mitigation they hope to achieve by 2010. These activities, which relate to reducing the impact of disasters whenever possible and to working toward better prediction and prevention methods, are becoming a fundamental component of local Red Cross/Red Crescent Society programs. The IFRC has recognized the following four points of action as most vital:

- Reducing the vulnerability of households and communities in disaster-prone areas and improving their ability to cope with the effects of disasters
- Strengthening the capacities of National Societies in disaster preparedness and postdisaster response
- Determining a role and mandate for National Societies in national disaster plans
- Establishing regional networks of National Societies that will strengthen the Federation's collective impact in disaster preparedness and response at the international level

They plan to increase local capacity to handle disasters, thus decreasing the magnitude of international assistance required on disaster onset. This increase in capacity eventually will result in a decreased loss of life and property, as each country becomes more developed and more able to prevent catastrophe. The IFRC aims to accomplish these results through their regular local capacity-building projects, performed in conjunction with research and analysis, which includes the following:

- Hazard prediction
- Risk and vulnerability assessment of individual groups or regions
- Assessment of local strength and capacity in disaster response
- Response network development
- Assessing of National Society disaster mitigation and response capacity
- Assessing national government preparedness and response plans

According to the Geneva Mandate on Disaster Reduction, which was adopted in 1999, the IFRC declared:

> We shall adopt and implement policy measures at the international, regional, sub-regional, national and local levels aimed at reducing the vulnerability of our societies to both natural and technological hazards through proactive rather than reactive approaches. These measures shall have as main objectives the establishment of hazard-resilient communities and the protection of people from the threat of disasters. They shall also contribute to safeguarding our natural and economic resources, and our social well being and livelihoods.

ASSISTANCE PROVIDED BY THE U.S. GOVERNMENT

U.S. Agency for International Development

The United States has several means by which it provides assistance to other nations requiring such aid in the aftermath of a disaster, accident (nuclear, biological, or chemical), or conflict. The U.S. agency tasked with providing development aid to other countries, the U.S. Agency for International Development (USAID), has also been tasked with coordinating the U.S. response to international disasters. USAID was created in 1961 through the Foreign Assistance Act, which was drafted to orga-nize U.S. foreign assistance programs and separate military and nonmilitary assis-tance. One branch of USAID, the Bureau for Humanitarian Response (BHR), manages the various mechanisms with which the United States can respond to humanitarian emergencies of all types. The office under BHR that most specifically addresses the needs of disaster and crisis victims by coordinating all nonfood aid provided by the government is the Office of U.S. Foreign Disaster Assistance (OFDA).

Office of Foreign Disaster Assistance

The OFDA is divided into four distinct subunits: Disaster Response Division (DRD); Prevention, Mitigation, Preparedness, and Planning (PMPP); Operations Support (OS); and Program Support (PS). The DRD handles the U.S. assistance provided to foreign disasters. The PMPP assists foreign nations with assistance to develop their ability to mitigate and prepare for disasters. The OS division handles the technical and logistical support of all OFDA projects, and the PS division works with the OFDA financial and accounting systems.

The administrator of USAID holds the title of President's Special Coordinator for International Disaster Assistance. When a disaster is declared in a foreign nation by the resident U.S. ambassador (or by the Department of State, if one does not exist), the USAID administrator is appealed to for help. This can be done when the magnitude of the disaster has overwhelmed a country's local response mechanisms, the government has requested assistance or will at least accept it, and it is in the interest of the U.S. government to assist. The OFDA is authorized to immediately disburse $25,000 in emergency aid to the U.S. Embassy to be spent at the discre-tion of the ambassador for immediate relief. The OFDA also can immediately send

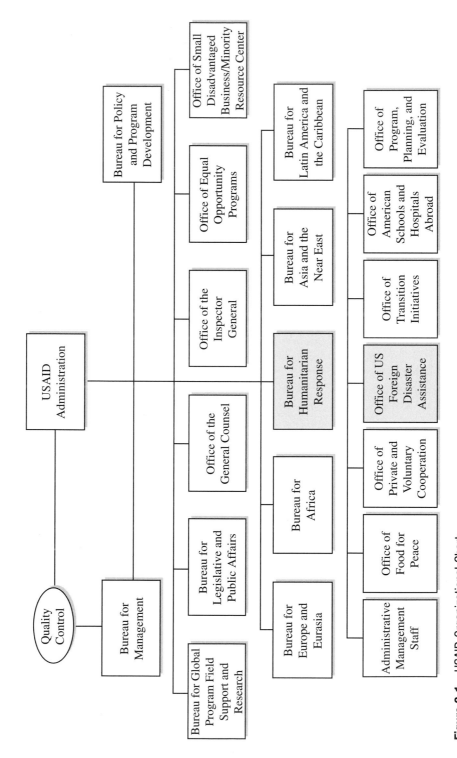

Figure 8-1   USAID Organizational Chart.

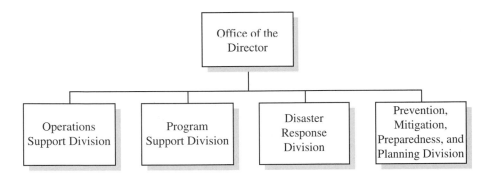

Figure 8-2 OFDA Organizational Chart.

regional advisors with temporary shelter and medical aid supplies from one of four OFDA stockpiles in Guam, Italy, Honduras, and the United States.

If the disaster is considerable in size, a Disaster Assistance Response Team (DART) is deployed to the country to assess the damages and recommend the level of assistance that should be made by the U.S. government. DARTs work quickly to develop a strategy to coordinate U.S. relief supplies; provide operational support; coordinate with other donor countries, UN agencies, NGOs, and the host government; and monitor and evaluate projects carried out with U.S. funds. In the largest of disasters, Response Management Teams (RMTs) may be established in both Washington, D.C. and the disaster site, to coordinate and offer administrative assistance and communication for the several DARTs that would be deployed.

The OFDA recently developed a Technical Assistance Group (TAG) to increase its capabilities in planning and programming. TAGs consist of scientists and specialists in agriculture and food security, emergency and public health, water and sanitation, geoscience, climate, urban planning, contingency planning, cartography, and so on. TAGs work with DARTS and RMTs in response, as well as USAID development missions in preparation and mitigation for future disasters.

In addition to the direct aid and logistical and operational support offered, the OFDA provides grants for relief assistance projects. These projects are carried out primarily by PVOs and NGOs, as well as IOs, the UN, and other various organizations (such as a pilots' club that is hired to transport supplies). Not all this monetary aid goes to response, however. The PMPP works to facilitate projects that aim to reduce the impact of disaster before they happen again. These types of projects seek to empower national governments to make them less likely to need international assistance in subsequent events. All these organizations are monitored carefully by the OFDA to ensure that they are working efficiently and are spending monetary resources sensibly.

Other USAID Divisions

Under the USAID BHR, several other offices provide humanitarian aid. The Office of Food for Peace (FFP) handles all the U.S. government's food assistance projects (U.S. food aid is categorized as Title II or Title III, with the first having no repayment obligations, and the second considered a bilateral loan). The Office of Transition Initiatives (OTI) works in postconflict situations to help sustain peace and establish democracy. The Department of State Bureau for Population, Refugees, and Migration (PRM) provides monetary grants to NGOs, PVOs, IOs, and the UN to respond to emergency refugee emergencies. A good portion of this assistance goes directly to the UNHCR. Lastly, the Department of Defense (DoD) responds through their Office of Peacekeeping and Humanitarian Affairs (PK/HA). It is important to note that the developed nations of the world are highly unlikely to receive U.S. assistance on the level that is provided to the developing nations.

The U.S. Military

The U.S. Military often is involved in relief efforts of natural and technological disasters and CHEs. The involvement of the military, a well-funded and equipped force whose primary function is national defense, brings about an entirely new perspective to the area of operations. It often is argued that nobody is better equipped to handle disasters than the military, with their wide assortment of heavy equipment, enormous reserve of trained personnel, and common culture of discipline and mission-oriented standard operation; however, it is also said that the military is a war agency, not a humanitarian assistance agency, and that these two organizational ideals are too fundamentally and diametrically opposed in practice to allow for effective military involvement.

The assistance of the military normally is requested by USAID/OFDA through the DoD Office of Political/Military Affairs. The chain of command for military operations begins with the President of the United States and the Secretary of Defense, collectively referred to as the National Command Authority (NCA). The NCA, which directs all functions of the U.S. Military, is advised by the Joint Chiefs of Staff (JCS) of the Army, Navy, Air Force, and Marines. The entire military force is divided into five geographic Areas of Responsibility (AORs) and two functional commands, as follows:

- U.S. Atlantic Command (USACOM): Norfolk, VA headquarters
- European Command (EUCOM): Stuttgart, Germany headquarters
- Pacific Command (PACOM): Honolulu, HI headquarters
- Central Command (CENTCOM): Tampa, FL headquarters
- Southern Command (SOUTHCOM): Miami, FL headquarters
- Special Operations Command (SOCOM): In command of special operations, including the Special Forces, Civil Affairs, and Psychological Operations; Tampa, FL headquarters
- Transportation Command (TRANSCOM): Provides management for all air/sea/land transportation; Scott Air Force Base, IL headquarters

The U.S. Military is heavily involved in the response to international disasters through organized operations termed Foreign Humanitarian Assistance (FHA) or Humanitarian Assistance Operations (HAO). FHAs are authorized by the DoD Office of Political/Military Affairs (DODPM) at the request of the OFDA (the President, as Commander-in-Chief, gives final authorization for any support operation).

Assistance may be provided in the form of physical or technical support, such as logistics, transportation, communications, relief distribution, security, and emergency medicine. In emergencies of natural or manmade origin that do not involve conflict, the role of the military is to provide support, rather than leadership, to the national government and the overall relief community.

The military is known for its self-contained operational abilities, arriving on-scene with everything they need, so to speak. Usually, they provide more than adequate personnel and supplies for the mission they were called to act upon. Once in-country, they work under the strict guidelines of Force Protection (enforced security of all military and civilian personnel, equipment, and facilities associated with their mission) and Rules of Engagement (ROE, a structured, preestablished guideline of "circumstances and limitations under which the military will initiate or continue combat engagement"). The ROE dictate military action in both peacekeeping and disaster operations.

If a particular command unit is tasked with assisting a relief operation, they may deploy a Humanitarian Assistance Survey Team (HAST) to conduct a needs assessment, which relates to the specific functions the military is suited to address. These assessments are occasionally much different than those generated by more humanitarian-based organizations, such as the UN or OFDA, because the military operates in such a fundamentally different fashion. The concerns of the HAST tend to focus on the military support requirements and the logistical factors involving deployment of troops. A Joint Task Force (JTF) will be established soon after to handle the management and coordination of military personnel activities, with a Commander for the JTF designated as the person in charge of the operation on-site; however, if an operation involves only one military service, or is minimal in size, a JTF may not be needed.

One of the main roles of the JTF is to establish a Civil Military Operations Center (CMOC). This center effectively functions to coordinate the military support capabilities in relation to the overall response structure involving all other players involved. The CMOC mobilizes requests for assistance from OFDA, the UN, NGOs, and the host government. All intermilitary planning is conducted through this center, including those operations involving cargo transportation and food logistics. This center is the primary node of information exchange to and from the JTF. CMOCs have taken on expanded responsibility in the past, including the reestablishment of government and civil society and the repair or rehabilitation of critical infrastructure.

THE INTERNATIONAL FINANCIAL INSTITUTIONS

The international financial institutions (IFIs) provide loans for development and financial cooperation throughout the world. They exist to ensure financial and market stability and to increase political balance. These institutions are made up of Member States, arranged on a global or regional basis, which work together to provide financial services to national governments through direct loans or projects. In the aftermath of disasters, it is common for nations with low capital reserve to request increased or additional emergency loans to fund the expensive task of reconstruction and rehabilitation. Without these IFIs, most developing nations would have no means with which to recover. The largest of these IFIs, The World Bank, and one of its subsidiaries, the International Monetary Fund (IMF), are detailed as follows. Other regional IFIs with similar functions include the Inter-American Development Bank (IDB), which works primarily in Central and South America, and the Asian Development Bank (ADB), based in Manila, Philippines, which works throughout the Asian continent.

The World Bank

The World Bank was conceived in 1944, during World War II, at Bretton Woods. Its inceptive purpose was to rebuild Europe, and France received the first World Bank loan of $250 million in 1947 for postwar reconstruction. This first lending arrangement reflects the standard of World Bank funding, and financial assistance for reconstruction has been provided regularly since that time in response to countless natural disasters and humanitarian emergencies. Most World Bank loans are provided to the Member States that are less developed.

Today, the World Bank is regarded as one of the largest sources of development assistance, and in the 2001 fiscal year, more than $17.3 billion in loans were provided in more than 100 countries. The World Bank is owned collectively by more than 180 countries and is based in Washington, D.C. It comprises several institutions referred to as the World Bank Group (WBG), which includes the International Bank for Reconstruction and Development (IBRD), the International Development Association (IDA), the International Finance Corporation (IFC), the Multilateral Investment Guarantee Agency (MIGA), and the International Centre for Settlement of Investment Disputes (ICSID). Its overall goal is to reduce poverty, and specifically to "individually help each developing country onto a path of stable, sustainable, and equitable growth, [focusing on] helping the poorest people and the poorest countries" (The World Bank). As disasters and CHEs are taking greater and greater toll on the economic stability of so many countries struggling financially, the Bank, as it is often referred to, is taking on a more central role in mitigation and reconstruction.

Developing nations, which are more likely to have no established mitigation or preparedness and therefore have little or no affordable access to disaster insurance, often sustain damage that is considered a total financial loss. In the periods of rehabilitation that follow the disaster, loans are essential to the success of programs and vital if any level of sustainability or increased disaster resistance is to be achieved. There are several points along this cycle in which the Bank lends assistance.

First, in regular financial assistance, the Bank has worked to ensure that borrowed funds are applied toward projects that give mitigation a central role during the planning phase. They utilize their privilege as financial advisor to guide financial planners who may forego these important measures to stretch the loaned capital as far as possible. They work to increase systems of prediction and risk analysis through their projects to adequately develop according to standards that account for recurrent disasters.

Once a disaster occurs, the Bank may be called on for help. Because it is not a relief agency, the Bank will not take on any role in the initial response; however, immediately upon agreeing to participate, it begins work on restoring damaged and destroyed infrastructure and restarting production capabilities. First, a team may be provided to assist in performing initial impact assessments, including an estimate of pure financial losses resulting from the disaster and an estimated cost of reconstruction including raised mitigation standards. Second, it could restructure the country's existing loan portfolio with the Bank in order to allow for expanded recovery projects. Third, projects that have not yet been approved (but are in the application process) can be redesigned to account for changes caused by the disaster. Finally, an Emergency Recovery Loan (ERL) can be granted, which would specifically address the issues of recovery and reconstruction.

ERLs are granted to restore affected economic and social institutions and to reconstruct physical assets such as essential infrastructure. It is important to note that ERLs are not designed for relief activities. They are most appropriate for disasters that have great adverse impact on the economy, are infrequent in nature (as recurrent disasters are accommodated by regular lending schemes), and create urgent needs. The loan is expected to eventually produce economic benefits to the borrowing government. The ERLs usually are implemented within three years and are flexible to accommodate for the specific needs of each unique scenario. Construction performed with the ERL must use disaster-resistant standards and include appropriate mitigation measures, thus providing an overall preparedness for the country affected. Once an ERL has been granted, the Bank coordinates with the IMF, the UNDP, NGOs, and several other international and local agencies to create a strategy that best utilizes granted funds in relation to the reconstruction effort as a whole.

The Bank works to increase resistance to repeat natural disasters through planning support in their regular lending programs that can be aimed toward mitigation and preparedness. The nature of their mission—to alleviate poverty—is in itself a mitigation measure. As part of its lending process, the Bank conducts vulnerability and risk assessments, which necessitate the subsequent consideration of any findings in planning of future development with Bank loans. The Bank is also a source of information on current hazard-resistant technology and provides the expertise for establishing more effective building codes and their enforcement. As countries develop, they increase their capacity to prepare and respond to disasters and establish the legal and political institutions that guide construction and settlement practices that ensure greater overall resilience. The Bank is arguably the most important player in attaining the means to do so.

The International Monetary Fund

The International Monetary Fund (IMF) was established in 1946 and has grown to a current membership of 183 countries. Its goals are to promote international monetary cooperation, exchange stability, and orderly exchange arrangements; to foster economic growth and high levels of employment; and to provide temporary financial assistance to countries to help ease balance of payments adjustment. It carries out these functions using loans, monitoring, and technical assistance.

In the event of an international disaster or CHE in a member country, the IMF utilizes its Emergency Assistance Specific Facility to provide rapid financial assistance. In these situations, it is not uncommon for a country to have severely exhausted its monetary reserves. The IMF's goals are to rebuild government capacity and to return stability to the local economy. In the event of a natural disaster, funding is directed toward local recovery efforts and for any economic adjustment that may be needed. If the situation is a postconflict one, its aim is to "reestablish macroeconomic stability and the basis for long-term sustainable growth" (IMF). The IMF will lend assistance only if a stable governing body is in place that has the capacity for planning and policy implementation and can ensure the safety of IMF resources. After stability has been sufficiently restored, increased financial assistance is offered, which will be used to develop the country in its postemergency status.

When a country wishes to request emergency assistance, it must submit a detailed plan for economic reconstruction and ensure that it will not create trade restrictions or intensify exchange. If the country is already working under an IMF loan, then assistance can come in the form of a reorganization within existing arrangements. Separate emergency assistance loans are also offered, which do not involve the regular criteria under which the countries must normally operate. These loans, although normally available only up to 25 percent of a country's preestablished lending quota, have been created in quantities reaching 50 percent of quota; however, this funding is provided only when the member country is "cooperating with the IMF to find a solution to its economic problems." These loans are required to be repaid within five years.

A country often requires technical assistance or policy advice because it is in a situation for which it has no experience or expertise. This is common in postconflict situations where a new government has been established and partnerships are being created for the first time. The IMF offers assistance in building capacity to implement macroeconomic policy. This can include tax and government expenditure capacity, the reorganization of fiscal, monetary, and exchange institutions, and guidance in the use of aid resources.

CONCLUSION

As global populations converge into more concentrated urban settlements, their collective hazard risks amplify. Loss of life and property caused by the realization of these hazard risks will overwhelm the response and recovery capacities of individual sovereign nations to an ever-increasing degree. Many of these disasters, particularly in the lesser-developed nations, will contribute to existing development obstacles and regional instability unless trends toward increased multilateral cooperation in disaster assistance are recognized more widely for their importance. The capabilities and organizational capacities of the international disaster management agencies listed in this chapter, namely national governments, nonprofit organizations, international organizations, and the international financial institutions, are vital for both the preparation and mitigation of hazard risks, and the response and recovery of actualized disasters.

CASE STUDY

THE GUJARAT, INDIA EARTHQUAKE

In Calcutta, India, as citizens were just starting to celebrate their country's 52nd Republic Day, highrise apartment buildings began to shake at a barely perceptible intensity. Little did anybody in that city know, they were not experiencing a weak local tremor but the far-reaching effects of the second most deadly earthquake to hit the country in recorded history taking place more than *1,900 kilometers away* in the state of Gujarat. In fact, the massive temblor, which struck at 8:46 A.M. on January 26, 2001, was also felt in Pakistan and Nepal.[1] This event, the worst earthquake to hit the state of Gujarat in 200 years and the most devastating disaster to hit the country of India in the past 50, struck an unprepared nation.

This case study discusses the origins and disaster history for the affected region and the damage inflicted by the Gujarat earthquake (also referred to as the Bhuj earthquake because of the epicenter's proximity to that city). Also examined is the response that followed by institutions including the national government of India and the state government of Gujarat, the government of the United States, the United Nations, and the multilateral lending institutions. Three nonprofits, the Red Cross, CARE, and Catholic Relief Services, are discussed in relation to their assistance, as a sample of the hundreds of agencies that responded.

The Earthquake

Origins, Geology, Disaster History

Gujarat's location in the west of India, bordering Pakistan, lies within the Himalayan collision zone where two surface plates (the Indo-Australian and the Eurasian) are slowly crashing together to form the world's youngest and tallest mountain chain at a pace of about two centimeters per year.[2] This movement is but one peril in a land that faces many natural challenges.

Cyclones, floods, drought, and earthquakes characterize Gujarat's history.[3] In the past 25 years, more than 3,000 people and 350 livestock have been killed and more than 1 million houses destroyed by almost yearly cyclones. Floods inundate an average of 300,000 hectares of land, damage an average of 37,000 houses, kill 135 people, and affect two million human lives in each average one-year span.[4] The district of Kuchchh, which is the largest in the state, is surrounded by a peculiar swamp called the Rann of Kuchchh, which floods annually and isolates the region from the rest of the Gujarat.[5] Drought is almost a yearly occurrence, with a particularly long three-year drought, which led up to and further complicated events discussed in this case.[6]

In terms of earthquakes, there have been many, with incidents measuring over 6.0 or greater on the Richter scale occurring in 1819 (8.3), 1903 (6.0), 1940 (6.0), and 1956 (7.0). Although the high vulnerability to these disasters has been long established as fact, there was no formalized government management plan to mitigate, prepare, or respond when the Gujarat quake struck. As a result, they were totally unprepared to handle the mass casualty events that ensued.[7] Ironically, this earthquake struck surprisingly close in location to the one that had occurred in 1819 along the same fault line in which many fewer lives were lost. A dramatic increase in development in that region with little or no building code enforcement is blamed for the much higher level of casualty even with a lower intensity of shaking.

Scope of the Quake

This was the largest earthquake to occur in India since an 8.5 magnitude event hit the state of Assam in 1950.[8] The Indian Meteorological Department (IMD) has recorded a Richter magnitude of 6.9 with location being northeast of Bhuj, although the U.S. Geological Survey (USGS) maintains that the magnitude was 7.9, and the epicenter lay north of Bachau in a location 50 kilometers from the IMD site.[9] The depth of the earthquake, also disputed, was eventually confirmed as approximately 20 kilometers, and resulting aftershocks with an unusual depth of 30 kilometers give the impression that the earthquake may have severed the lithosphere.[10] There was little surface deformation because of the depth, with no clearly discernible cracks on the surface such as those seen with more shallow quakes; however, the liquefaction phenomenon was widespread because of the intensity,[11] and in some cases, rivers that had been dry for more than a century became activated.[12]

Most of the communication infrastructure was immediately destroyed, and a good portion of the transportation infrastructure was damaged. The local government had no immediate means to alert the central government of their imminent needs. This resulted in the lack of an initial assessment, and urban search-and-rescue teams were not sent in time to be fully effective in their missions. The bulk of the initial rescue missions were carried out by neighbors helping neighbors, digging with their bare hands and personal tools.[13] Nobody outside the state could have guessed the magnitude of damage they would find in the coming days, and the character of the first response reflected this knowledge gap; however, when the rescue teams reached the relatively easily accessible city of Ahmedabad and observed the damage, they immediately knew they were going to confront worse conditions in Kuchchh, where the epicenter was located. They moved relief material and volunteers to that region without preassessment.[14]

The earthquake caused damage in 7,904 villages in 21 of the state's 25 districts.[15] The district of Kuchchh sustained the bulk of the damage, with more than 400 villages affected. The towns that suffered most significantly were Bhuj, Bachhau, Anjar, Rapar, and Gandhidham, where virtually 100 percent of the buildings were damaged.[16] This district sustained 90 percent of the deaths and 78 percent of the injuries reported overall, and contained 257,000 of the houses damaged or destroyed.[17] Three hundred kilometers from the epicenter, however, in the city of Ahmedabad, 179 buildings were destroyed.[18]

In many of the areas that were isolated, there was no food or medical relief for up to five days, and people began looting what they could in desperation.[19] In Bachhau, where 30,000 people of 40,000 were cut off from the relief, armed gangs formed and began attacking survivors for money or food.[20] These problems ceased almost immediately upon the arrival of assistance, illustrating the effect a timely response can have on the security of an affected region.

Damage Caused

The damage resulting from this earthquake is a good indicator of the extent to which megahazards will affect nations financially in the twenty-first century because sustained losses repeatedly exceeded $1 billion. The following list summarizes these damages:

- In pure asset losses, the World Bank and Asian Development Bank estimate that India's losses will exceed $2.1 billion.
- The official government death toll, based on family registration of death and most likely severely underestimated as result, is 20,005 people; 166,812 were injured, about 20,000 seriously.[21]
- Almost 16 million people, or 1 in 3 in the state of Gujarat, were affected in some way by the January 26 events.[22]
- About 400,000 structures collapsed, and an additional 500,000 to 800,000 were damaged. In the Kuchchh district alone, 300 primary healthcare centers and 1,300 child nutrition centers were lost.
- The damage to the state's infrastructure, administration, and communications was extensive and remained a major burden on resources in the reconstruction phase.[23] Several of the sustained damages are listed below:
  - The main telecommunications link with Kuchchh snapped and 147 exchanges were damaged in the initial tremor, confounded further by 82,000 damaged phone lines.[24] The remaining open lines were quickly flooded.
  - Most power facilities were damaged to some extent, and 925 villages lost power.[25]
  - Drinking water and irrigation systems were affected in 1,340 villages, with 1,100 of those villages reporting severe damages.
  - Of 240 damaged reservoir dams that supply the water for these irrigation and domestic needs, 20 need to be completely rebuilt.[26]
  - More than 100 kilometers of roads were severely damaged, several railroad lines needed repair, and 5 of 10 piers at Kandla Port (the major shipping port in the state) were destroyed.[27]
  - Approximately 9,600 primary schools, 2,040 secondary schools, and 140 technical institutions will need to be rebuilt.[28]
  - The handicraft industry in Kuchchh suffered the loss of more than 3,000 artisans, and in Dhamadka village, almost 70 percent of the workers in this industry were lost. More than 3,000 small-scale and cottage industries and 20 medium to large-scale enterprises were affected.[29]

Because the impact of this event was not initially communicated to the Government of India, a resulting underestimation of its severity was conveyed to the world community of responders.[30] Much of the initial response was further hampered by the fact that many of the responders (e.g., fire, police, health) were either dead, injured, or attending to family emergencies, which diverted their attention away from the greater relief effort.[31] The scope of rehabilitation required is close to inconceivable, and 18 months later many anxious people are still sleeping out in the open or under plastic sheeting.

The Participants

The State- and National-Level Governments of India

This event was "the biggest challenge Gujarat has ever faced."[32] By most accounts, the initial response by the state government was nonexistent, primarily because of the complete lack of emergency preparedness and resultant chaos that ensued. Police and fire brigades, the personnel that traditionally respond first in these situations, were occupied with duties related to security and logistics for flag-raising ceremonies and parades.[33] Most government personnel were taking advantage of the long weekend and were not prepared to suddenly return to work.[34]

The Government of Gujarat immediately airlifted a team of five officials headed by the Additional Chief Secretary, which arrived in Bhuj within six hours of the first tremors.[35] This team, although experienced in the management of engineering and medical response, was much too small to handle an event of this magnitude. To increase the rescue staff available, all government officials were officially called off vacation, and an appeal for volunteerism was made to doctors, engineers, retired government officials, and others with applicable skills.[36] Schools and colleges were uniformly closed to ensure that students would be available for the relief and rescue efforts.[37] A state control room was made functional on the first day, and its effectiveness increased once the communication lines were repaired on day two.[38] Ham radio, satellite, and cellphone stations were established for both public and private use.[39] It was not until the third day, however, that the state government diverted heavy equipment used for irrigation, roads, and construction to the search, rescue, and food distribution operations.[40]

The Government of India, on the other hand, took charge almost immediately and responded to an event that would have challenged even the most developed nations. It is important to note that even though Prime Minister Vajpayee never formally requested international assistance, he did let it be known that offers of aid were welcome and would gladly be accepted.[41]

Because of the communication infrastructure problems, the initial Government of India response was small and mounted only in Ahmedabad, where reports of damage could be broadcast.[42] The Government of India had no formal disaster management plan that defined the responsibilities of the separate government agencies, so the approach was centralized. Assets had not been inventoried, and their mobilization was not as rapid as it could have been.[43] Other than these initial issues, the government response was one to be commended.

The Krishi Control Room was set up to coordinate the central government response and provide constant communications and updates.[44] The Chief Secretary began holding twice-daily meetings to review the progress and planning of the relief efforts, and a hotline was set up between the Prime Minister and the

State Governor to facilitate communication.[45] Local Emergency Operation Control Rooms of varying capability and equipment were set up in tents or structures that had not collapsed, in the localities that suffered the worst damage. These centers acted as information nodes and assisted in the central government coordination to the sites.[46] Two major locations were established as collection, tracking, and distribution centers at Gujarat College and at a town hall, for the tremendous flow of donated goods.[47]

Doctors and nurses were sent to each region with appropriate medical equipment and vehicles.[48] Fifteen thousand Gujarat Electricity Board personnel and 30 truckloads of equipment were dispatched to repair the electrical power in the affected regions.[49] A government survey team was created and examined the status of the buildings that remained standing to determine their safety.[50] Fifteen thousand Indian military service personnel and significant heavy equipment were deployed to provide transportation and distribution support to relief operations, and to repair the airports and bridges that had been damaged.[51,52] The government sent out a request to businesses that operate cranes, gas cutters, and construction equipment to volunteer their services.[53]

When the temperature began to fall at night, temporary shelters were provided as quickly as possible.[54] The water supply, which was already deficient because of the drought, was supplemented by tankers in Kuchchh. Various foods and cooking supplies were distributed, including the allotment of 20 kilograms of wheat and 5 kilograms of rice for each family. For the many families who lost food ration cards (prequake government subsidies), replacements were given.[55] One month's worth of grass was distributed to cattle owners in the region's hardest hit areas.[56] Public service announcements were taped and announced on radio and television instructing people not to enter damaged buildings that could collapse.[57] Customs and excise taxes on all goods imported or manufactured for the relief efforts was waived, and the ban on foreign technology and foreign aid that was in effect was suspended as well.[58] To show government support and sympathy, Prime Minister Vajpayee visited the area and is said to have foregone regular security and stayed longer than originally planned to convey his message.[59]

The Empowered Group of Central Ministers was created to coordinate the emergency response and met for the first time on January 30, 2001. It consists of representatives from the departments of Home Affairs, Railways, Textiles, Consumer Affairs, Information & Broadcasting, Defense, Finance, Civil Supplies, Health, Rural Development, Housing, Agriculture, Communications, and Power.[60] Their purpose was to do the following:

1. Consider the report of the Crisis Management Group and give such directions as considered appropriate.
2. Decide on all action necessary to provide immediate relief to the victims.
3. Consider measures necessary for relief and rehabilitation of the affected.
4. Consider long-term institutional and organizational measures that are necessary for management and mitigation of such natural disasters.[61]

For the restoration efforts, the Gujarat Earthquake Rehabilitation Fund was created to raise money. Grants were distributed according to the extent of financial and physical damage. The Government of Gujarat State Disaster Management Authority was created to better enable these reconstruction efforts, heralding $1 billion in aid to assist more than 300,000 families according to level of village damage, distance from the epicenter, and the original house value.[62] A national Department of Earthquake Relief was also created, as part of the department of General Administration.[63] Finally, a plea was made to ban all public celebrations until February 28, and to ask that "those celebrating marriage and other social programs [are] modest and austere."[64]

The U.S. Government

The U.S. Government, one of the largest donors in the relief effort, provided aid through the Department of Defense (DoD) and the U.S. Agency for International Development (USAID) Office of Foreign Disaster Assistance (OFDA). Between the two agencies, the United States contributed $13.1 million to the response effort. The DoD provided airlifts for all the donated goods, a 2.5-ton truck, two forklifts, two 400-gallon tankers, 10,000 blankets, 1,500 sleeping bags, and 92 50-person tents. A six-person military assessment team consisting of experts in communications, logistics, and technical support, was provided to advise the government responders.[65] The OFDA provided assistance through donated commodities and through grants via organizations such as CARE, Catholic Relief Services, and the World Food Program. Three airlifts by OFDA (valued at $2,426,463), which carried technical equipment, shelters, blankets, sleeping bags, water and sanitation equipment, and other goods, supplied relief to more than 450,000 people.[66] Grant programs that OFDA dollars facilitated included water sanitation, disease surveillance, emergency shelters, relief distribution, medical support, trauma counseling, and food assistance. In addition to these projects, a USAID Disaster Assistance Response Team (DART) of 11 people was dispatched to conduct emergency needs assessments and coordinate the distribution of all relief supplies donated by the U.S. Government.[67] Finally, $100,000 was given to the Prime Minister's Gujarat Rehabilitation Relief Fund.

The United States has remained active in the recovery and rehabilitation of Gujarat. USAID developed the Gujarat Earthquake Recovery Initiative, which is aimed at families in the poorest communities. An allocation of $10 million has been granted, and the funds will come from existing USAID budget resources to be used by various NGOs and multilateral organizations like the UN. There are four established areas for which the funds will be used:

1. Cash, for work and other NGO programs to help repair roads, wells, water systems, homes, workplaces, and other infrastructure needed to restart economic activities.
2. Cash, for work and other NGO programs to clear away debris and rubble and repair public facilities such as health clinics and child nutrition centers.
3. Survey support, to assess damaged (but still standing) buildings, to determine whether they can be repaired and retrofitted or if they need to be demolished and rebuilt.
4. Support, to municipalities and local NGOs to develop community renewal plans that will help reconstruct devastated communities.[68]

The United States was just one of many countries that responded, providing a total of about $90 million to support the relief and reconstruction of Gujarat.[69] Other nations that assisted include Australia, Austria, Belgium, Canada, Denmark, Finland, France, Germany, Greece, Hungary, Ireland, Israel, Italy, Japan, Luxembourg, Monaco, Nepal, the Netherlands, New Zealand, Norway, Oman, Poland, Russia, Singapore, Spain, Sweden, Switzerland, Taiwan, Turkey, and the UK.[70]

The Nongovernmental Organizations

More than 200 NGOs engaged in the response and relief effort in India, creating a daunting task of cooperation and coordination.[71] Initially, there was no built-in government mechanism to organize the relief. Under these chaotic circumstances, the organizations worked out a system of coordination on their own, which attempted to create an optimal working arrangement for the disaster and increase the effectiveness of response to the greatest number of those in need. It is reported that this was the first time coordination efforts such as these had taken place in India, and they were primarily successful.[72] Three of these organizations' responses, those of CARE, Catholic Relief Services, and the Red Cross, are described as follows.

Care. CARE mobilized the morning after the quake to perform an initial assessment of the Kuchchh district. They provided an immediate supply of medical equipment, food, blankets, tarps, tents (10,000 family-sized), and water-purification tablets. CARE emergency medical teams provided treatment and trauma counseling to survivors in the hard-to-reach areas of Anjar, Bachau, Rapar, and Bhuj.[73] With the help of a USAID grant, they were able to provide food and survival kits to assist 50,000 people, to encourage them to remain in their home areas rather than become displaced.[74]

CARE's work in India lasted through the end of February, helping more than 175,000 people in the remote villages where they felt need was the greatest.[75] In this time, they helped build at least 118 community service facilities (e.g., schools, health centers, government offices) and 105 water systems (locally managed for sustainability), increased access to employment and training for 6,000 people, and rebuilt damaged irrigation systems and watershed management schemes. Overall, their goal was to increase the general capacity of the earthquake victims through many self-help initiatives.

Catholic Relief Services. Catholic Relief Services (CRS), like CARE, was the recipient of a large portion of the USAID grants. In addition, they committed $650,000 in private funds, as well as their Africa-based emergency technical unit and staff from various locations including Bosnia.[76] Initial cash resources were designated for the installation of temporary shelter and to meet the personal hygiene needs of more than 65,000 people in 73 villages. Mental Health units were established to provide trauma counseling for the injured, their families, and the most vulnerable groups (e.g., women, children, lower-caste members, elderly, and minorities).[77] One year later, CRS was still working on follow-up projects to increase the likelihood of program success and are creating village resource maps to maximize the overall target population size.[78]

American Red Cross. The American Red Cross is one of the most experienced organizations in responding to international disasters of every type. They were one of the first organizations on the ground in Gujarat, working with a team of 11 American experts trained in logistics, communication, mental health, and family tracing. This team supported the overall International Red Cross team of more than 120 people. The Red Cross distributed almost $2 million in supplies to nearly 100,000 victims. Included in this aid were 13,000 five-gallon buckets, 550 rolls of plastic sheeting, 15,000 kitchen sets, 25,000 tarps, 15,000 blankets, and 5,000 tents. They purchased and distributed emergency health kits, from the World Health Organization, which included medicine, intravenous fluids, surgical tools, and other medical supplies.[79]

The Red Cross plans to assist the state of Gujarat in the reconstruction as well. Their current projects aim to do the following:

1. Help rebuild community infrastructure to provide safe, clean water, including the repair and installation of water collection, storage, and sanitation.
2. Develop a trained network of Indian mental health professionals who will provide mental health counseling for this and other disasters.
3. Provide community health education programs to improve access to basic healthcare and prevent the spread of communicable diseases.[80]

These efforts complement the $15 million in aid provided by the International Federation of Red Cross/Red Crescent Societies (IFRC), of which a portion was used to construct a 310-bed, high-tech emergency hospital in Bhuj.[81,82]

The United Nations

The UN agencies responded immediately, having access to all government information through their established in-country presence. The UN Development Programme (UNDP) was coordinator, assisting in responses of the World Food Programme (WFP), the UN Children's Fund (UNICEF), the Office of the Coordination of Humanitarian Affairs (OCHA), the International Labor Organization (ILO), and the World Health Organization (WHO). The accomplishments of each of these agencies is described as follows:

UN Development Programme. The UNDP deployed its Disaster Response Team, whose responsibility was to coordinate the entire emergency response until the UNDP could formally assume that role.[83] In addition, supported by $2.75 million from the governments of the United States, Britain, and Italy, the UNDP coordinated the UN body needs assessments, activity identification, project proposal design and implementation, monitoring, and quality control.[84] The UNDP and the UN volunteers they oversee worked to address the issue of the houses destroyed in the quake. Using "roaming teams," they worked with local communities to develop and fund projects for the distribution of building materials and the construction of temporary shelters. These teams also monitored the progress of the projects. The UNDP provided $100,000 for immediate relief through a project in partnership with two of the leading women's organizations in Gujarat: the Self-Employed Women's Association (SEWA) and the NGO Kutch Mahila Vikas Sangthan, who put together survival kits for families in addition to helping with general housing issues.[85] The UNDP sent 35 UN volunteers into several regions where no other NGOs had initiated work or provided assistance, and plans to eventually have 5,000 volunteers working on the overall recovery effort.[86]

The UNDP continues to be the UN coordinating body for reconstruction, and this is to be a long and arduous task. They have continued to work with the central Government of India and the state Government of Gujarat in implementing plans to provide permanent housing to the homeless, using construction design that is resistant to the many risks encountered in that region.[87] All these projects are merely in addition to those the UNDP already is conducting throughout India.

The World Food Programme. The WFP launched a $4.14 million project that provided relief food rations to 300,000 people for four months. Most of these people received packages of wheat flour and lentils, to help them survive the months following the earthquake. They specifically targeted a group of 178,000 children below the age of five and pregnant and nursing mothers, and provided them with highly nutritious biscuits and a fortified blended food called Indiamix. A special Joint Logistics Center was initiated in Bhuj on February 11, with a $2.3 million budget, to coordinate the overall relief efforts for the victims and airlift the relief material from a UN Humanitarian Response Depot (UNHRD) in Brindisi, Italy to Bhuj.[88]

UN Children's Fund. Just two days after the earthquake struck, UNICEF sent a team of 15 members based in Gujarat to distribute 15,000 blankets, 1 million chloroquine tablets to purify drinking water, and medical supplies that could help 30,000 people for three months. In the next 72 hours, they provided an additional $600,000 in medical equipment. Over the course of the next few weeks, during the response phase, UNICEF supplied 83 mobile water tankers, countless medical supplies of every type, 75,000 blankets, measles vaccines to more than 400,000 children, water supply systems, 700 large tents (to act as temporary classrooms and healthcare centers), school supplies, vitamin A for one million children, one million oral rehydration packets, refrigerators, generators, and 106,000 family survival kits.[89] UNICEF is continuing to work with the Government of Gujarat to rebuild many of the schools that were damaged or destroyed, and is helping the communities in the state prepare emergency preparedness plans. UNICEF contributed more than $21 million to relief and reconstruction.

Office of the Coordination of Humanitarian Affairs. OCHA sent a five-member UN disaster assessment and coordination team on January 27 to assist the UNDP in the response phase of the disaster.[90] It provided an emergency grant of $150,000 from its own resources and from prepositioned funds from the Governments of Denmark and Norway to purchase tents and blankets. Together with the WFP, OCHA organized the three relief flights from the UNHRD in Brindisi, Italy. Periodically during the response phase, OCHA issued Situation

Reports in order to keep the international community informed and to raise support for the affected population.[91]

International Labor Organization. The ILO's activities were aimed at creating short-term work opportunities in cleanup, rebuilding the infrastructure and housing, and "protecting vulnerable groups such as young women and children."[92] They established programs that addressed aspects of disaster recovery relating to their main concern of labor issues. These projects sought to gather statistics relating to the effect on the job market from losses in employees and employment, migration flows, and the skills of the victims. Using what they refer to as "labor-intensive methods," they provided immediate employment opportunities to stimulate local markets and provide people with self-reliance. They concentrated on the most vulnerable groups, such as women and children, and worked with other agencies (such as UNICEF) to curb the disaster effects that lead to child labor, child trafficking, and sexual exploitation.[93]

World Health Organization. WHO sent a team of nine public health experts to Gujarat to perform a rapid health assessment of the region. A disease surveillance desk was established in the main emergency operations center in Bhuj to monitor the possible outbreak of disease (which often appears in mass-casualty events).[94] Experts from WHO provided technical advice to the state government and health officials on public health issues. They also provided emergency health materials, including trauma kits, emergency health kits, and other essential medical supplies, all within the first days of the disaster. What was most needed, however, was the rehabilitation of the damaged and destroyed healthcare facilities, and they were working with the experience they had acquired in the same region after the 1999 cyclone that caused similar destruction.[95]

The International Development Banks

It is important to mention the international development banks that worked with the Government of India to finance reconstruction loans that are essential to the recovery of the state. Although these institutions played a vital role in establishing the preliminary and final assessments of the damages and reconstruction needs resulting from the quake, they do not perform any duties related specifically to the response. Their involvement in the reconstruction is essential because they are providing the capital, without which nothing could be rebuilt, and are working with the Government of India in developing a reconstruction plan that will be able to better sustain the types of natural disasters that afflict the area on a regular basis.[96]

References

1. Government of Gujarat, India. *Earthquake in Gujarat*, (n.d.) http://gujaratindia.com/preliminary.html, retrieved October 30, 2001, p. 1.
2. World Bank and Asian Development Bank. *Gujarat Earthquake Recovery Program: Assessment Report*, March 14, 2001, p. 10.
3. *Ibid*, p. 6.
4. *Ibid*, p. 9.
5. Krishna Vatsa. *The Bhuj Earthquake, District of Kutch, State of Gujarat (India) January 26, 2001*, March 16, 2001, DRM-World Institute for Disaster Risk Management, p. 1.

6. U.S. House of Representatives. *The Earthquake in India: The American Response*, March 1, 2001, Washington D.C., U.S. Government Printing Office, p. 10.

7. EERI. *EERI Preliminary Government Response Report, Earthquake in Gujarat, India*, January 26, 2001, (n.d.) www.eeri.org/earthquakes/reconn/bhuj_india/pgovtrest.html, retrieved October 29, 2001, p. 1.

8. USAID. *India—Earthquake, Fact Sheet #14 (FY 2001)*, February 8, 2001, www.usembassy.state.gov/posts/in1/wwwhguj.html, p. 1.

9. Ravi Sinha and Rajib Shaw. *The Bhuj Earthquake of January 26, 2001: Consequences and Future Challenges*, April 26, 2001, p. 1.

10. CIRES. *26 January 2001 Bhuj Earthquake, Gujarat, India*, August 2001, University of Colorado, http://cires.Colorado.edu/~bilham/gujarat2001.html, p. 3.

11. *See* note 2, p. 2.

12. *See* note 10.

13. *See* note 9, p. 4.

14. *Ibid*, p. 11.

15. *Ibid*, p. 4.

16. *See* note 5, p. 6.

17. *See* note 6.

18. *See* note 1.

19. *Ibid*, p. 3.

20. "Food Riots Hamper Quake Relief Work in Gujarat," *Times of India*, January 30, 2001.

21. *See* note 6.

22. Government of India. *The Government of India's Official Website on the Gujarat Earthquake*, (n.d.) http://gujarat-earthquake.gov.in/mainpage.htm, p. 3.

23. *Ibid*, p. 2.

24. *Ibid*, pp. 1–9.

25. *Ibid*, p. 14.

26. *See* note 2, p. 1.

27. *See* note 22, pp. 11–13.

28. *See* note 2, p. 1.

29. *Ibid*.

30. *See* note 1, p. 2.

31. *See* note 22, p. 1.

32. *See* note 1, p. 11.

33. *See* note 7.

34. *See* note 1.

35. *Ibid*.

36. *Ibid*, p. 1.

37. *Ibid*, p. 3.

38. *Ibid*, p. 6.

39. *Ibid*, p. 3.

40. *See* note 6, p. 11.

41. *See* note 9, p. 9.

42. *Ibid*.

43. *See* note 22, p. 5.

44. *See* note 1, pp. 5, 7.

45. *See* note 9, p. 10.

46. *See* note 1, p. 7.

47. *Ibid*, p. 6.

48. *See* note 5, p. 7.

49. *See* note 1, p. 7.

50. *See* note 8, p. 8.

51. *See* note 1, p. 6.

52. *Ibid*, p. 6.

53. *Ibid*, p. 4.

54. *Ibid*, p. 9.
55. *Ibid*, p. 11.
56. *Ibid*, p. 6.
57. *Ibid*, pp. 33, 36.
58. *Ibid*, p. 4.
59. *See* note 22, p. 5.
60. Government of India, "The Central Government's Relief Efforts," 2001, http://www.gujarat-earthquake.gov.in/ministry-wise.htm, last accessed May 2003.
61. *See* note 5, p. 9.
62. *See* note 1, p. 11.
63. *Ibid*.
64. *See* note 8, p. 3.
65. *Ibid*.
66. *Ibid*.
67. USAID. *Summary of U.S. Government Assistance for the Victims of the Gujarat Earthquake*, February 12, 2001, p. 1.
68. *See* note 6, pp. 14–15.
69. *Ibid*, p. 39.
70. *See* note 8, p. 3.
71. "Gujarat Tragedy: Could Be a Blessing in Disguise," *The Statesman* (India), April 23, 2001, http://web.lexis-nexis.com.
72. *See* note 9, p. 4.
73. *See* note 6, p. 23.
74. USAID. *USAID Awards $2.2 Million in Contracts for Indian Earthquake*, February 5, 2001, USAID Press Office, p. 1.
75. *See* note 6, p. 23.
76. *Ibid*, p. 16.
77. *Ibid*, pp. 13–19.
78. Interaction. *Earthquake in India*, June 2001 www.interaction.org/india00/index.html, p. 6.
79. American Red Cross. *India Quake Relief 2001*, (n.d.) www.redcross.org/news/in/0101bhuj/india.html, p. 2.
80. *Ibid*, p. 3.
81. *See* note 9, p. 12.
82. *See* note 2, p. 3.
83. *Ibid*, p. 1.
84. Madhu Nainan. "Quake-Proof Houses Set the Model for Gujarat Reconstruction," *Agence France-Presse*, March 13, 2001, p. 5.
85. UNDP Press Release. *UN Steps Up Relief Operations in Gujarat*, United Nations, February 3, 2001, p. 2.
86. *Ibid*, p. 6.
87. *See* note 2, p. 45.
88. UN India. *The UN System Response to the Gujarat Earthquake*, August 2, 2001, www.un.org.in/dmt/guj/unresguj80201c.htm, retrieved October 9, 2001, p. 6.
89. Unicef, *India: Earthquake, Complete Emergency Report*, (n.d.) www.unicefusa.org/alert/emergency/indiaeq/archive.html, p. 6.
90. *See* note 2, p. 3.
91. *See* note 88, p. 7.
92. UNDP Press Release. "UN Steps Up Relief Operations in Gujarat," United Nations, February 3, 2001, p. 2.
93. *See* note 88, p. 7.
94. *See* note 2, p. 3.
95. *See* note 92, p. 2.
96. *See* note 2, p. 4.

9. Emergency Management and the New Terrorist Threat

INTRODUCTION

The terrorist attacks of September 11 prompted dramatic changes in emergency management in the Untied States. These attacks and the subsequent Anthrax scare in Washington, D.C., in October 2001 have been the impetus for a reexamination of the nation's emergency management system, its priorities, funding, and practices. These changes are ongoing and will continue for the foreseeable future.

Prior to September 11, the Nunn-Lugar legislation provided the primary authority and focus for domestic federal preparedness activities for terrorism. Several agencies, the Federal Emergency Management Agency (FEMA), Department of Justice (DOJ), Department of Health and Human Resources (HHS), Department of Defense (DoD), and the National Guard were all involved, and jockeying for leadership of the terrorism issue. There were some attempts at coordination but, in general, agencies pursued their own agendas. The biggest difference among the agencies was the level of funding available, with DoD and DOJ controlling the most funds. State and local governments were confused, felt unprepared and complained of the need to recognize their vulnerability and needs should an event happen. The TOPOFF exercise, held in 1999, reinforced these concerns and vividly demonstrated the problems that could arise in a real event. The events of September 11, unfortunately, validated their concerns and visibly demonstrated the need for changes in the federal approach to terrorism.

The changes fall into five general categories, including (1) first responder practices and protocols, (2) preparing for terrorist acts, (3) funding the war on terrorism, (4) creation of the Department of Homeland Security, and (5) the shift in focus of the nation's emergency management system to the war on terrorism. This chapter explores these categories, identifies issues, and discusses the implications of this new direction for emergency management. Where appropriate, a historic perspective to these changes is provided.

CHANGES IN EMERGENCY MANAGEMENT AND THE WAR ON TERRORISM

There are five groups that must be fully engaged in the nation's war on terrorism: the diplomats, the intelligence community, the military, law enforcement, and emergency management. The principal goal of the diplomats, intelligence, military, and law enforcement is to reduce if not eliminate the possibility of future terrorist attacks on American citizens inside our borders and abroad.

The goal of emergency management should be to be prepared and to reduce the future impacts in terms of loss of life, injuries, property damage, and economic disruption caused by the next terrorist attack. As President Bush and many of his advisors have repeatedly informed the nation, it is not a question of if but rather when the next terrorist attack occurs. Therefore, it is incumbent upon emergency managers to apply the same diligence to preparing for the next bombing or biochemical event as they do for the next hurricane or flood or tornado. The focus of emergency management in the war on terrorism must be on reducing the danger to first responders, to the public, the business community, the economy, and our way of life from future terrorist attacks. This change must occur at all levels of the emergency management system, federal, state and local.

Figure 9-1   Oklahoma City, OK, April 26, 1995—Scene of the devastation following the Oklahoma City bombing. FEMA News Photo.

The war on terrorism has resulted in unprecedented funding resources being made available to the emergency management community. The federal government has recognized the role that state and local first responders played in limiting further harm from the September 11 and other previous terrorist attacks, and now, for the first time in memorable history, vast sums of money from the federal government are available for first responder equipment and training, for planning and exercises, and for the development of new technologies. Funding for the Federal Emergency Management Agency (FEMA) has increased, as has the amount of funds FEMA delivers to state and local emergency management organizations.

Historically, FEMA had distributed about $175 million annually to its state and local emergency management partners. Since 2001, the amount of money granted to these agencies is measured in the billions of dollars, with the FY2006 budget request for such items and activities set at over $3.5 billion. New federal funding sources also have opened up for emergency managers from the Department of Defense, the Department of Justice, and the Department of Health and Human Resources to fund contingency plans, technology assessment and development, and bioterror equipment and training. These changes in funding for emergency management has been felt most significantly at the state and local levels.

The creation of the Department of Homeland Security (DHS) represented a landmark change for the federal community, especially for emergency management. The consolidation of all federal agencies involved in fighting the war on terrorism follows the same logic that first established FEMA in 1979. At that time, then-President Carter, at the request and suggestion of the nation's governors, consolidated all the federal agencies and programs involved in federal disaster relief, preparedness, and mitigation into one single federal agency, FEMA.

The director of the new agency, FEMA, reported directly to the President as will the OHS secretary. However, now that FEMA is a component of DHS, the FEMA director no longer reports directly to the President but rather to the DHS secretary. The impact of this change will be known only after the initial preparedness and response focus, brought about by the spectacular nature of the September 11 attacks, wanes, and the need for a strong voice leading emergency management once again returns.

At the request of President George W. Bush, FEMA established the Office of National Preparedness in 2001 to focus attention on the then undetermined terrorist threat and other national security issues. This was the first step in the refocusing of FEMA's mission and attention from an all-hazards approach to emergency management embraced by the Clinton administration. The shift in focus was accelerated by the events of September 11 and has been embraced by state and local emergency management operations across the country. A similar shift of focus in FEMA occurred in 1981 at the beginning of the Reagan administration. Then the shift of focus was from disaster management to planning for a nuclear war. For the remaining years of the Reagan administration and the four years of President George H. W. Bush's administration, FEMA resources and personnel focused their attention of ensuring continuity of government operations in the event of a nuclear attack. Little attention was paid to natural hazard management and FEMA was left unprepared to deal with a series of catastrophic natural disasters starting with Hurricane Hugo in 1989 and culminating with Hurricane Andrew in 1992.

If history repeats itself, the current change in focus away from the all-hazard approach of the 1990s could result in a weakening of FEMA's natural disaster management capabilities in the future. Once the final assessments emerge from the response and recovery from the devastating 2004 hurricane season, we will have a better idea about whether or not this has occurred.

SUMMARY OF SEPTEMBER 11 EVENTS

Measuring the far-reaching impacts of the events of September 11 on emergency management can be done in a wide variety of ways. In the following sections of this chapter, we will discuss some of the organizational, funding, technology, and operational changes that these events initiated. We will also examine how the focus of the emergency management in this country shifted because of these events. In this section we will examine the size and breadth of these events through an examination of some of the financial costs, principally spending by FEMA and other federal government agencies, in responding to and assisting in the recovery from these events.

When considering the impacts of the events of September 11, the first impact that must be considered is the horrific loss of life in New York City, Virginia, and Pennsylvania. After years of researching lists of who may or may not have been involved at each of the locations, a final tally of 2,985 killed was determined. Of this amount, 184 died at the Pentagon, 40 in Pennsylvania, and the remaining 2,761 at the World Trade Center (*USA Today* 2002).

The attacks on the World Trade Center and the Pentagon together could arguably be considered the first national disaster event, outside of wartime, in the history of the United States. It is the first disaster event in this country that impacted all citizens of this country and left all citizens and communities with an uneasy sense of vulnerability. However, it is the economic consequences of these attacks that were felt in all parts of our country and, in fact, around the world that make this disaster event truly national in scope. Measuring the economic impacts of such an event is daunting and some measures will take years to complete, but a quick review of some the economic impacts measured to date clearly illustrate the breadth and width of this disaster's impact on the economic well-being of the people of the United States.

Figure 9-2   New York, NY, September 27, 2001—The remaining section of the World Trade Center is surrounded by a mountain of rubble following the September 11 terrorist attacks. Photo by Bri Rodrguez/FEMA News Photo.

PARTIAL LIST OF ECONOMIC LOSSES CAUSED BY SEPTEMBER 11 EVENTS AS OF DECEMBER 2001

The Airline Industry

Prior to September 11, 2001, the airline industry was facing financial troubles due to internal organizational difficulties, low ridership, rising labor costs, a failing economy, and subsequent unexpectedly low profits. The September 11 attacks, which directly targeted the airline industry, was a blow that appeared to be potentially fatal. To avoid a full collapse of the industry, Congress passed the Air Transportation Safety and Stabilization Act (P.L. 107-42) in the weeks immediately following the attacks. This legislation provided $5 billion in direct compensation to the airlines, over 90 percent of which was disbursed in the first year, and $10 billion in guaranteed loans.

A major reduction in the demand levels for the industry spurred the airlines to reduce their operations by 20 percent and eliminate almost 100,000 jobs. Unfortunately, passenger demand dropped faster than even these conservative estimates could accommodate, falling to 66 percent of capacity by the end of 2001's fourth quarter. The immediate financial losses were grave, with the industry as a whole losing a net $7.7 billion, despite the benefits they had received from the Stabilization Act. In 2002, the trend continued, with losses exceeding

continues

even the most pessimistic estimates at over $10 billion. 2002 was the worst in the industry's history. 2003 showed a moderate gain in business, with approximately $8.6 billion in losses. However, 2004 showed the best returns since 2000, with net losses of only around $5 billion, blamed mostly on the rising costs of fuel experienced during the military operations in Iraq.

The outlook for 2005 has been sour, but is actually a marked improvement over the previous four years with estimates of net losses totaling between $1 and $3 billion by various analysts. There are signs that ridership has risen to pre-September 11, 2001 levels. The key factor to all this information, though, is that there are obvious winners and losers within the industry emerging. Although these industry-wide numbers are negative, there are some airlines that are posting high positive profits, which indicates that there are others posting great net losses, resulting in an overall negative figure. The result is expected to come in the form of several bankruptcies, layoffs, and buyouts as the trend toward recovery continues.

Sources: Ramstack, Tom. 2004. "Airline Traffic at Pre-September 11th Totals," *The Washington Times*, July 29.
U.S. Subcommittee on Aviation. 2002. Hearing on Financial Condition of the Airline Industry.

The Insurance Industry

The September 11 terrorist attacks resulted, at the time, in the second greatest amount of insured losses on record, at a total of $37 billion (the greatest being 1992, when hurricanes Andrew and Iniki caused $38 billion in insured losses). The events led to a full reorganization of the industry, and the passage of the Terrorism Risk Insurance Act of 2002. The Act, which requires that private insurance companies operating in the United States offer insurance against acts of international terrorism events, reinsures the industry against the losses that might occur from such events. In the years that have followed the 2001 attacks, there have been no attacksz on U.S. soil, and as a result, no additional losses. With increased premiums, and new premiums collected on terrorism-based policies, the events eventually will cause industry-wide profits in the absence of future attacks.

In an unexpected turn of events, 2004 brought about insurance claims that broke the 1992 record for the internationally-based insurance industry's payouts by $4 billion, with $42 billion in losses posted. This time, however, it was natural disasters that resulted in the bulk of losses, including the succession of hurricanes that struck the United States and the 10 typhoons that struck Asia. The catastrophic tsunami events will not have a great impact on the industry as most of the structures in the affected regions were uninsured.

Source: Poon, Arthur. 2004. "Insurers worldwide pay out record S$69 billion this year," *The Straits Times* (Singapore), December 29.

Costs Associated with Federal/State Disaster Assistance

The cost to the federal government for the response and recovery of the World Trade Center was formally estimated to be $20 billion, though other informal measures are more difficult to assess and likely will raise this figure significantly. FEMA provided 42 percent of the federal share, with $8.8 billion in aid. HUD gave the second largest share, $2.48 billion, or 17 percent of the total share, and DOT ranked third at $2.37 billion (11.5 percent). All other federal agencies contributed a total of $820 million, which amounted to 4 percent of the total federal share. Also included in the federal figures of aid are the tax benefits associated with the New York City Liberty Zone—an area of the city where new tax incentives have resulted in over $5 billion in indirect economic aid to the city and its residents. The exact amount of aid resulting from this program will never be known, and as such, the exact figure of federal aid will remain an estimate.

Since September 11, the costs associated with securing the nation from future acts of terrorism have eclipsed this $20 billion figure, through the creation of the Department of Homeland Security, the costs of airport security, police and fire department overtime, special events security, equipment and training grants, technology grants, port security—the list is extensive. The cost of all these measures combined, across the four years since the terrorist attacks, amounts to several hundred billion dollars, and will likely continue to rise for many years to come.

U.S. Unemployment

Immediately following the September 11 attacks, specific industries laid off hundreds of thousands of workers, including the food and beverage industry (42,000 workers), hotels (46,000 workers), and the airline industry (over 100,000 workers). The economic downturn that existed before the attacks but increased in severity following them, resulted in economy-wide losses in jobs, with the unemployment rate rising from 4.0 percent in 2000, to 4.8 percent in 2001, 5.8 percent in 2002, 6 percent in 2003, and finally falling to 5.5 percent in 2004. Although many factors affect the rise and fall of unemployment, it was irrefutably proven that the attacks had a direct, significant impact on these numbers, and affected specific industries more severely than others. However, the rise in security-related jobs has offset these negative figures, and moreover, as of the beginning of 2005, unemployment rates appear to be falling further away from their recent 2003 low, to 5.2 percent.

Sources: DRI-WEFA. 2001. Greatest U.S. Employment Loss in 20 Years. DRI-WEFA Economic Briefings. November 2.
CBS News. 2005. Unemployment Rate Dips, But . . . CBSNews.com, February 4.

Another measure of the size of these events is the costs to the federal government in providing disaster relief. As of the end of 2004, FEMA had disbursed over $8.8 billion to the city and state of New York for emergency and recovery work. This total represents only FEMA's expenditures on this disaster and does not include expenditures by other federal agencies, insurance companies, and the private sector. According to FEMA records this total would place the World Trade Center disaster above all disasters on FEMA's list of Top Ten Natural Disasters presented in Table 9-1. (FEMA does not have a comparable list for technological disasters that could be used to compare the events of September 11, so the natural disaster list is used.)

Table 9-1   Top Ten Natural Disasters (Ranked By FEMA Relief Costs)

Event	Year	FEMA Funding
Northridge earthquake (CA)	1994	$6.967B
Hurricane Georges (AL, FL, LA, MS, PR, VI)	1998	$2.255B
Hurricane Andrew (FL, LA)	1992	$1.814B
Tropical Storm Allison (FL, LA, MS, PA, TX)	2001	$1.375B
Hurricane Hugo (NC, SC, PR, VI)	1989	$1.307B
Midwest floods (IL, IA, KS, MN, MO, NE, ND, SD, WI)	1993	$1.140B
Hurricane Floyd (CT, DE, FL, ME, MD, NH, NJ, NY, NC, PA, SC, VT, VA)	1999	$1.054B
Loma Prieta earthquake (CA)	1989	$865.8M
Red River Valley floods (MN, ND, SD)	1997	$741.2M
Miami floods (FL)	2000	$623.1M

Source: www.fema.gov

Summaries of selected FEMA costs associated with the World Trade Center disaster are as follows.

FEDERAL DISASTER EXPENDITURES FOR THE WORLD TRADE CENTER DISASTER

Federal Disaster Assistance Committed to New York City Following the September 11th Terrorist Attacks

Initial Response to the Attacks: $2.55 billion

- FEMA—$2.2 billion
- DOT—$100 million
- HUD—$250 million

Numerous assistance programs are included in this grouping, such as search and rescue operations, debris removal operations, emergency transportation measures, and emergency utility service repair. FEMA provided the bulk of the federal funds for initial response efforts—$2.20 billion—and DOT and HUD provided the bulk of the remaining funds.

Highlights:

- Search and rescue—$22 million (largest in U.S. history)
- Debris removal—$1.7 billion ($1 billion of which will be used to establish an insurance company to cover the city and any contractors from potential claims that may arise)
- Emergency transportation measures—$299 million
- Other response assistance (health monitoring, EPA cleanup, etc.)—$285 million
- Emergency and temporary utility service—$250 million (to be disbursed)

Compensation for Losses: $4.81 billion

- FEMA—$3.84 billion
- HUD—$960 million

This funding, provided by FEMA and HUD, compensated state and local organizations, individuals, and businesses for disaster-related costs, such as mortgage and rental assistance to individuals and grants to businesses to cover economic losses.

Highlights:

- NYPD and NYFD benefits, wages, and other reimbursement—$643 million
- Other public assistance to NYC, New York State, and other organizations—$847 million
- Nontraditional assistance—$1 billion
- FEMA Hazard Mitigation Grant Program—$377 million
- Mortgage and Rental Assistance Program—$200 million
- Crisis counseling—$99 million
- Individual and Family Grant Program—$110 million
- Other FEMA assistance—$34 million
- HUD Residential Grant Program—$106 million
- HUD Business Assistance Program—$510 million

Infrastructure Restoration: $5.57 billion

- FEMA—$2.75 billion
- DOT—$2.24 billion
- HUD—$580 million

The majority of this funding is a combination of FEMA and DOT funds to rebuild and enhance the lower Manhattan transportation system, including the construction or repair of roads, subways, ferries, and railroads. HUD is funding efforts to improve utility infrastructure.

continues

Highlights:

- Projects planned to restore and enhance the lower Manhattan transportation system—$4.55 billion
- Permanent utility infrastructure repairs and improvements—$750 million
- Short-term capital projects—$68 million

Economic Revitalization: $5.54 billion

- HUD—$520 million
- Liberty Zone—$5.03 billion

Efforts to revitalize the economy in lower Manhattan include the Liberty Zone tax benefit plan (a congressionally created tax benefit plan in lower Manhattan)—an estimated benefit of $5.03 billion—and $515 million in HUD funding for business attraction and retention programs. Once the city, state, and HUD finalize plans for the remaining $1.16 billion, these funds will most likely be directed to infrastructure restoration and improvements and/or economic revitalization:

Highlights:

- Liberty Zone tax benefits—approximately $5 billion, but expected to grow
- HUD Business Assistance Programs and planning for rebuilding and permanent memorial—$515 million

Assistance by Agency to New York City:

- FEMA—$8.8 billion (42.9%)
- Liberty Zone Tax Benefits—$5.03 billion (24.5%)
- HUD—$3.48 billion (17%)
- DOT—$2.37 billion (11.5%)
- Other agencies—$820 million (4%)

Source: U.S. General Accounting Office Report GAO-03–1174T

On May 1, 2003, FEMA closed its Application Assistance Center located in Lower Manhattan. In its one year and seven months of operation, the center's staff assisted over 190,000 individuals and small business owners who were applying for grants and assistance with temporary housing, mortgage and rental assistance, and low-interest disaster assistance loans.

FIRST RESPONDER EVALUATION

In July and August 2002, two September 11–related after-action reports were released: "Improving NYPD Emergency Preparedness and Response" prepared by McKinsey & Company for the New York City Police Department and "Arlington County After-Action Report on the Response to the September 11 Terrorist Attack on the Pentagon" prepared for Arlington County, Virginia, by Titan Systems Corporation. Both reports are based on hundreds of interviews with event participants and reviews of organizational plans. These reports provide lessons learned and present hundreds of recommendations.

The NYPD report did not pass judgment on the success or failure of the NYPD on September 11 but rather assessed the NYPD's response objectives and instruments in order to identify 20 improvement opportunities for the NYPD of which six merited immediate action:

- Clearer delineation of roles and responsibilities of NYPD leaders
- Better clarity in the chain of command
- Radio communications protocols and procedures that optimize information flow
- More effective mobilization of members of the service
- More efficient provisioning and distribution of emergency and donated equipment
- A comprehensive disaster response plan, with a significant counter-terrorism component (McKinscy & Company 2002)

The Arlington County After-Action Report did declare the response by the county and others to the Pentagon terrorist attack a success that "can be attributed to the efforts of ordinary men and women performing in extraordinary fashion" (Titan Systems Corporation, 2002). The terrorist attack on the Pentagon sorely tested the plans and skills of responders from Arlington County, Virginia, other jurisdictions, and the federal government. "Notable Facts about Sept. 11 at the Pentagon" compiled in the report are provided as follows.

NOTABLE FACTS ABOUT SEPT. 11 AT THE PENTAGON

- The first Arlington County emergency response unit arrived at the crash site less than three minutes after impact.
- Lieutenant Robert Medarios was the first Arlington County Police Department command-level official on-site. He made a verbal agreement with a representative of the Defense Protective Service that Arlington County would lead the rescue efforts of all local and federal agencies.
- Over 30 urban search and rescue teams, police departments, fire departments, and federal agencies assisted Arlington's police and fire in the rescue. Some of these important partners included the FBI, the Federal Emergency Management Agency, U.S. Park Police, Defense Protective Service, the Military District of Washington, the Metropolitan Washington Airport Authority, the Virginia Department of Emergency Management, and USAR teams from Albuquerque, N.M., Fairfax County, Va., Montgomery County, Md., and Memphis, Tenn.

continues

- Captain Dennis Gilroy and the team on Foam Unit 161 from the Fort Meyer Fire Station were on-site at the Pentagon when Flight 77 crashed into the building. Firefighters Mark Skipper and Alan Wallace who were next to the unit received burns and lacerations but immediately began helping Pentagon employees, who were trying to escape from harm's way, out of the first-floor windows.
- Capt. Steve McCoy and the crew of Engine 101 were on their way to fire staff training in Crystal City when they saw the plane fly low overhead and an explosion from the vicinity of the Pentagon. McCoy was the first person to call Arlington County's Emergency Communications Center to report the plane crash.
- The Arlington County American Red Cross Chapter coordinated support from the Red Cross. The chapter had 80 trained volunteers at the time of the attack, but the organization's mutual-aid arrangements with other chapters garnered nearly 1,500 volunteers who helped support the emergency services personnel, victims, and their families.
- Business supporters set up temporary food service on the Pentagon parking lot for rescue workers. Over 187,940 meals were served to emergency workers. Many other businesses brought phones for rescuers to call home, building materials, and other vital necessities.
- Over 112 surgeries on nine burn victims were performed in three weeks. One of the nine burn victims died after having over 60 percent of her body burned. There were 106 patients that reported to area hospitals with various injuries.
- 189 people died at the Pentagon—184 victims and five terrorists.
- On the morning of Sept. 11, 1941, exactly 40 years before the terrorist attacks of 2001, the original construction on the Pentagon began.

Source: "Arlington County After-Action Report on the Response to the September 11 Terrorist Attack on the Pentagon," prepared for Arlington County, Virginia, by Titan Systems Corporation.

The Arlington County report contains 235 recommendations and lessons learned. Of these many recommendations, the report highlights examples of lessons learned in two categories: things that worked well and contributed to the overall success of the response and challenges encountered and overcome by responders that could serve as examples for other jurisdictions in the future. These lessons learned are presented next.

LESSONS LEARNED AT THE PENTAGON

The Arlington County After Action Report contains 235 recommendations and lessons learned, each of which must be understood within the context and setting of the Pentagon response. Some specifically apply to a particular response element or activity. Others address overarching issues that apply to Arlington County and other jurisdictions, particularly those in large metropolitan areas. They have not been weighted or prioritized. This is a task best left to those with operational responsibilities and budgetary authority.

Capabilities Others Should Emulate

1. **ICS and Unified Command.** The primary response participants understood the ICS, implemented it effectively, and complied with its provisions. The ACFD, an experienced ICS practitioner, established its command presence literally within minutes of the attack. Other supporting jurisdictions and agencies, with few exceptions, operated seamlessly within the ICS framework. For those organizations and individuals unfamiliar with the ICS and Unified Command, particularly the military, which has its own clearly defined command and control mechanisms, the Incident Commander provided explicit information and guidance early during the response and elicited their full cooperation.

2. **Mutual Aid and Outside Support.** The management and integration of mutual-aid assets and the coordination and cooperation of agencies at all government echelons, volunteer organizations, and private businesses were outstanding. Public safety organizations and chief administrative officers (CAOs) of nearby jurisdictions lent their support to Arlington County. The response to the Pentagon attack revealed the total scope and magnitude of support available throughout the Washington metropolitan area and across the nation.

3. **Arlington County CEMP.** The CEMP proved to be what its title implies. It was well thought out, properly maintained, frequently practiced, and effectively implemented. Government leaders were able to quickly marshal the substantial resources of Arlington County in support of the first responders, without interfering with tactical operations. County board members worked with counterparts in neighboring jurisdictions and elected federal and state officials to ensure a rapid economic recovery, and they engaged in frequent dialogue with the citizens of Arlington County.

4. **Employee Assistance Program (EAP).** At the time of the Pentagon attack, Arlington County already had in place an aggressive, well-established EAP offering critical incident stress management (CISM) services to public safety and other county employees. In particular, the ACFD embraced the concept and encouraged all its members to use EAP services. Thus, it is not surprising that the EAP staff was well received when they arrived at the incident

continues

site within three hours of the attack. During the incident response and in follow-up sessions weeks afterward, the EAP proved invaluable to first responders, their families, and the entire county support network. This is a valuable resource that must be incorporated in response plans.

5. **Training, Exercises, and Shared Experiences.** The ACED has long recognized the possibility of a weapons of mass destruction (WMD) terrorist attack in the Washington metropolitan area and has pursued an aggressive preparedness program for such an event, including its pioneering work associated with the MMRS. In preparation for anticipated problems associated with the arrival of Y2K, Arlington County government thoroughly exercised the CEMP. In 1998, the FBI Washington Field Office (WFO) established a fire liaison position to work specifically with area fire departments. Washington metropolitan area public safety organizations routinely work together on events of national prominence and shared jurisdictional interests, such as presidential inaugural celebrations, Heads of State visits, international conferences such as the periodic International Monetary Fund (IMF) conference, and others. They also regularly participate in frequent training exercises including those hosted by the Pentagon and MDW. All this and more contributed to the successful Pentagon response.

Challenges That Must Be Met

1. **Self-Dispatching.** Organizations, response units, and individuals proceeding on their own initiative directly to an incident site, without the knowledge and permission of the host jurisdiction and the Incident Commander, complicate the exercise of command, increase the risks faced by bonafide responders, and exacerbate the challenge of accountability. WMD terrorist event response plans should designate preselected and well-marked staging areas. Dispatch instructions should be clear. Law enforcement agencies should be familiar with deployment plans and quickly establish incident site access controls. When identified, self-dispatched resources should be immediately released from the scene, unless incorporated into the Incident Commander's response plan.

2. **Fixed and Mobile Command and Control Facilities.** Arlington County does not have a facility specifically designed and equipped to support the emergency management functions specified in the CEMP. The conference room currently used as the EOC does not have adequate space and is not configured or properly equipped for that role. The notification and recall capabilities of the Emergency Communications Center are constrained by equipment limitations and there are no protected telephone lines for outside calls when the 9-1-1 lines are saturated. The ACED does not have a mobile command vehicle and relied on the use of vehicles belonging to other organizations and jurisdictions. The ACPD mobile command unit needs to be replaced or extensively modernized.

3. **Communications.** Almost all aspects of communications continue to be problematic, from initial notification to tactical operations. Cellular telephones were of little value in the first few hours and cellular priority access service (CPAS) is not provided to emergency responders. Radio channels were initially over-saturated and interoperability problems among jurisdictions and agencies persist. Even portable radios that are otherwise compatible were sometimes preprogrammed in a fashion that precluded interoperability. Pagers seemed to be the most reliable means of notification when available and used, but most firefighters are not issued pagers. The Arlington County EOC does not have an installed radio capacity and relied on portable radios coincidentally assigned to staff members assigned duties at the EOC.

4. **Logistics.** Arlington County, like most other jurisdictions, was not logistically prepared for an operation of the duration and magnitude of the Pentagon attack. The ACED did not have an established logistics function, a centralized supply system, or experience in long-term logistics support. Stock levels of personal protective equipment (PPE), critical high-demand items (such as batteries and breathing apparatus), equipment for reserve vehicles, and medical supplies for EMS units were insufficient for sustained operations. These challenges were overcome at the Pentagon with the aid of the more experienced Fairfax County Fire and Rescue Department logistics staff. A stronger standing capacity, however, is needed for a jurisdiction the size of Arlington County.

5. **Hospital Coordination.** Communications and coordination were deficient between EMS control at the incident site and area hospitals receiving injured victims. The coordination difficulties were not simple equipment failures. They represent flaws in the system present on September 11. Regional hospital disaster plans no longer require a clearinghouse hospital or other designated communications focal point for the dissemination of patient disposition and treatment information. Thus, hospitals first learned of en route victims when contacted by transporting EMS units, and EMS control reconstructed much of the disposition information by contacting hospitals after the fact. Although the number of victims of the Pentagon attack were fewer than many anticipated, they were not insignificant. An incident with more casualties would have seriously strained the system.

Source: "Arlington County After-Action Report on the Response to the September 11 Terrorist Attack on the Pentagon," prepared for Arlington County, Virginia, by Titan Systems Corporation.

The events at the World Trade Center and the Pentagon varied significantly in size and impact but, from a responder's perspective, they were similar in terms of surprise and challenges. There are striking similarities between the "improvement opportunities" listed in the NYPD report and the "lessons learned" in the Arlington County report.

Although the specifics vary, both responses identified issues in five key areas:

- Command
- Communications
- Coordination
- Planning
- Dispatching Personnel

Many of the actions taken after September 11 by government officials and emergency managers at the federal, state and local levels reflect the need for changes in order to prepare for the next terrorist event.

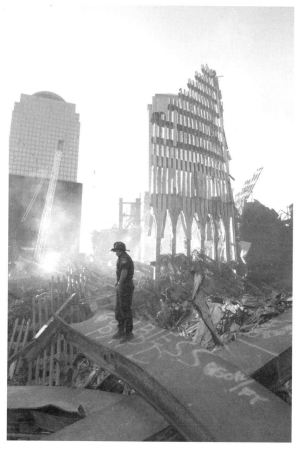

Figure 9-3   New York, NY, October 4, 2001—NY Firefighter chief at the site of the World Trade Center. Photo by Andrea Booher/FEMA News Photo.

FEDERAL GOVERNMENT TERRORISM ACTIVITY

For FEMA and its partner agencies in the former Federal Response Plan (FRP—now the National Response Plan—see Chapter 4), the most significant actions taken by the federal government to combat terrorism were the creation of the Department of Homeland Security (DHS) and the Global War on Terrorism (which has involved direct military action in both Afghanistan and Iraq in addition to the diplomatic and other nonmilitary actions throughout the rest of the world).

For state and local emergency managers, the most significant result of federal government actions since September 11 has been the increased funding and additional funding agencies providing support for first responders and emergency management terrorism planning and prevention activities, and the fundamental shift in funding from more traditional hazard management to management of the terrorist threat.

For the American people the most significant impacts of federal government activities to combat terrorism is the confusion resulting from the terrorism threat warnings being issued by public officials, an uncertainty regarding individual risk presented by the terrorist threat, and the effects resulting from the nation participating in a major overseas conflict (alteration in social program funding, increased security measures at public events and in transportation, and the displacement of reservist family members, employees, and business owners—many of whom are first responders—and changes in social programs that the federal government has influenced).

All three perspectives will be discussed in this section.

The Department of Homeland Security (DHS)

On November 25, 2002, President Bush signed into law the Homeland Security Act of 2002 (HS Act) (Public Law 107-296), and announced that former Pennsylvania Governor Tom Ridge would become secretary of a new Department of Homeland Security (DHS) to be created through this legislation. This act, which authorized the greatest federal government reorganization since President Harry Truman joined the various branches of the armed forces under the Department of Defense, was charged with a threefold mission of protecting the United States from further terrorist attacks, reducing the nation's vulnerability to terrorism, and minimizing the damage from potential terrorist attacks and natural disasters.

The sweeping reorganization into the new department, which officially opened its doors on January 24, 2003, joined together over 179,000 federal employees from 22 existing federal agencies under a single, cabinet-level organization. The legislation also included several changes within other federal agencies that were only remotely affiliated with DHS.

The creation of DHS was the culmination of an evolutionary legislative process that began largely in response to criticism that increased federal intelligence interagency cooperation could have prevented the September 11 terrorist attacks. The White House and Congress both had recognized that a Homeland Security czar would require both a staff and a large budget in order to succeed, and thus began deliberations to create a new cabinet-level department that would fuse many of the security-related agencies dispersed throughout the federal government.

Figure 9-4   New York, NY, November 1, 2001—FEMA's Disaster Field Office in New York has been ground zero for the agency's operations in the aftermath of the World Trade Center tragedy. Photo by Larry Lerner/FEMA News Photo.

For several months during the second half of 2002, Congress jockeyed between different versions of the Homeland Security bill in an effort to establish legislation that was passable yet effective. Lawmakers were particularly mired on the issue of the rights of employees—an issue that prolonged the legal process considerably. Furthermore, efforts to incorporate many of the intelligence-gathering and investigative law enforcement agencies, namely the National Security Agency (NSA), the FBI, and the CIA, into the legislation failed.

Despite these delays and setbacks, after the 2002 midterm elections, the Republican seats gained in both the House and Senate gave the President the leverage he needed to pass the bill without further deliberation (H.R., 299-121 on November 13, 2002; Senate, 90-9 on November 19, 2002). Although the passage of this act represented a significant milestone, the implementation phase presented a tremendous challenge; a concern expressed by several leaders from the agencies that were to be absorbed. On November 25, 2002, President Bush submitted his Reorganization Plan (as required by the legislation), which mapped out the schedule, methodology, and budget for the monumental task.

Beginning March 1, 2003, almost all the federal agencies named in the Act began their move, whether literally or symbolically, into the new department. Those remaining followed on June 1, 2003, with all incidental transfers completed by September 1, 2003. Although a handful of these agencies remained intact after the move, most were fully incorporated into one of four new directorates; Border and Transportation Security (BTS), Information Analysis and Infrastructure Protection (IAIP), Emergency Preparedness and Response (EP&R), and Science and Technology

(S&T). A fifth directorate, Management, incorporated parts of the existing administrative and support offices within the merged agencies.

Secretary Ridge was given exactly one year to develop a comprehensive structural framework for DHS, and to name new leadership for all five directorates and other offices created under the legislation.

In addition to the creation of the Department of Homeland Security, the HS Act made several changes to other federal agencies and their programs, and created several new programs. A list of the most significant follows.

- Established a National Homeland Security Council within the Executive Office of the President, which assesses U.S. objectives, commitments, and risks in the interest of Homeland Security, oversees and reviews federal homeland security policies, and makes recommendations to the President.
- Transferred the Bureau of Alcohol, Tobacco, and Firearms (ATF) from the Department of the Treasury to the Department of Justice (DOJ).
- Explicitly prohibited both the creation of a national ID card and the proposed Citizen Corps "Terrorism Information and Prevention System" (Operation TIPS, which encouraged transportation workers, postal workers, and public utility employees to identify and report suspicious activities linked to terrorism and crime). The Act also reaffirmed the Posse Comitatus Act, which prohibits the use of the Armed Forces in law enforcement activities except under Constitutional or Congressional authority (the Coast Guard is exempt from this Act).
- The "Arming Pilots Against Terrorism Act," incorporated into the HS Act, allows pilots to defend aircraft cockpits with firearms or other "less-than-lethal weapons" against acts of criminal violence or air piracy, and provides anti-terrorism training to flight crews.
- The Critical Infrastructure Information Act (2002), incorporated in the HS Act, exempts certain components of critical infrastructure from Freedom of Information Act (FOIA) regulations.
- The "Johnny Michael Spann Patriot Trusts" was created to provide support for surviving spouses, children, or dependent parents, grandparents, or siblings of various federal employees who die in the line of duty as result of terrorist attacks, military operations, intelligence operations, or law enforcements operations.

On November 30, 2004, following the presidential elections, DHS Secretary Ridge announced his resignation. After an initial nomination of NYPD commissioner Bernard Kerik for the position, which was withdrawn due to questions about an undocumented immigrant he employed at his home, Federal Judge Michael Chertoff was named to lead the agency.

Homeland Security Department Subcomponents and Agencies

The Department of Homeland Security is a massive agency, with many responsibilities in a staggeringly wide range of program areas, approximately 180,000 employees, a massive multibillion dollar budget, and an ambitious list of tasks and goals. The department leverages resources within federal, state, and local governments, coordinating the transition of multiple agencies and programs into a single, integrated agency focused on protecting the American people and their homeland. More than 87,000 different governmental jurisdictions at the federal, state, and local level have homeland security responsibilities.

The following list comprises the major components that make up the Department of Homeland Security.

Office of the Secretary. The staff functions in the Office of the Secretary oversee activities with other federal, state, local, and private entities as part of a collaborative effort to strengthen our borders, provide for intelligence analysis and infrastructure protection, improve the use of science and technology to counter weapons of mass destruction, and create a comprehensive response and recovery division. Within the Office of the Secretary there are multiple offices that contribute to the overall Homeland Security mission:

- Office of the Chief Privacy Officer
- Office of Civil Rights and Civil Liberties
- Office of Counter Narcotics
- Office of General Counsel
- Office of the Inspector General
- Office of Legislative Affairs
- Office of National Capital Region Coordination
- Office of the Private Sector
- Office of Public Affairs
- Office of State and Local Government Coordination and Preparedness

Border and Transportation Security (BTS). Border and Transportation Security (BTS) secures our nation's borders and transportation systems and enforces the nation's immigration laws. The directorate includes the following components:

- Transportation and Security Administration (TSA)—www.tsa.gov
- Customs and Border Protection—www.cbp.gov
- Immigration and Customs Enforcement—www.ice.gov

Emergency Preparedness and Response (EP&R). Emergency Preparedness and Response (EP&R), building on the long and solid track record of FEMA, ensures that the nation is prepared for incidents, whether natural disasters or terrorist attacks, and oversees the federal government's national response and recovery strategy.

Information Analysis and Infrastructure Protection (IAIP). Information Analysis and Infrastructure Protection (IAIP) helps deter, prevent, and mitigate acts of terrorism by assessing vulnerabilities in the context of continuously changing threats. IAIP strengthens the nation's protective posture and disseminates timely and accurate information to our federal, state, local, private, and international partners. The components of this directorate include:

- Homeland Security Operations Center (HSOC)
- Information Analysis (IA)
- Infrastructure Protection (IP)

Science and Technology (S&T). The Directorate of Science and Technology (S&T) serves as the primary research and development arm of Homeland Security, using the nation's scientific and technological resources to provide federal, state, and local officials with the technology and capabilities to protect the homeland. The focus is on catastrophic terrorism—threats to security that could result in large-scale loss of life and major economic impact. S&T's work is designed to counter those threats, both by evolutionary improvements to current technological capabilities and development of revolutionary, new technological capabilities.

- Office of National Laboratories
- Homeland Security Laboratories
- Homeland Security Advanced Research Projects Agency (HSARPA)

Office of Management. The Under Secretary for Management is responsible for the budget, appropriations, expenditure of funds, accounting and finance, procurement, information technology systems, facilities, property, equipment, other material resources, and the identification and tracking of performance measurements relating to the responsibilities of Homeland Security.

U.S. Citizenship and Immigration Services (USCIS). U.S. Citizenship and Immigration Services (USCIS) ensures that America "continues to welcome visitors, refugees, immigrants, asylum seekers, and new citizens while protecting the nation from terrorism, unlawful entrants, and illegal residents." USCIS is made up of 15,000 employees in 250 offices worldwide and is subsidized largely by revenue generated from fees paid for immigration benefits. Within USCIS, the Office of Citizenship was established to develop and implement public outreach and education initiatives to promote U.S. citizenship. The components of this directorate include:

- U.S. Citizenship and Immigration Services
- Office of Citizenship
- National Customer Service Center

U.S. Coast Guard. The U.S. Coast Guard protects the public, the environment, and U.S. economic interests in the nation's ports and waterways, along the coast, on international waters, or in any maritime region as required to support national security.

U.S. Secret Service (USSS). The U.S. Secret Service (USSS) is responsible for the protection of the President, the nation's leaders, as well as the country's financial and critical infrastructures. USSS is a crucial component of Homeland Security. USSS is organized into two major components, one focused on protection and the other focused on investigation.

SELECT STRATEGIC GOALS FOR PROTECTION AND RESPONSE FROM THE U.S. DEPARTMENT OF HOMELAND SECURITY STRATEGIC PLAN

Strategic Goal 1: Awareness. Identify and understand threats, assess vulnerabilities, determine potential impacts, and disseminate timely information to our homeland security partners and the American public.

- Objective 1.1. Gather and fuse all terrorism-related intelligence; analyze, and coordinate access to information related to potential terrorist or other threats.

    Intelligence and information analysis is an integral component of our nation's overall efforts to protect against and reduce our vulnerability to terrorism. We will receive, assess, and analyze information from law enforcement, the intelligence community and nontraditional sources (e.g., state and local, private sector) to increase situational awareness of terrorist threats and specific incidents. We will review and, as necessary, work to improve policies for law enforcement and intelligence information sharing within the federal government and between state and local authorities. Data collection and analysis capabilities will be supported through investment in, and development of, leading-edge information analysis, data mining, data warehousing, and threat/vulnerability mapping applications and tools, and recruiting, training, and retaining human analysts.

- Objective 1.2. Identify and assess the vulnerability of critical infrastructure and key assets.

    We will conduct and sustain a complete, current, and accurate assessment of our nation's infrastructure sectors and assets. We will use modeling, simulation, and risk-based analytic tools to prioritize our work with an emphasis on critical infrastructure and key resources that could be catastrophically exploited. By establishing this understanding of the full array of critical infrastructure facilities and assets, how they interact, and the interdependencies across infrastructure sectors, we will be in a position to anticipate the national security, economic, and public safety implications of terrorist attacks and will prioritize protective measures accordingly.

- Objective 1.3. Develop timely, actionable, and valuable information based on intelligence analysis and vulnerability assessments.

    We will integrate intelligence, threat, and infrastructure vulnerability information to provide our national leaders, decision makers, and the owners and operators of our critical infrastructure and key assets with the increasingly targeted and actionable information necessary in the post 9/11 threat environment. We will build an intelligence analysis structure that coordinates with the rest of the federal government, as well as state, local, and tribal governments; the private sector; and our international partners. Our national imperative is to improve the sharing, analysis, integration of all-source threat, risk, and infrastructure vulnerability information so appropriate preventative and protective actions can be taken.

- Objective 1.4. Ensure quick and accurate dissemination of relevant intelligence information to homeland security partners, including the public.

  Securing the homeland is a joint effort of the federal government; state, local, and tribal governments; the private sector; our international partners; and the public. Therefore we will work to empower those partners by disseminating relevant intelligence and threat information to them accurately and as quickly as possible. We will work with our partners to remove roadblocks to information sharing. We will administer the Homeland Security Advisory System, including the issuance of public advisories and coordination of warning information with other agencies. We will deploy and operate tools and secure communications channels to analyze and disseminate information to relevant agencies as quickly and efficiently as possible.

Strategic Goal 2: Prevention. Detect, deter, and mitigate threats to our homeland.

- Objective 2.1. Secure our borders against terrorists, means of terrorism, illegal drugs, and other illegal activity.

  We interdict terrorist activities by targeting unlawful migration of people, cargo, drugs, and other contraband, while facilitating legitimate migration and commerce. The department will enforce border security in an integrated fashion at ports of entry, on the borders, on the seas, and before potential threats can reach our borders. Through the continued deployment of the appropriate balance of personnel, equipment and technology we will create "smart borders." Not only will we create more secure United States borders but, in conjunction with international partners, we will extend our zones of security beyond our physical borders identifying, prioritizing, and interdicting threats to our nation before they arrive. We will develop and provide resources for a cohesive, unified enforcement capability that makes our border security effective, smarter, and stronger.
- Objective 2.2. Enforce trade and immigration laws.

  We will enforce all applicable laws in an integrated fashion while facilitating free commerce and the flow of legal immigration and travel into the United States. We will interdict smuggling and stop other illegal activities that benefit terrorists and their supporters. We will build a unified, cohesive enforcement capability to actively conduct and coordinate law enforcement operations.
- Objective 2.3. Provide operational end users with the technology and capabilities to detect and prevent terrorist attacks, means of terrorism, and other illegal activities.

  The nation's technical superiority in science and technology is key to securing the homeland. We will use, leverage, and enhance the vast resources and expertise of the federal government, private sector, academic community, nongovernmental organizations, and other scientific bodies. We will

continues

develop new capabilities to facilitate the sharing of information and analysis; test and assess threats and vulnerabilities; counter various threats, including weapons of mass destruction and illegal drugs; and mitigate the effects of terrorist attacks. We will also focus our efforts on developing technology to detect and prevent the illicit transport of chemical, biological, radiological, and nuclear materials. We will develop and deploy the capabilities, equipment, and systems needed to anticipate, respond to, and recover from attacks on the homeland.

- Objective 2.4. Ensure national and international policy, law enforcement, and other actions to prepare for and prevent terrorism are coordinated.

  We will effectively coordinate and communicate with other federal agencies; and with state, local, and tribal governments; the private sector; and the American people. Increasing and coordinating information sharing between law enforcement, intelligence, and military organizations will improve our ability to counter terrorists everywhere. We will coordinate training and education across multiple levels, both national and international, ensuring common standards and approaches to recognizing key indicators of future terrorist actions.

- Objective 2.5. Strengthen the security of the nation's transportation systems.

  Transportation systems have the unique ability to be either a means of delivering weapons of terror or the target of a direct terrorist attack. Our domestic transportation system is intertwined inextricably with the global transportation infrastructure. Safety and security are two sides of the same coin. We will strengthen the security of the transportation network while we work to remove all threats or barriers to the safe movement of commerce and people. We will coordinate with federal, state, local, and tribal agencies, as well as our international and private sector partners, to ensure the transportation system remains a safe and vital economic link, while preventing terrorists from using transportation conveyances or systems to deliver implements of destruction.

- Objective 2.6. Ensure the security and integrity of the immigration system.

  We will ensure that immigrants and nonimmigrants comply with laws and security mandates to prevent persons who seek to exploit the economic and social benefits of immigration or engage in illegal activities from obtaining lawful status. We will strengthen legal protections and design programs appropriately to create a more secure immigration system. We will make decisions in a timely and efficient manner by applying technology and allocating our resources to provide actionable and accurate information. We will ensure that those persons entitled to benefits receive them through verification services and encouraging employers to verify status. We will refer illegal aliens to enforcement entities for prosecution or removal from the United States.

Strategic Goal 3: Protection. Safeguard our people and their freedoms, critical infrastructure, property, and the economy of our nation from acts of terrorism, natural disasters, or other emergencies.

- Objective 3.1. Protect the public from acts of terrorism and other illegal activities.

  We must not let the threat of terrorism alter the American way of life. We will identify and disrupt terrorists and criminals before they threaten the well-being of American citizens. Our investigative efforts will focus on identifying the tools and conveyances used by terrorists and criminals, and apprehending suspect individuals. Through our partnerships with other agencies and through our own efforts we will coordinate and apply knowledge and skills acquired through years of practical use in drug interdiction and airspace security to remain at the forefront of global law enforcement and counterterrorism efforts. We will ensure that our nation's shipping routes do not become avenues of entry for terrorists, their weapons, or supplies. We will conduct national and international investigations to gather evidence of violations of United States laws, and prevent terrorist groups from obtaining sensitive weapons of United States origin.
- Objective 3.2. Reduce infrastructure vulnerability from acts of terrorism.

  We will lead and coordinate a national effort to secure America's critical infrastructure. Protecting America's critical infrastructure is the shared responsibility of federal, state, local, and tribal governments, in active partnership with the private sector, which owns approximately 85 percent of our nation's critical infrastructure. Using the results of modeling, simulation, and analytic tools to prioritize our efforts, we will implement standardized and tiered protective measures that are rapidly adjustable to counter various levels of threat. We will coordinate the implementation of a comprehensive integrated national plan to protect both our physical and cyber infrastructure and significantly reduce vulnerabilities, while ensuring that government at all levels enables, and does not inhibit, the private sector's ability to carry out its protection responsibilities.
- Objective 3.3. Protect against financial and electronic crimes, counterfeit currency, illegal bulk currency movement, and identity theft.

  A principal component of homeland security is economic security, including protection of the nation's currency and financial payment systems. The Department of Homeland Security participates in task forces and other joint operations with the financial community and with federal, state, local, and tribal law enforcement partners to investigate crimes targeting the stability, reliability, and security of financial systems. To prevent, detect, and investigate various forms of electronic crimes, we will operate a nationwide network of Electronic Crimes Task Forces. We will maintain an overseas investigative presence where criminal groups engage in the counterfeiting of United States currency and other financial crimes targeting our homeland. International drug traffickers steal $20 to $30 billion annually from the United States economy. Much of these illegal funds are shipped out of the United States as bulk currency. This weakens our economy and strengthens the ability of the international drug traffickers to destabilize the governments

continues

of their countries by bribery or to finance terrorist activities. We will investigate, identify, and seize outbound shipments to take away this ability to fund illegal activities.

- Objective 3.4. Secure the physical safety of the President, Vice President, visiting world leaders, and other protectees.

  We will protect our nation's leaders and visiting dignitaries from all threats, including terrorists and other criminals; natural, technological, and man-made emergencies; and preventable accidents. We will coordinate with military, federal, state, local, and tribal law enforcement organizations to ensure their safety. We will evaluate information received from law enforcement and intelligence agencies and other sources to investigate, apprehend, and prosecute, if appropriate, those who pose a threat. We will ensure that protectees have a safe environment in which to continue their operations in the event of any threat contingency.

- Objective 3.5. Ensure the continuity of government operations and essential functions in the event of crisis or disaster.

  We will partner with other federal departments and agencies to ensure the continuous operation of the federal government and to secure the survival of an enduring constitutional government in times of attack, national emergency, or disaster. We will provide alternative facilities, equipment, and communications capabilities to ensure that the federal government is capable of performing its essential functions, and that the nation will continue to be governed as set forth in the United States Constitution.

- Objective 3.6. Protect the marine environment and living marine resources.

  We will partner with other nations; federal agencies; state, local, and tribal governments; and responsible sectors of the maritime industry, to ensure the quality of our marine resources are protected. We will encourage, pursue, and enforce bilateral and regional agreements with other governments to ensure that the world's living marine resources are properly maintained and managed. The ability to use unpolluted waters for transportation and recreation is vital to the safety of our citizens and the economy of our nation; we will work to ensure compliance with existing regulations and consider others that may be required to protect our marine environment. We will maintain an uncompromising commitment to the stewardship of our national living marine resources through the highest caliber enforcement of fisheries laws and regulations supporting the national policy.

- Objective 3.7. Strengthen nationwide preparedness and mitigation against acts of terrorism, natural disasters, or other emergencies.

  The best way to protect against the effects of harmful incidents is to be prepared. Preparedness and mitigation are important elements in reducing the impacts of acts of terror and other disasters. We will ensure all levels of public safety and emergency management are capable of rapid and effective response by establishing a unified, capabilities-based preparedness strategy incorporating all-hazard assessments, training, exercises, and assistance for federal, state, tribal, and local governments, first responders, and communities. We will establish, implement, and evaluate capabilities through a system of national standards, mutual aid systems, and credentialing protocols, and supply technologies for rapid and interoperable communications, personal protection, and incident management. We will implement and sustain a national citizen preparedness movement that includes private sector involvement. We will expand the nation's community risk management capabilities and reduce the nation's vulnerability to acts of terrorism and other disasters through effective vulnerability assessments and risk management programs.

Strategic Goal 4: Response. Lead, manage, and coordinate the national response to acts of terrorism, natural disasters, or other emergencies.

- Objective 4.1. Reduce the loss of life and property by strengthening nation-wide response readiness.

  The nation must have a vigorous capability to respond when disaster strikes. We will strengthen the national capability to respond to disasters of all types, including terrorism, through the integration of Department of Homeland Security response systems and teams and the completion of catastrophic all-hazard plans for the nation's most vulnerable communities and geographic areas, including tactical elements to ensure coordinated response operations, logistics, and support. We will provide health and medical response readiness through integrated planning, surge capacity to address health and medical emergencies or acts of terrorism, and will develop the logistical capacity to provide intermediate emergency housing to large displaced populations following major disasters.
- Objective 4.2. Provide scalable and robust all-hazard response capability.

  The nation will know it can rely on us to respond in time of need. We will provide and coordinate a quick and effective response when state, local, and tribal resources are overwhelmed by disasters and emergencies. We will bring the right people and resources to bear where and when they are needed most, including medical, urban search and rescue, and incident management capabilities, and will assist all mariners in peril. We will provide integrated logistical support to ensure a rapid and effective response and coordinate among Department of Homeland Security and other federal, state, and local operations centers consistent with national incident command protocols. We will work with our partners to create and implement a National Incident Management System and a single, all-discipline National Response Plan that will strengthen the nation's ability to respond to catastrophic events of all types, including terrorism.
- Objective 4.3. Provide search and rescue services to people and property in distress.

  Mariners operate in an unforgiving and often remote environment that increases the risk of injury, loss of life, and property. We will continue to use our maritime expertise, assets, and around-the-clock, on-call readiness to conduct search and rescue missions to save lives and property. We will also

continues

continue to partner with other nations, federal, state, local agencies, maritime industry, professional mariners, commercial providers, and volunteer organizations to assist mariners in distress and protect property in imminent danger. A number of projects are under way that will improve our ability to respond to maritime distress incidents. Recapitalization of aviation, surface, command, and control architecture and supporting logistic and personnel systems, as well as the procurement of specialized boats and attainment of additional search planning tools, will greatly enhance our ability to assist mariners in distress.

Strategic Goal 5: Recovery. Lead national, state, local, and private sector efforts to restore services and rebuild communities after acts of terrorism, natural disasters, or other emergencies.

- Objective 5.1. Strengthen nationwide recovery plans and capabilities.
  We will work with our partners to ensure the nation's capability to recover from multiple or simultaneous disasters, including terrorist use of weapons of mass destruction, other man-made hazards and natural disasters, through the development and maintenance of short- and long-term plans and capabilities.
- Objective 5.2. Provide scalable and robust all-hazard recovery assistance.
  We will lead the nation's recovery from the impacts of disasters and emergencies. We will deliver timely and appropriate assistance to individuals and families following acts of terrorism, natural disasters, and other emergencies, acknowledging the unique requirements of recovery from catastrophic disasters and weapons of mass destruction events. We will provide help to restore services and public facilities, and provide states and other partners with professional, readily deployable, trained and certified leaders and staff to manage all levels and types of disasters. We will make assistance available to states and local governments for the management, mitigation, and control of local hazards and emergencies that threaten to become major disasters.

Source: U.S. Department of Homeland Security Strategic Plan

On July 13th, 2005, DHS Secretary Michael Chertoff released a six-point agenda that would be used to guide a reorganization of the Department aimed at streamlining its efforts. The agenda followed an initial review that Chertoff initiated immediately upon assuming the leadership position. The review was designed to closely examine the Department in order to discover ways in which leadership could better manage risk in terms of threat, vulnerability and consequence; prioritize policies and operational missions according to this risk-based approach; and establish a series of preventive and protective steps that would increase security at multiple levels. According to the six-point agenda, changes that will occur at DHS will focus on:

- Increasing overall preparedness, particularly for catastrophic events;
- Creating better transportation security systems to move people and cargo more securely and efficiently;

- Strengthening border security and interior enforcement and reforming immigration processes;
- Enhancing information sharing (with partners);
- Improving financial management, human resource development, procurement and information technology within the department; and
- Realigning the department's organization to maximize mission performance.

Secretary Chertoff announced several new policy initiatives that will be included in the overhaul of the department, including:

- A new approach to securing borders through additional personnel, new technologies, infrastructure investments, and interior enforcement – coupled with efforts to reduce the demand for illegal border migration by channeling migrants seeking work into regulated legal channels;
- Restructuring the current immigration process to enhance security and improve customer service;
- Reaching out to State homeland security officials to improve information exchange protocols, refine the Homeland Security Advisory System (HSAS), support State and regional data fusion centers, and address other topics of mutual concern; and
- Investing in DHS personnel by providing professional career training and other development efforts.

One of the most significant changes that is to occur as result of the six-point agenda is an organizational restructuring of the Department. Chertoff asserts that these changes are being made to increase the Department's ability to prepare, prevent, and respond to terrorist attacks and other emergencies. Changes to be performed include:

- A new Directorate of Policy will be created to centralize and improve policy development and coordination. This directorate will be led by an Under Secretary upon enactment of legislation, and will serve as the primary Department-wide coordinator for policies, regulations, and other initiatives. This Directorate is being created to ensure the consistency of policy and regulatory development across various parts of the Department, as well as to perform long-range strategic policy planning. It will assume the policy coordination functions previously performed by the Border and Transportation Security (BTS) Directorate. It will also create a single point of contact for internal and external stakeholders by consolidating or co-locating similar activities from across the department. This new Directorate will include:
  - Office of International Affairs;
  - Office of Private Sector Liaison;
  - Homeland Security Advisory Council;
  - Office of Immigration Statistics; and
  - Senior Asylum Officer
- A new Office of Intelligence and Analysis will be created to strengthen intelligence functions and information sharing. This office will ensure that information is gathered from all relevant field operations and other parts of the

intelligence community; analyzed with a mission-oriented focus; informative to senior decision-makers; and disseminated to the appropriate Federal, State, local, and private sector partners. Led by a Chief Intelligence Officer who reports directly to the Secretary, this office will be comprised of analysts within the former Information Analysis directorate and draw on expertise of other DHS components with intelligence collection and analysis operations.

- A new Director of Operations Coordination will be created to improve coordination and efficiency of operations. This official will work to enable DHS to more effectively conduct joint operations across all organizational elements; coordinate incident management activities; and utilize all resources within the Department to translate intelligence and policy into immediate action. The Homeland Security Operations Center, which serves as the nation's nerve center for information sharing and domestic incident management on a full-time basis, will be a critical part of this new office.

- The Information Analysis and Infrastructure Protection Directorate will be renamed the Directorate for Preparedness and consolidate preparedness assets from across the department. The Directorate for Preparedness will facilitate grants and oversee nationwide preparedness efforts supporting first responder training, citizen awareness, public health, infrastructure and cyber security, and ensure proper steps are taken to protect high-risk targets. The directorate will be managed by an Under Secretary and include:
  - A new Assistant Secretary for Cyber Security and Telecommunications, responsible for identifying and assessing the vulnerability of critical telecommunications infrastructure and assets; providing timely, actionable and valuable threat information; and leading the national response to cyber and telecommunications attacks;
  - A new Chief Medical Officer, responsible for carrying out the Department's responsibilities to coordinate the response to biological attacks and to serve as a principal liaison between DHS and the Department of Health and Human Services, the Centers for Disease Control, the National Institutes of Health, and other key parts of the biomedical and public health communities;
  - Assistant Secretary for Infrastructure Protection;
  - Assets of the Office of State and Local Government Coordination and Preparedness responsible for grants, training and exercises;
  - U.S. Fire Administration; and
  - Office of National Capitol Region Coordination.

- FEMA will report directly to the Secretary of Homeland Security in order to improve national response and recovery efforts by focusing FEMA on its core functions. Under the new DHS proposed by the agenda, FEMA will focus on response and recovery.

- The Federal Air Marshal Service will be moved from the Immigration and Customs Enforcement (ICE) bureau to the Transportation Security Administration to increase operational coordination and strengthen efforts to meet this common goal of aviation security.

- A new Office of Legislative and Intergovernmental Affairs will be created, which will merge certain functions among the Office of Legislative Affairs and the Office of State and Local Government Coordination in order to stream-line intergovernmental relations efforts and better share homeland security information with members of Congress, as well as state and local officials.

- The Office of Security will be moved such that it will be under the direction of the Under Secretary for Management in order to better manage informa-tion systems, contractual activities, security accreditation, training, and resources.

Funding for First Responders and Emergency Management

For state and local government, the events of September 11 (see Table 9-2) resulted in an extraordinary increase in funding for first responders—fire, police, and emer-gency medical technicians—and emergency management activities. Also, the number of federal government agencies and programs now providing funds for these activities has increased significantly. In the first responder community, historically only the police have received significant funding from the federal government. Fire departments across the country traditionally have raised the majority of their funding from local sources. Emergency medical technicians are often private contractors paid for by local and state government sources.

Proper training and equipping of firefighters responding to a biochemical terror-ist attack has been a concern among the fire services community and FEMA since the early 1990s. Passage of the Fire Prevention and Assistance Act in 2000 was the first effort by Congress to support the nation's paid and volunteer fire departments. In Spring 2001, FEMA initiated a new Fire Grant program that provided $100 million in small grants to local fire departments for equipment, protective gear, train-ing, and prevention programs. In 2002, the amount available for FEMA fire grants increased to $300 million. By 2004, that amount had risen to over $700,000, and to over $1.8 million by 2006. In addition to the annual fire grants, the bulk of the $3.5 to $4 billion spent on first responders each year has been designated for equipping and training of first responders for future terrorist events (see proposed 2006 budget figures in Table 9-2).

Table 9-2 Local First Responder Funding Figures: 2004–2006 (Dollars in Thousands)

Funding Area	FY2004 Enacted	FY2005 Enacted	FY2006 Proposed	Change between 2005 & 2006
State and local grants, Citizen Corps; Other grant programs	$2,405,722	$1,665,856	$1,240,000	−$425,856
Urban Areas Security Initiative; Targeted Infrastructure Protection programs	$720,723	$1,157,857	$1,620,000	+$462,143
Fire Act Grants	$745,575	$715,000	$500,000	−$215,000
State and local Training, Exercise and technical assistance	$320,100	$446,133	$204,756	−$241,377
Total	$4,192,119	$3,984,846	$3,564,756	−$420,090

Source: Homeland Security Budget in Brief: FY 2006

FEMA is not the only source of terrorism funding for state and local government. The Department of Justice, through a variety of programs, is making funding available for the acquisition of equipment and technology. The Department of Health and Human Resources is making available substantial funding to state and local government to address the threat of biochemical terrorist attacks. The Centers for Disease Control is providing funding for public health planning and capacity building and bolstering the national pharmaceutical stockpile. The Department of Defense currently provides funding for emergency management training for military personnel and community officials.

Communicating Threat Information to the American People

As noted earlier, in Objective 1.4 of the DHS Strategic Plan,

> Securing the homeland is a joint effort of the federal government; state, local and tribal governments; the private sector; our international partners; and the public. Therefore we will work to empower those partners by disseminating relevant intelligence and threat information to them accurately and as quickly as possible. We will work with our partners to remove roadblocks to information sharing. We will administer the Homeland Security Advisory System, including the issuance of public advisories and coordination of warning information with other agencies. We will deploy and operate tools and secure communications channels to analyze and disseminate information to relevant agencies as quickly and efficiently as possible.

The Homeland Security Advisory System was borne out of Homeland Security Presidential Directive—3 (HSPD-3), issued on March 11, 2002, which stated that:

> The nation requires a Homeland Security Advisory System to provide a comprehensive and effective means to disseminate information regarding the risk of terrorist acts to federal, state, and local authorities and to the American people. Such a system would provide warnings in the form of a set of graduated "Threat Conditions" that would increase as the risk of the threat increases. At each Threat Condition, federal departments and agencies would implement a corresponding set of "Protective Measures" to further reduce vulnerability or increase response capability during a period of heightened alert.

> This system is intended to create a common vocabulary, context, and structure for an ongoing national discussion about the nature of the threats that confront the homeland and the appropriate measures that should be taken in response. It seeks to inform and facilitate decisions appropriate to different levels of government and to private citizens at home and at work.

There are three components of the system, which is designed to combine threat information with vulnerability assessments and provide communications to public safety officials and the public. They are as follows:

- **Homeland Security Threat Advisories.** Contain actionable information about an incident involving, or a threat targeting, critical national networks or infrastructures or key assets. They could, for example, relay newly developed procedures that, when implemented, would significantly improve security or protection. They could also suggest a change in readiness posture, protective actions, or response. This category includes products formerly named alerts, advisories, and sector notifications. Advisories are targeted to federal, state, and local governments, private sector organizations, and international partners.

- **Homeland Security Information Bulletins.** Communicate information of interest to the nation's critical infrastructures that do not meet the timeliness, specificity, or significance thresholds of warning messages. Such information may include statistical reports, periodic summaries, incident response or reporting guidelines, common vulnerabilities and patches, and configuration standards or tools. It also may include preliminary requests for information. Bulletins are targeted to federal, state, and local governments, private sector organizations, and international partners.
- **Color-coded Threat Level System.** Used to communicate with public safety officials and the public at-large through a threat-based, color-coded system so that protective measures can be implemented to reduce the likelihood or impact of an attack. Raising the threat condition has economic, physical, and psychological effects on the nation; so, the Homeland Security Advisory System can place specific geographic regions or industry sectors on a higher alert status than other regions or industries, based on specific threat information.

Figure 9-5 provides suggestions for public action in accordance with the five color codes of the Homeland Security Advisory System.

The following information, based upon the same color-coded chart, provides DHS recommendations to federal departments and agencies.

Guidance for Federal Departments and Agencies

The following Threat Conditions each represent an increasing risk of terrorist attacks. Beneath each Threat Condition are some suggested Protective Measures, recognizing that the heads of federal departments and agencies are responsible for developing and implementing appropriate agency-specific Protective Measures:

1. **Low Condition (Green).** Declared when there is a low risk of terrorist attacks. Federal departments and agencies should consider the following general measures in addition to the agency-specific Protective Measures they develop and implement:
   - Refining and exercising as appropriate preplanned Protective Measures
   - Ensuring personnel receive proper training on the Homeland Security Advisory System and specific preplanned department or agency Protective Measures
   - Institutionalizing a process to assure that all facilities and regulated sectors are regularly assessed for vulnerabilities to terrorist attacks, and all reasonable measures are taken to mitigate these vulnerabilities
2. **Guarded Condition (Blue).** Declared when there is a general risk of terrorist attacks. In addition to the Protective Measures taken in the previous Threat Condition, federal departments and agencies should consider the following general measures in addition to the agency-specific Protective Measures that they will develop and implement:

Citizen Guidance on the Homeland Security Advisory System

Risk of Attack	Recommended Actions for Citizens
GREEN Low Risk	→ Develop a family emergency plan. Share it with family and friends, and practice the plan. Visit www.Ready.gov for help creating a plan. → Create an "Emergency Supply Kit" for your household. → Be informed. Visit www.Ready.gov or obtain a copy of "Preparing Makes Sense, Get Ready Now" by calling 1-800-BE-READY. → Know how to shelter-in-place and how to turn off utilities (power, gas, and water) to your home. → Examine volunteer opportunities in your community, such as Citizen Corps, Volunteers in Police Service, Neighborhood Watch or others, and donate your time. → Consider completing an American Red Cross first aid or CPR course, or Community Emergency Response Team (CERT) course.
BLUE Guarded Risk	→ *Complete recommended steps at level green.* → Review stored disaster supplies and replace items that are outdated. → Be alert to suspicious activity and report it to proper authorities.
YELLOW Elevated Risk	→ *Complete recommended steps at levels green and blue.* → Ensure disaster supply kit is stocked and ready. → Check telephone numbers in family emergency plan and update as necessary. → Develop alternate routes to/from work or school and practice them. → Continue to be alert for suspicious activity and report it to authorities.
ORANGE High Risk	→ *Complete recommended steps at lower levels.* → Exercise caution when traveling, pay attention to travel advisories. → Review your family emergency plan and make sure all family members know what to do. → Be Patient. Expect some delays, baggage searches, and restrictions at public buildings. → Check on neighbors or others that might need assistance in an emergency.
RED Severe Risk	→ *Complete all recommended actions at lower levels.* → Listen to local emergency management officials. → Stay tuned to TV or radio for current information/instructions. → Be prepared to shelter-in-place or evacuate, as instructed. → Expect traffic delays and restrictions. → Provide volunteer services only as requested. → Contact your school/business to determine status of work day.

Developed with input from the <u>American Red Cross</u>.

Figure 9-5   Homeland Security Advisory System.
Source: www.dhs.gov

- Checking communications with designated emergency response or command locations
- Reviewing and updating emergency response procedures
- Providing the public with any information that would strengthen its ability to act appropriately

3. **Elevated Condition (Yellow).** Declared when there is a significant risk of terrorist attacks. In addition to the Protective Measures taken in the previous Threat Conditions, federal departments and agencies should consider the following general measures in addition to the Protective Measures that they will develop and implement:
   - Increasing surveillance of critical locations
   - Coordinating emergency plans as appropriate with nearby jurisdictions
   - Assessing whether the precise characteristics of the threat require the further refinement of preplanned Protective Measures
   - Implementing, as appropriate, contingency and emergency response plans

4. **High Condition (Orange).** Declared when there is a high risk of terrorist attacks. In addition to the Protective Measures taken in the previous Threat Conditions, federal departments and agencies should consider the following general measures in addition to the agency-specific Protective Measures that they will develop and implement:
   - Coordinating necessary security efforts with federal, state, and local law enforcement agencies or any National Guard or other appropriate armed forces organizations
   - Taking additional precautions at public events and possibly considering alternative venues or even cancellation
   - Preparing to execute contingency procedures, such as moving to an alternate site or dispersing their workforce
   - Restricting threatened facility access to essential personnel only

5. **Severe Condition (Red).** Reflects a severe risk of terrorist attacks. Under most circumstances, the Protective Measures for a Severe Condition are not intended to be sustained for substantial periods of time. In addition to the Protective Measures in the previous Threat Conditions, federal departments and agencies also should consider the following general measures in addition to the agency-specific Protective Measures that they will develop and implement:
   - Increasing or redirecting personnel to address critical emergency needs
   - Assigning emergency response personnel and prepositioning and mobilizing specially trained teams or resources
   - Monitoring, redirecting, or constraining transportation systems
   - Closing public and government facilities

The Department of Homeland Security also helps citizens and business owners to prepare for future acts of terrorism through their Ready.gov campaign. The Web-based public education campaign provides a "common sense framework designed to launch a process of learning about citizen preparedness."

DHS urges citizens to stay informed about how to react to various disaster scenarios. These include biological, chemical, explosive, nuclear, radiological, and natural disasters. Ready.gov states:

> Terrorists are working to obtain biological, chemical, nuclear, and radiological weapons, and the threat of an attack is very real. Here at the Department of Homeland Security, throughout the federal government, and at organizations across America we are working hard to strengthen our nation's security. Whenever possible, we want to stop terrorist attacks before they happen. All Americans should begin a process of learning about potential threats so we are better prepared to react during an attack. While

there is no way to predict what will happen, or what your personal circumstances will be, there are simple things you can do now to prepare yourself and your loved ones.

Some of the things you can do to prepare for the unexpected, such as assembling a supply kit and developing a family communications plan, are the same for both a natural or man-made emergency. However, as you will see throughout the pages of **Ready.gov**, there are important differences among potential terrorist threats that will impact the decisions you make and the actions you take. With a little planning and common sense, you can be better prepared for the unexpected.

Appendix D illustrates the recommendations for citizens to stay prepared provided by Ready.gov. More detailed recommendations for each step are provided at www.Ready.gov.

Accomplishments of the Department of Homeland Security

The Department of Homeland Security (DHS) has undergone two full budget cycles since it was established in March 2003. The Fiscal Year 2006 President's Budget includes critical programs and activities that build on the department's many accomplishments.

- The Department has significantly improved security measures at every entry point into the United States. Whether by land, sea, or air, it is substantially more difficult for terrorists to enter our country now than ever before.
- We have created a new communications network to make sure that those who most need threat-related and operational security information get it on a real-time basis. Prior to September 11, 2001, there were few formal channels in place to push timely, threat-related information down to state and local governments or the private sector—the people who actually have the responsibility for implementing much of the nation's security measures. Within the federal government, there was also a need for additional information sharing among federal partners.
- The Department has made it harder to attack the physical and cyber systems that support key business, government, and community activities, facilities, and networks. The Department identified and prioritized the nation's critical infrastructures and then took immediate action by implementing new security measures. From banks to bridges, nuclear plants to the water supplies, key assets are better protected now than they ever were before.
- DHS has provided unprecedented levels of funding and resources to state, local, and private sector partners to protect and prepare America's communities and individual citizens. We continue to improve ways for first responders across the nation to be better equipped, better trained, and more capable of communicating across the public safety community.
- United States-Visitor and Immigrant Status Indicator Technology (US-VISIT) was successfully implemented at all 115 U.S. international airports and 14 seaports, and immediately demonstrated results by preventing individuals with criminal records and immigration violations from entering the United States.

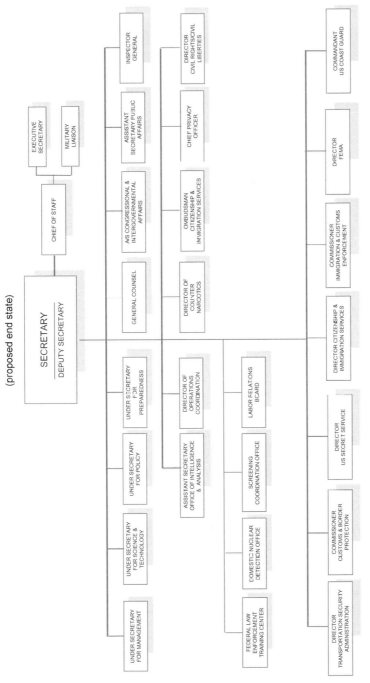

Department of Homeland Security Organization Chart

(proposed end state)

Figure 9-6   Organizational Chart for the Department of Homeland Security. *Source:* www.dhs.gov.

In addition, US-VISIT successfully deployed initial capability to the 50 busiest land border ports of entry in December 2004.

- The U.S. Coast Guard (USCG) developed, reviewed, and approved 9,000 domestic vessel security plans; 3,200 domestic facility plans; 48 area maritime security plans and committees; and verified security plan implementation on 8,100 foreign vessels.
- USCG interdicted nearly 11,000 undocumented migrants attempting to enter the country illegally by sea.
- USCG saved the lives of nearly 5,500 mariners in distress and responded to more than 32,000 calls for rescue assistance.
- The Department established "the One-Stop-Shop" for first responder grants, which allows a single point of entry to the federal government for homeland security preparedness resources.
- The Homeland Security Operations Center's (HSOC) Homeland Security Information Network (HSIN) Secret level connectivity has been expanded to state level Emergency Operations Centers in all 50 states.
- Working closely with importers, carriers, brokers, freight forwarders, and others, Customs and Border Protection (CBP) has developed a seamless, security-conscious trade environment resistant to the threat of international terrorism. The Customs-Trade Partnership Against Terrorism (C-TPAT) has become the largest government/private partnership to arise from September 11.
- Federal Emergency Management Agency (FEMA) provided $4.9 billion in aid including hurricane relief efforts for victims and communities affected by disasters. DHS responded to 65 major disaster declarations and seven emergencies in FY 2004.
- Passenger screening kept 6,501,193 prohibited items from coming on board aircrafts during FY 2004.
- A total of 428 million people, including 262 million aliens, were processed at land, air, and sea ports of entry. Of that number 643,000 aliens were deemed inadmissible under U.S. law.
- Immigration and Customs Enforcement (ICE) officers achieved a 112 percent increase over the prior year for fugitive apprehensions, resulting in more than 7,200 arrests. More than 150,000 aliens were removed in 2004, 53 percent of whom were criminals. This is an all-time record.
- Border Patrol agents apprehended almost 1.2 million illegal aliens between our official ports of entry.
- The Automated Biometric Identification System (IDENT)/Integrated Automated Fingerprint System (IAFIS), which allows the rapid identification of individuals with outstanding criminal warrants through electronic comparison of ten-print digital fingerscans against DHS and FBI systems, is now operational at all Border Patrol stations, every air and seaport of entry, and the 50 busiest land ports of entry.
- The Container Security Initiative (CSI), which involves prescreening shipping containers to detect and interdict terrorists' weapons and other illegal material, was expanded to include 21 countries. CSI is now operational in 34 foreign ports in Europe, Asia, and Africa.

- Approximately 600 million checked bags were screened using advanced explosive technologies in 2004.
- 15,560 federal workers were engaged in response and recovery operations for the declared disasters of 2004, including more than 11,000 FEMA personnel and 1,900 disaster medical specialists.
- More than 2,500 criminal investigations were conducted involving the illegal export of U.S. arms and strategic technology, including weapons of mass destruction (WMD).
- The Federal Law Enforcement Training Center (FLETC) provided basic and advanced law enforcement training to more than 44,750 students, representing 81 federal agencies, as well as state, local, and international law enforcement organizations.
- The U.S. Secret Service (USSS) planned, designed, and implemented security for five events designated as *National Security Special Events* (State of Union Address, G-8 Economic Summit, Former President Ronald Reagan Funeral, Democratic National Convention and Republican National Convention).
- USSS arrested 30 individuals involved in global cyber organized crime, domestically and internationally. Industry experts estimate that $1 billion in total fraud loss was prevented.
- The Science and Technology (S&T) Directorate has implemented initiatives in chemical, biological, radiological, nuclear, and explosive (CBRNE) countermeasures, cargo security, border and transportation security, interoperability, standards for emergency responders, and cyber security. These initiatives have resulted in improved security of U.S. borders, transportation systems, and critical infrastructure, and resulted in the greater preparedness of our nation.

Source: www.dhs.gov

DHS Budget

The proposed FY 2006 budget request includes a total of $41.1 billion for the Department of Homeland Security. This is an increase of seven percent over the enacted FY 2005 funding (excluding Project BioShield, a $5.6 billion program aimed at the development of vaccines and other medical responses to biological, chemical, nuclear, and radiological weapons). The FY 2006 budget request includes several key initiatives that are expected to allow the Department to integrate and consolidate existing security functions to more effectively serve the DHS overall mission of making America safer.

Overall FY 2006 Budget Highlights

Among the entities with significant budgetary increases are Immigration and Customs Enforcement with a 13.5 percent increase and the U.S. Coast Guard with an increase of more than nine percent.

The budget includes the establishment of the Domestic Nuclear Detection Office (DNDO). The DNDO will develop, acquire, and support the deployment and improvement of a domestic system to detect and report attempts to import, assemble, or transport a nuclear explosive device, fissile material, or radiological material intended for illicit use. The DNDO will be located within DHS and will be jointly staffed with representatives from DHS, the Department of Energy, the Department of Defense, and the FBI, with coordination between the Department of Justice, Department of State, intelligence community, and other departments as needed.

The budget proposes to consolidate the various DHS screening activities with the formation of the Office of Screening Coordination and Operations (SCO) within the Border and Transportation Security (BTS) directorate. This new organization will attempt to enhance terrorist-related screening through comprehensive, coordinated procedures that detect, identify, track, and interdict people, cargo, and other entities and objects that pose a threat to homeland security. This effort brings together several similar ongoing screening efforts under one office, including: United States-Visitor and Immigrant Status Indicator Technology (US-VISIT); Secure Flight and Crew Vetting; Free and Secure Trade (FAST); NEXUS/Secure Electronic Network for Travelers Rapid Inspection (SENTRI); Transportation Worker Identification Credential (TWIC); Registered Traveler; Hazardous Materials Trucker Background Checks; and Alien Flight School Checks.

The awarding of over $2 billion in grants for states and urban areas will now be based upon assessments of risk and vulnerability, as well as the needs and priorities identified in state and regional homeland security plans, rather than having a standard base rate. The proposed Targeted Infrastructure Protection program would provide $600 million in integrated grants, enabling DHS to supplement state, local and regional government efforts in their protection of critical national infrastructures such as seaports, mass transit, railways, and energy facilities.

In FY 2006, DHS seeks to consolidate the research, development, test, and evaluation (RDT&E) activities within the DHS Science and Technology (S&T) directorate. This consolidation, in the amount of $127 million, will bring the scientific and engineering personnel and other RDT&E resources of the Department under a single accountable authority. The Department requests $49.9 million to begin to establish a regional structure for DHS and integrate and identify efficiencies within information technology, facilities, and operations centers across DHS. Of the 22 agencies that were brought together to form the Department, 12 have regional and field structures ranging in size from three to 30 offices distributed throughout the nation.

Aviation security is a shared responsibility of the federal government, airports, airlines, and traveling public. Airport screening, one element of aviation security, benefits passengers and air carriers by protecting them from threats. These costs should be borne primarily by the beneficiaries of screening services. The budget proposes raising the fee on a typical one-leg ticket from $2.50 one way to $5.50. For passengers traveling multiple legs on a one-way trip, that fee would increase from the current maximum of $5.00 to $8.00. Fees cover nearly the full cost of aviation screening operations.

FY 2006 DHS BUDGET HIGHLIGHTS

The FY 2006 DHS budget revolves around five major themes: Revolutionizing the Borders, Strengthening Law Enforcement, Improving National Preparedness and Response, Leveraging Technology, and Creating a Twenty-first Century Department.

Revolutionizing the Borders

- **Weapons of Mass Destruction (WMD) Detection Technology** is an integral part of the DNDO comprehensive strategy to address the threat of nuclear and radiological terrorism. The budget includes $125 million to purchase additional Radiation Portal Monitors (RPMs) and pilot advanced next generation RPMs to detect both gamma and neutron radiation at our borders.
- The **Container Security Initiative (CSI)**, which focuses on prescreening cargo before it reaches our shores, will have a preventative and deterrence effect on the use of global containerized shipping of WMD and other terrorist equipment. Egypt, Chile, India, Philippines, Venezuela, Bahamas, and Honduras have been identified as pilots for screening in FY 2006. An increase of $5.4 million over FY 2005 is included in CBP's budget for CSI, for a total request of $138.8 million.
- **CBP Targeting Systems** aid in identifying high-risk cargo and passengers. The budget includes a total of $28.3 million for these system initiatives, which includes a $5.4 million increase.
- **America's Shield Initiative (ASI)** enhances electronic surveillance capabilities along the northern and southern land borders of the U.S. by improving the sensor and video surveillance equipment deployed to guard against the entry of illegal aliens, terrorists, WMDs, and contraband into the United States. The budget includes $51.1 million, an increase of $19.8 million.
- **US-VISIT**, which is proposed for consolidation within the SCO, increases from $340 million to $390 million. The increase will provide for the accelerated deployment of US-VISIT at the land borders and enhance access for border personnel to immigration, criminal, and terrorist information.
- The **Customs Trade Partnership Against Terrorism (C-TPAT)** focuses on partnerships all along the entire supply chain, from the factory floor, to foreign vendors, to land borders and seaports. The budget includes an increase of $8.2 million, for a total amount of $54.3 million. The increase will enhance our ability to conduct additional supply chain security validations.
- **Border Patrol Staffing** would increase along the southwest border and coastal areas, in part to replace some Border Patrol agents shifted to the northern border as required by the Patriot Act. An increase of 210 agents and $36.9 million is included in the budget for the Border Patrol. This increases the Border Patrol Agents to 10,949.

continues

- **Long Range Radar** technology is used by the Office of Air and Marine Operations to detect and intercept aircraft attempting to avoid detection while entering the United States. CBP and the Department of Defense will assume responsibility for operating and maintaining these systems from the FAA beginning in FY 2006. CBP's share is $44.2 million in the budget.

Strengthening Law Enforcement

- The **Armed Helicopter for Homeland Security Project** increases by $17.4 million in the budget. These funds will provide equipment and aircraft modifications to establish armed helicopter capability at five USCG Air Stations. This will provide the USCG and DHS with the tools needed to respond quickly and forcefully to emergency maritime threats. A total of $19.9 million is included in the budget for this project.
- The **Integrated Deepwater System** increases by $242 million to a total of $966 million in FY 2006 to continue the acquisition of the USCG's Maritime Security Cutter–Large, complete design of the Maritime Security Cutter–Medium, promote completion of the Multi-Mission Cutter Helicopter (reengineered and electronically upgraded HH-65 helicopter) and significantly improve fixed and rotary wing aircraft capabilities. These upgrades will increase awareness and are crucial for an integrated, interoperable border and port security system.
- The **Response Boat-Medium Project** increases the effort to replace the USCG's 41-foot utility boats and other large nonstandard boats with assets more capable of meeting all of the USCG's multimission operational requirements by $10 million. A total of $22 million is proposed in the budget for this effort.
- The **Federal Air Marshal Service (FAMS)** seeks a total of $688.9 million. This funding will allow ICE to protect air security and promote public confidence in our civil aviation system.
- **Detention and Removal** within ICE increases by $176 million for detention and removal activities. Total increases for this program are approximately 19 percent above the FY 2005.
- **Temporary Worker Worksite Enforcement** increases will more than double the resources available for worksite enforcement including employer audits, investigations of possible violations, and criminal case presentations. An increase of $18 million is in the budget.
- **Federal Flight Deck Officers (FFDO)/Crew Member Self-Defense (CMSD) Training** is increased by $11 million in FY 2006 for a total of $36.3 million. This allows for the expansion of the semi-annual firearm requalification program for FFDO personnel and to fund the first full year of the CMSD training program.

Improving National Preparedness and Response

- **Federal assistance for our nation's first responder community**. The budget includes $3.6 billion for grants, training, and technical assistance administered by the Office of State and Local Government Coordination and Preparedness (SLGCP). This funding will support state and local agencies as they equip, train, exercise, and assess preparedness for emergencies regardless of scale or cause.
- **Enhanced Catastrophic Disaster Planning** is budgeted at $20 million for FEMA to work with states and localities, as well as other federal agencies, to develop and implement plans that will improve the ability to respond to, and to recover from, catastrophic disasters.
- The **Office of Interoperability and Compatibility (OIC)** within the S&T Directorate will allow the Department to expand its leadership role in interoperable communications that could be used by every first responder agency in the country. The OIC has currently identified three program areas: communications, equipment, and training. With $20.5 million in FY 2006, the OIC will plan and begin to establish the training and equipment programs, as well as continue existing communication interoperability efforts through the SAFECOM Program.
- Replacement of the USCG's **High Frequency (HF) Communications System**, funded at $10 million in the budget, will replace unserviceable, shore-side, high-power, high-frequency transmitters, significantly improving long-range maritime safety and security communications.
- The **Rescue 21** project is funded at $101 million in the budget to continue recapitalizing the Coast Guard's coastal zone communications network. This funding will complete system infrastructure and network installations in 14 regions and begin development of regional designs for the remaining 11 regions.

Leveraging Technology

- **Low Volatility Agent Warning System** is a new FY 2006 initiative totaling $20 million. Funding is included to develop a system that will serve as the basis for a warning and identification capability against a set of chemical agents whose vapor pressure is too low to be detected by conventional measures.
- **Counter-MAN Portable Air Defense Systems (C-MANPADS)** funding is increased by $49 million to a total of $110 million in the budget. This program will continue to research the viability of technical countermeasures for commercial aircraft against the threat of shoulder-fired missiles.
- **Cyber Security** is enhanced in the budget to augment a 24/7 cyber threat watch, warning, and response capability that would identify emerging threats and vulnerabilities and coordinate responses to major cyber security inci-

continues

dents. An increase of $5 million is proposed in the budget for this effort, bringing the program total to $73.3 million.

- **Secure Flight/Crew Vetting** requests an increase of $49 million to field the system developed and tested in FY 2005. The funds will support testing, information systems, connectivity to airlines and screen systems, and daily operations. This also includes an increase of $3.3 million for crew vetting.
- The budget includes $174 million to complete installation of **High Speed Operational Connectivity (Hi-SOC)** to passenger and baggage screening checkpoints to improve management of screening system performance.
- **Emerging Checkpoint Technology** is enhanced by $43.7 million in FY 2006 to direct additional resources to improve checkpoint explosives screening. This assures that TSA is on the cutting edge, ahead of the development of increasingly well-disguised prohibited items. This proposed increase will result in investing more than $100 million invested in FY 2005 and FY 2006 for new technology to ensure improved screening of all higher risk passengers.
- **Homeland Secure Data Network (HSDN)** includes $37 million in the budget. These funds will streamline and modernize the classified data capabilities in order to facilitate high-quality and high-value classified data communication and collaboration.
- The **Homeland Security Operations Center (HSOC)** funding is increased by $26.3 million bringing its FY 2006 funded level to $61.1 million. This includes an increase of $13.4 million for the Homeland Security Information Network (HSIN) and an increase of $12.9 million to enhance HSOC systems and operations.

Creating a Twenty-first Century Department

- **Electronically managing enterprise resources for government effectiveness and efficiency (eMerge2)** funding of $30 million in the budget to continue implementation of a DHS-wide solution that delivers accurate, relevant, and timely resource management information to decision makers. By delivering access to critical information across all components, the Department will be able to better support its many front-line activities.
- **MAX HR** funding of $53 million is to continue the design and deployment of a new human resources system. As outlined in final regulations, issued jointly on February 1, 2005, by Secretary Ridge and the Director of the Office of Personnel Management (OPM) Kay Coles James, the MAXHR system provides greater flexibility and accountability in the way employees are paid, developed, evaluated, afforded due process, and represented by labor organizations. The goal is a twenty-first century personnel system that enhances mission-essential flexibility and preserves core civil service principles and the merit system.

> • The **Information Sharing and Collaboration (ISC)** program will affect the policy, procedures, technical, process, cultural, and organizational aspects of information sharing and collaboration, including coordinating ISC policy with other federal agencies, drafting technical and operational needs statements, performing policy assessments, and analyzing new requirements. The total funding for FY 2006 will be $16.5 million.
>
> *Source*: DHS Proposed 2006 Budget, www.DHS.gov

The 911 Commission

In late 2002, in an effort to "prepare a full and complete account of the circumstances surrounding the" terrorist attacks that occurred on September 11, 2001, the National Commission on Terrorist Attacks Upon the United States (more commonly known as the 911 Commission) was formed. This commission set out to determine the shortfalls and the lessons learned from the preparedness for and response to international terrorism within the United States, and to formulate recommendations for activities that would help improve these systems in case of future threats and attacks.

The Commission, which consisted of five Republicans and five Democrats, interviewed over 1200 people from 10 countries, including several past and present government officials at the federal, state, and local levels, and studied millions of pages of documentation, to accurately assess the events. On July 22 of 2004, the 911 Commission released its long-awaited report. Although there was initial criticism of earlier Commission reports and its members, including claims of bias, difficulty in attaining cooperation from White House officials, partisanship, among others, the final report's findings generally have been met with approval and acceptance for their recommendations.

The report found many opportunities that could have been exploited by the federal government to stop the terrorists who attacked in 2001, including:

- Not watchlisting future hijackers Hazmi and Mihdhar, not trailing them after they traveled to Bangkok, and not informing the FBI about one future hijacker's U.S. visa or his companion's travel to the United States
- Not sharing information linking individuals in the *Cole* attack to Mihdhar
- Not taking adequate steps in time to find Mihdhar or Hazmi in the United States
- Not linking the arrest of Zacarias Moussaoui, described as interested in flight training for the purpose of using an airplane in a terrorist act, to the heightened indications of attack
- Not discovering false statements on visa applications
- Not recognizing passports manipulated in a fraudulent manner
- Not expanding no-fly lists to include names from terrorist watchlists
- Not searching airline passengers identified by the computer-based CAPPS screening system

- Not hardening aircraft cockpit doors or taking other measures to prepare for the possibility of suicide hijackings

The report also identified failures on the part of U.S. government policy that could have prevented the attacks, including:

- **Imagination.** The Commission saw this as the most important failure. They do not believe leaders understood the gravity of the threat, or that terrorist danger from Bin Ladin and al Qaeda was a major topic for policy debate among the public, the media, or in the Congress. Al Qaeda's new brand of terrorism presented challenges to U.S. governmental institutions that they were not well-designed to meet. Though top officials all told the Commission that they understood the danger, the Commission believed there was uncertainty among them as to whether this was just a new and especially venomous version of the ordinary terrorist threat the United States had lived with for decades, or it was indeed radically new, posing a threat beyond any yet experienced.
- **Policy.** The Commission felt that terrorism was not the overriding national security concern for the U.S. government under either the Clinton or the pre-9/11 Bush administration. The policy challenges were linked to this failure of imagination. Officials in both the Clinton and Bush administrations regarded a full U.S. invasion of Afghanistan as practically inconceivable before 9/11.
- **Capabilities.** Before 9/11, the United States tried to solve the al Qaeda problem with the capabilities it had used in the last stages of the Cold War and its immediate aftermath. The Commission claims these capabilities were insufficient. The CIA had minimal capacity to conduct paramilitary operations with its own personnel, and it did not seek a large-scale expansion of these capabilities before 9/11. The CIA also needed to improve its capability to collect intelligence from human agents.

  At no point before 9/11 was the Department of Defense fully engaged in the mission of countering al Qaeda, even though this was perhaps the most dangerous foreign enemy threatening the United States. NORAD itself was barely able to retain any alert bases at all. Its planning scenarios occasionally considered the danger of hijacked aircraft being guided to American targets, but only aircraft that were coming from overseas.

  The Commission saw the most serious weaknesses in agency capabilities in the domestic arena. The FBI did not have the capability to link the collective knowledge of agents in the field to national priorities. Other domestic agencies deferred to the FBI. FAA capabilities were weak. Any serious examination of the possibility of a suicide hijacking could have suggested changes to fix glaring vulnerabilities—expanding no-fly lists, searching passengers identified by the CAPPS screening system, deploying federal air marshals domestically, hardening cockpit doors, alerting air crews to a different kind of hijacking possibility than they had been trained to expect. Yet the FAA did not adjust either its own training or training with NORAD to take account of threats other than those experienced in the past.

- **Management.** The Commission reported that the missed opportunities to thwart the 9/11 plot were also symptoms of a broader inability to adapt the way government manages problems to the new challenges of the twenty-first century. Action officers should have been able to draw on all available knowledge about al Qaeda in the government. Management should have ensured that information was shared, and duties were clearly assigned across agencies and across the foreign-domestic divide. There were also broader management issues with respect to how top leaders set priorities and allocated resources. The U.S. government did not find a way of pooling intelligence and using it to guide the planning and assignment of responsibilities for joint operations involving entities as disparate as the CIA, the FBI, the State Department, the military, and the agencies involved in homeland security.

In addition to these general findings, the Commission also reported a description of several specific findings they claim resulted in the inability of the government to thwart the attacks, and its ability to respond once they occurred, including:

- Unsuccessful diplomacy
- Lack of military operations
- Problems with the intelligence community
- Problems in the FBI
- Permeable borders and immigration controls
- Permeable aviation security
- Terrorist financing
- The lack of an improved homeland defense
- Problems with emergency response systems
- The poor response of Congress to the terrorist threat

The 911 Commission made recommendations that fell into two general categories: What To Do? and How To Do It? The following information comes directly from the Executive Summary of the 911 Commission Report.

911 COMMISSION REPORT RECOMMENDATIONS

What To Do? A Global Strategy

The enemy is not just "terrorism." It is the threat posed specifically by Islamist terrorism, by Bin Ladin and others who draw on a long tradition of extreme intolerance within a minority strain of Islam that does not distinguish politics from religion, and distorts both.

The enemy is not Islam, the great world faith, but a perversion of Islam. The enemy goes beyond al Qaeda to include the radical ideological movement, inspired in part by al Qaeda, that has spawned other terrorist groups and violence. Thus our strategy must match our means to two ends: dismantling the al Qaeda network and, in the long term, prevailing over the ideology that contributes to Islamist terrorism.

The first phase of our post-9/11 efforts rightly included military action to topple the Taliban and pursue al Qaeda. This work continues. But long-term success demands the use of all elements of national power: diplomacy, intelligence, covert action, law enforcement, economic policy, foreign aid, public diplomacy, and homeland defense. If we favor one tool while neglecting others, we leave ourselves vulnerable and weaken our national effort.

What should Americans expect from their government? The goal seems unlimited: Defeat terrorism anywhere in the world. But Americans have also been told to expect the worst: An attack is probably coming; it may be more devastating still.

Vague goals match an amorphous picture of the enemy. Al Qaeda and other groups are popularly described as being all over the world, adaptable, resilient, needing little higher-level organization, and capable of anything. It is an image of an omnipotent hydra of destruction. That image lowers expectations of government effectiveness.

It lowers them too far. Our report shows a determined and capable group of plotters. Yet the group was fragile and occasionally left vulnerable by the marginal, unstable people often attracted to such causes. The enemy made mistakes. The U.S. government was not able to capitalize on them.

No president can promise that a catastrophic attack like that of 9/11 will not happen again. But the American people are entitled to expect that officials will have realistic objectives, clear guidance, and effective organization. They are entitled to see standards for performance so they can judge, with the help of their elected representatives, whether the objectives are being met.

We propose a strategy with three dimensions: (1) attack terrorists and their organizations, (2) prevent the continued growth of Islamist terrorism, and (3) protect against and prepare for terrorist attacks.

- Attack Terrorists and Their Organizations
  - Root out sanctuaries. The U.S. government should identify and prioritize actual or potential terrorist sanctuaries and have realistic country or

regional strategies for each, utilizing every element of national power and reaching out to countries that can help us.

- Strengthen long-term U.S. and international commitments to the future of Pakistan and Afghanistan.
- Confront problems with Saudi Arabia in the open and build a relationship beyond oil, a relationship that both sides can defend to their citizens and includes a shared commitment to reform.

- Prevent the Continued Growth of Islamist Terrorism. In October 2003, Secretary of Defense Donald Rumsfeld asked if enough was being done "to fashion a broad integrated plan to stop the next generation of terrorists. "As part of such a plan, the U.S. Government should
  - Define the message and stand as an example of moral leadership in the world. To Muslim parents, terrorists like Bin Ladin have nothing to offer their children but visions of violence and death. America and its friends have the advantage—our vision can offer a better future.
  - Where Muslim governments, even those who are friends, do not offer opportunity, respect the rule of law, or tolerate differences, then the United States needs to stand for a better future.
  - Communicate and defend American ideals in the Islamic world, through much stronger public diplomacy to reach more people, including students and leaders outside of government. Our efforts here should be as strong as they were in combating closed societies during the Cold War.
  - Offer an agenda of opportunity that includes support for public education and economic openness.
  - Develop a comprehensive coalition strategy against Islamist terrorism, using a flexible contact group of leading coalition governments and fashioning a common coalition approach on issues like the treatment of captured terrorists.
  - Devote a maximum effort to the parallel task of countering the proliferation of weapons of mass destruction.
  - Expect less from trying to dry up terrorist money and more from following the money for intelligence, as a tool to hunt terrorists, understand their networks, and disrupt their operations.
- Protect against and Prepare for Terrorist Attacks
  - Target terrorist travel, an intelligence and security strategy that the 9/11 story showed could be at least as powerful as the effort devoted to terrorist finance.
  - Address problems of screening people with biometric identifiers across agencies and governments, including our border and transportation systems, by designing a comprehensive screening system that addresses common problems and sets common standards. As standards spread, this necessary and ambitious effort could dramatically strengthen the world's ability to intercept individuals who could pose catastrophic threats.

continues

- Quickly complete a biometric entry-exit screening system, one that also speeds qualified travelers.
- Set standards for the issuance of birth certificates and sources of identification, such as driver's licenses.
- Develop strategies for neglected parts of our transportation security system. Since 9/11, about 90 percent of the nation's $5 billion annual investment in transportation security has gone to aviation, to fight the last war.
- In aviation, prevent arguments about a new computerized profiling system from delaying vital improvements in the "no-fly" and "automatic selectee" lists. Also, give priority to the improvement of check-point screening.
- Determine, with leadership from the President, guidelines for gathering and sharing information in the new security systems that are needed, guidelines that integrate safeguards for privacy and other essential liberties.
- Underscore that as government power necessarily expands in certain ways, the burden of retaining such powers remains on the executive to demonstrate the value of such powers and ensure adequate supervision of how they are used, including a new board to oversee the implementation of the guidelines needed for gathering and sharing information in these new security systems.
- Base federal funding for emergency preparedness solely on risks and vulnerabilities, putting New York City and Washington, D.C., at the top of the current list. Such assistance should not remain a program for general revenue sharing or pork-barrel spending.
- Make homeland security funding contingent on the adoption of an incident command system to strengthen teamwork in a crisis, including a regional approach. Allocate more radio spectrum and improve connectivity for public safety communications, and encourage wide-spread adoption of newly developed standards for private-sector emergency preparedness—since the private sector controls 85 percent of the nation's critical infrastructure.

How to Do It? A Different Way of Organizing Government

The strategy we have recommended is elaborate, even as presented here very briefly. To implement it will require a government better organized than the one that exists today, with its national security institutions designed half a century ago to win the Cold War. Americans should not settle for incremental, ad hoc adjustments to a system created a generation ago for a world that no longer exists.

Our detailed recommendations are designed to fit together. Their purpose is clear: to build unity of effort across the U.S. government. As one official now serving on the front lines overseas put it to us: "One fight, one team."

We call for unity of effort in five areas, beginning with unity of effort on the challenge of counterterrorism itself:

- Unifying strategic intelligence and operational planning against Islamist terrorists across the foreign-domestic divide with a National Counterterrorism Center
- Unifying the intelligence community with a new National Intelligence Director
- Unifying the many participants in the counterterrorism effort and their knowledge in a network-based information sharing system that transcends traditional governmental boundaries
- Unifying and strengthening congressional oversight to improve quality and accountability
- Strengthening the FBI and homeland defenders

Unity of Effort: A National Counterterrorism Center. The 9/11 story teaches the value of integrating strategic intelligence from all sources into joint operational planning—with *both* dimensions spanning the foreign-domestic divide.

- In some ways, since 9/11, joint work has gotten better. The effort of fighting terrorism has flooded over many of the usual agency boundaries because of its sheer quantity and energy. Attitudes have changed. But the problems of coordination have multiplied. The Defense Department alone has three unified commands (SOCOM, CENTCOM, and NORTHCOM) that deal with terrorism as one of their principal concerns.
- Much of the public commentary about the 9/11 attacks has focused on "lost opportunities." Though characterized as problems of "watch-listing," "information sharing," or "connecting the dots," each of these labels is too narrow. They describe the symptoms, not the disease.
- Breaking the older mold of organization stovepiped purely in executive agencies, we propose a National Counterterrorism Center (NCTC) that would borrow the joint, unified command concept adopted in the 1980s by the American military in a civilian agency, combining the joint intelligence function alongside the operations work.
- The NCTC would build on the existing Terrorist Threat Integration Center and would replace it and other terrorism "fusion centers" within the government. The NCTC would become the authoritative knowledge bank, bringing information to bear on common plans. It should task collection requirements both inside and outside the United States.
- The NCTC should perform joint operational planning, assigning lead responsibilities to existing agencies and letting them direct the actual execution of the plans.
- Placed in the Executive Office of the President, headed by a Senate-confirmed official (with rank equal to the deputy head of a cabinet department) who reports to the National Intelligence Director, the NCTC would

continues

track implementation of plans. It would be able to influence the leadership and the budgets of the counterterrorism operating arms of the CIA, the FBI, and the departments of Defense and Homeland Security.

- The NCTC should *not* be a policymaking body. Its operations and planning should follow the policy direction of the President and the National Security Council.

Unity of Effort: A National Intelligence Director. Since long before 9/11—and continuing to this day—the intelligence community is not organized well for joint intelligence work. It does not employ common standards and practices in reporting intelligence or in training experts overseas and at home. The expensive national capabilities for collecting intelligence have divided management. The structures are too complex and too secret.

- The community's head—the Director of Central Intelligence—has at least three jobs: running the CIA, coordinating a 15-agency confederation, and being the intelligence analyst-in-chief to the President. No one person can do all these things.
- A new National Intelligence Director should be established with two main jobs: (1) to oversee national intelligence centers that combine experts from all the collection disciplines against common targets—like counterterrorism or nuclear proliferation; and (2) to oversee the agencies that contribute to the national intelligence program, a task that includes setting common standards for personnel and information technology.
- The national intelligence centers would be the unified commands of the intelligence world—a long-overdue reform for intelligence comparable to the 1986 Goldwater-Nichols law that reformed the organization of national defense. The home services—such as the CIA, DIA, NSA, and FBI—would organize, train, and equip the best intelligence professionals in the world, and would handle the execution of intelligence operations in the field.
- This National Intelligence Director (NID) should be located in the Executive Office of the President and report directly to the President, yet be confirmed by the Senate. In addition to overseeing the National Counterterrorism Center described earlier (which will include both the national intelligence center for terrorism and the joint operations planning effort), the NID should have three deputies:
  - For foreign intelligence (a deputy who also would be the head of the CIA)
  - For defense intelligence (also the undersecretary of defense for intelligence)
  - For homeland intelligence (also the executive assistant director for intelligence at the FBI or the undersecretary of homeland security for information analysis and infrastructure protection)
- The NID should receive a public appropriation for national intelligence, should have authority to hire and fire his or her intelligence deputies, and should be able to set common personnel and information technology policies across the intelligence community.

- The CIA should concentrate on strengthening the collection capabilities of its clandestine service and the talents of its analysts, building pride in its core expertise.
- Secrecy stifles oversight, accountability, and information sharing. Unfortunately, all the current organizational incentives encourage overclassification. This balance should change; and as a start, open information should be provided about the overall size of agency intelligence budgets.

Unity of Effort: Sharing Information. The U.S. government has access to a vast amount of information. But it has a weak system for processing and using what it has. The system of "need to know" should be replaced by a system of "need to share."

- The President should lead a government-wide effort to bring the major national security institutions into the information revolution, turning a mainframe system into a decentralized network. The obstacles are not technological. Official after official has urged us to call attention to problems with the unglamorous "back office" side of government operations.
- But no agency can solve the problems on its own—to build the network requires an effort that transcends old divides, solving common legal and policy issues in ways that can help officials know what they can and cannot do. Again, in tackling information issues, America needs unity of effort.

Unity of Effort: Congress. Congress took too little action to adjust itself or to restructure the executive branch to address the emerging terrorist threat. Congressional oversight for intelligence—and counterterrorism—is dysfunctional. Both Congress and the executive need to do more to minimize national security risks during transitions between administrations.

- For intelligence oversight, we propose two options: either a joint committee on the old model of the Joint Committee on Atomic Energy or a single committee in each house combining authorizing and appropriating committees. Our central message is the same: the intelligence committees cannot carry out their oversight function unless they are made stronger, and thereby have both clear responsibility and accountability for that oversight.
- Congress should create a single, principal point of oversight and review for homeland security. There should be one permanent standing committee for homeland security in each chamber.
- We propose reforms to speed up the nomination, financial reporting, security clearance, and confirmation process for national security officials at the start of an administration, and suggest steps to make sure that incoming administrations have the information they need.

Unity of Effort: Organizing America's Defenses in the United States. We have considered several proposals relating to the future of the domestic intelligence and counterterrorism mission. Adding a new domestic intelligence agency will

continues

not solve America's problems in collecting and analyzing intelligence within the United States. We do not recommend creating one.

- We propose the establishment of a specialized and integrated national security workforce at the FBI, consisting of agents, analysts, linguists, and surveillance specialists who are recruited, trained, rewarded, and retained to ensure the development of an institutional culture imbued with a deep expertise in intelligence and national security.

At several points we asked: Who has the responsibility for defending us at home? Responsibility for America's national defense is shared by the Department of Defense, with its new Northern Command, and by the Department of Homeland Security. They must have a clear delineation of roles, missions, and authority.

- The Department of Defense and its oversight committees should regularly assess the adequacy of Northern Command's strategies and planning to defend against military threats to the homeland.
- The Department of Homeland Security and its oversight committees should regularly assess the types of threats the country faces, in order to determine the adequacy of the government's plans and the readiness of the government to respond to those threats.

We call on the American people to remember how we all felt on 9/11, to remember not only the unspeakable horror but how we came together as a nation—one nation. Unity of purpose and unity of effort are the way we will defeat this enemy and make America safer for our children and grandchildren. We look forward to a national debate on the merits of what we have recommended, and we will participate vigorously in that debate.

Source: 911 Commission Final Report

STATE GOVERNMENT TERRORISM ACTIVITY

Governors, and the states they govern, are recognized for the critical role they play in Homeland Security. State and local law enforcement and health personnel provide the first line of defense in protecting critical infrastructure and public health and safety. Should an incident occur, state and local personnel are the first to respond to an emergency and the last to leave the scene. Governors, with the support of the federal government, are responsible for coordinating state and local resources to effectively address natural disasters, accidents, and other types of major emergencies, including terrorist incidents.

The national effort to protect the nation from acts of terrorism has been conducted with equal strength at the state level as has been seen at the federal level. As the recipients of a bulk of the homeland security funding that has been distributed by the Department of Homeland Security and other federal agencies, the states have had the ability to administer new statewide programs aimed at bringing preparedness and prevention to each and every community.

State Homeland Security entities were created to ensure that the states are preparing for the wide range of terrorist attacks that have been identified by DHS and other entities. These state offices accomplish this by facilitating the interaction and coordination that is needed among each state's governor's office, the homeland security director, the state emergency management office, other state agencies, local governments, the private sector, volunteer organizations, and the federal government.

Following the attacks of September 11, the governors designated individuals from various backgrounds in state government to serve as their state homeland security directors. Among the states and territories, there is no common model; however, in several states, the homeland security director serves as an advisor to the governor in addition to coordinating state emergency management, law enforcement, health, and related public safety functions. In other models, governors designated the state's Adjutant General as homeland security advisor. Although governors generally have opted not to create unique cabinet-level positions with oversight over all state agencies, they did form homeland security task forces. The task forces typically consist of executive office staff and agency heads from law enforcement, fire and rescue, public health, National Guard, transportation, public works, and information technology.

STATE OFFICES OF HOMELAND SECURITY HAVE BEEN PLACED IN ALL OF THE FOLLOWING STATE GOVERNMENT AGENCIES SINCE 2001, IN ORDER OF MOST TO LEAST COMMON:

- Governor's office
- Military/Adjutant General
- Emergency Management
- Public Safety
- Law Enforcement
- Attorney General
- Lt. Governor
- Land Commissioner

Source: National Emergency Management Association, National Governors Association

---

In August 2002, the NGA Center for Best Practices of the National Governors Association released "States' Homeland Security Priorities." A list of 10 "major priorities and issues" was identified by the NGA center through a survey of states' and territories' state homeland security offices. (NGA Center for Best Practices, 2002). A list of these priorities follows.

LIST OF STATES' HOMELAND SECURITY PRIORITIES

- Coordination must involve all levels of government.
- The federal government must disseminate timely intelligence information to the states.
- States must work with local governments to develop interoperable communications between first responders and adequate wireless spectrum must be set aside to do the job.
- State and local governments need help and technical assistance to identify and protect critical infrastructure.
- Both the states and federal government must focus on enhancing bioterrorism preparedness and rebuilding the nation's public health system to address twenty-first century threats.
- The federal government should provide adequate federal funding and support to ensure that homeland security needs are met.
- The federal government should work with states to protect sensitive security information, including restricting access to information available through "freedom of information" requests.
- An effective system must be developed that secures points of entry at borders, airports, and seaports without placing an undue burden on commerce.
- The National Guard has proven itself to be an effective force during emergencies and crises. The mission of the National Guard should remain flexible, and Guard units should remain primarily under the control of the governor during times of crises.
- Federal agencies should integrate their command systems into existing state and local incident command systems (ICS) rather than requiring state and local agencies to adapt to federal command systems.

Source: NGA Center for Best Practices, Issue Brief, August 19, 2002

LOCAL GOVERNMENT TERRORISM ACTIVITIES

The Counties

Emergency preparedness, mitigation, response, and recovery all occur at the local community level. This is true for terrorism preparedness, mitigation, response, and recovery activities. It is at the local level that the critical planning, communications, technology, coordination, command, and spending decisions matter the most. The priorities of groups such as the National Conference of Mayors and the National Associations of Counties (NACo) represent what matters at the local community level in the fight against terrorism. The fight against terrorism has spawned a series of new requirements in preparedness and mitigation planning at the local level.

NACo has created a "Policy Agenda to Secure the People of America's Counties." This policy paper states: "Counties are the first responders to terrorist attacks, natural disasters and major emergencies" (NACo, 2004). NACo has established a 43-member NACo Homeland Security Task Force that in July of 2004 reaffirmed a set of 21 recommendations concerning homeland security issues. The 21 NACO recommendations are presented here.

NACO HOMELAND SECURITY NATIONAL OBJECTIVES AND FUNDING RECOMMENDATIONS FOR COUNTIES AND HOMELAND SECURITY

1. **National Strategies for the Nation**
   A national long-term strategy for homeland security should be developed to guide federal, state, and local preparedness efforts. Input from local and state stakeholders must be included in the development of this strategy and funding must be consistent with national goals and objectives.

2. **Sustained Funding for Homeland Security**
   Congress must provide sustained funding for homeland security to enhance the ability of local governments to protect their communities. Funding for federal public safety programs that existed prior to September 11 should not be supplanted by recent homeland security funding.

3. **Base Level of Preparedness for All Communities**
   Federal funding allocations for homeland security should ensure a base level of preparedness to all states and regions to ensure that all citizens are protected from the threat of terrorism.

4. **High Threat Funding to Most Critical Areas**
   Federal funding to the nations high threat urban areas (cities, counties, contiguous counties, and mutual aid partners) must be provided. These areas of high visibility, heightened threat and risk, vulnerable critical infrastructure and population density have a heightened sense of vulnerabilities to terrorist attacks.

5. **Expediting Assistance at All Levels of Government**
   Federal and state assistance for homeland security, public health, "all hazards," and safety must reach first responders in an expedited fashion. As a result, all levels of government should work together to ensure the timely distribution of assistance to first responders. In the event that federal, state, and local government legal, procedural, and/or procurements processes delay the expenditure of funds, efforts must be made to establish an expedited authorization and appropriation process.

continues

Public Health

6. **Fund Local Public Health Emergency Preparedness**

   Congress should continue to provide adequate funding for HHS cooperative agreements with the states for public health emergency preparedness
   and give strong direction to the states to ensure that: 1) no less than 80%
   of the funds are used to improve local preparedness and local infrastructure; and 2) county public health agencies are consulted and concur with
   the state plans for expenditures of these funds.

7. **Ensure an Adequate Supply of Vaccines and Antibiotics**

   The federal government should ensure an adequate supply of appropriate
   antibiotics, vaccines, and other relevant medications, and medical supplies
   are made available to counties and other local communities in a timely
   manner as part of the stockpiled push packages administered by the CDC.
   Also, the federal government must continue to build an advanced surveillance system for the detection and identification of biological and other
   harmful agents.

8. **Train Health Personnel**

   Public and private sector health personnel should receive adequate training
   to manage public health emergencies, in cooperation with federal, state, and
   local governments. Although specific training relative to bioterrorism is
   needed, general competency building in public health is also needed to
   assure that the workforce is fully prepared.

9. **Ensure That Adequate Medical Surge Capacity Exists**

   The federal government, in cooperation with state and local governments, should ensure that the medical surge capacity needs associated
   with events of mass casualties and large outbreaks of infectious diseases
   can be met, particularly in communities that serve as regional medical
   centers.

Information Sharing and Critical Infrastructure Security

10. **Sharing of Intelligence**

    The federal government must develop an efficient and comprehensive
    system for the sharing, analysis, and dissemination of intelligence between
    federal, state, and local public safety agencies in concert with local gov-
    ernments.

11. **Balance Heightened Border Security with Economic Activity**

    Improve border security operations to enhance the nation's ability to restrict
    the movement of weapons, weapons components, or potential terrorists into
    the country and eliminate their ability to operate within our borders, in such
    a way that heightened security does not impede with the ability to continue
    active cross-border commerce.

12. **Securing Critical Infrastructure**

    The federal government should provide assistance to counties for enhancing critical infrastructure and key resources. Enhanced coordination between local governments and the private sector is critical for ensuring the preparedness of states and localities and for protecting vital physical and economic infrastructure. State and local intelligence information should be utilized in the development and continued refinements of DHS's national critical infrastructure protection list.

13. **Help Localities Secure Public Utilities and a Safe Water Supply**

    Congress should authorize funds for drinking water systems and other public utilities (large and small) to conduct physical vulnerability assessments, emergency planning, and security enhancements. Additional research should be conducted into the threats to water and sewer systems and other public utilities and the development of methods and technologies to prevent and respond to such attacks.

14. **Reimburse Counties for Costs Incurred on Behalf of the Federal Government**

    The federal government should reimburse counties for the local public safety and law enforcement costs associated with requests to provide security to federal installations and federally owned infrastructure within their jurisdictions, and for the federal use of county facilities and other federally mandated expenses incurred during an emergency and/or a heightened sense of alert.

Emergency Planning and Public Safety

15. **Assist Counties to Develop Evacuation Capacity**

    Support assistance to counties for the evaluation of transportation and other infrastructure systems and evacuation planning, including developing capacity at the local level to facilitate evacuations.

16. **Train County Elected and Appointed Officials to Prepare for and Respond to Acts of Terror**

    Federal, state, and local governments should collaborate to train first responders to respond to acts of terror, utilizing and expanding upon existing training facilities and opportunities to their fullest extent. Curricula also should be established for the specific purpose of training elected county officials and other representatives of general-purpose local governments. A standard, core set of competencies should be developed and cross-discipline training must be encouraged.

17. **One-Stop Clearinghouse**

    The federal government should create a "one-stop" clearinghouse for grants, training programs, and other disaster preparedness assistance for state and local governments and public safety agencies.

continues

18. **Assist Public Safety Communications Interoperability and Interference Issues**

    The federal government should assist counties to provide the broadest possible interoperability between public safety agencies across voice, data, and geo-data and wireless technologies. The federal government also should assist counties in obtaining additional spectrum as soon as possible to address interoperability and dead zone problems created by congestion and interference with commercial services. In the event of a disaster or terrorist attack, all first responders should have access to a common set of frequencies that can be used to communicate between agencies. Manufacturers should expand their commitment to producing standards-compliant communications infrastructure. Equally important, the public safety community should be made aware of standards-compliant equipment, and the importance of public safety participation in standards development efforts should be emphasized. Working with the first responder's community, a common standard "language" for interoperability communication needs to be established so that responders from various agencies can act on specific instructions without mistake or delay.

19. **Establish a Public Communication Network**

    A communication network capable of delivering information in a timely manner between the federal government, state and local governments, and the general public should be established.

20. **Urge the Release of Federal Research to Assist Counties**

    The federal government should make its research and information available to counties at the earliest possible time—including declassifying such information as appropriate—to facilitate their use by counties to prepare for and respond to acts of terrorism and other emergencies.

21. **Provide Immunity to Encourage Mutual Aid and Support**

    The federal government and state governments, where applicable, should provide legal immunity from civil liability for counties and other local governments responding collaboratively to emergencies outside their primary jurisdiction. Also, the federal and state governments should allow reimbursement under public assistance for assistance rendered by mutual aid partners.

Source: NACo Homeland Security Task Force, July 2004

As previously mentioned, the Global War on Terrorism has caused various hardships at the state and local levels. One particular hardship that has been endured is the loss of critical employees serving as military reservists on deployment in Afghanistan, Iraq, and elsewhere. NACo performed a survey of county governments entitled "How Has the Deployment of Reserves Affected Your County?" to assess these hardships, the results of which are summarized next.

EFFECTS OF MILITARY RESERVISTS' DEPLOYMENT ON COUNTY GOVERNMENTS

Counties were asked if county employees who are members of the reserves been called up for duty. Of the 164 responding counties, 43 percent report that employees have been called up. Of these counties that have had employees called up, 76 percent have had less than five employees called to the military. Twelve percent have had between 5 to 10 employees called up and 8 percent have had more than 20.

Departmental Distribution

Counties were asked to list the departments that were most affected by the call up. Seventy-four percent report that police/sheriff departments were affected. This was followed by 28 percent stating other departments and 18 percent reporting fire and emergency medical departments and public works departments were also affected. Nine percent report that transit and transportation as well as administration departments were among those affected.

Benefits for the Military

Counties were asked about the benefits their county employees received while serving on active duty in the military. Forty-three percent report that benefits stopped in accordance with the time period required by federal law. However, 35 percent of the responding counties indicate that they have established policy that continues benefits to the military beyond those required by federal law. Sixteen percent of responding counties report continuing to extend benefits to the military based on state law.

Hardships Caused by Deployment

Counties are coping with missing employees in several ways. Fifty-nine percent are reallocating other staff to fill the positions of missing employees, and 46 percent have hired temporary staff. More than 14 percent indicate that they have had to cut back on service delivery while these employees are deployed.

Counties are making do, with 52 percent reporting that their counties have not experienced a hardship while these employees are on active duty. Examining this response by population size however, paints a different picture. Sixty-nine percent of counties with populations below 10,000 report the deployment has created a hardship for them.

Of the 48 percent of counties reporting hardships caused by the current deployment, several provided the following anecdotes:

- *We have a very tight budget and hiring temporary help has placed an additional burden on the county. We have had a large murder trial in the*

continues

> *last month that has taxed the Sheriff's Office and they have needed all personnel.*
> - *Temporary employees are not certified as police officers so they are still understaffed.*
> - *It is difficult to recruit, hire, and train Juvenile Probation Counselors when you don't know for how long they'll be hired. Training is expensive and takes about two years.*
> - *With one of seven deputies on staff, the other six had to take up his shifts because we couldn't find another deputy since we were paying his salary and benefits in his absence.*
> - *Especially for 24/7 operations such as sheriff deputies, we have to pay overtime to backfill the shifts while picking up the pay difference for the employee.*
> - *The Sheriff's Department has had to reduce service in some instances when part-time staff could not fill the empty slot.*
> - *We have been forced to use overtime to compensate for the absent staff. Additionally, this has caused us to prioritize duties and not accomplish some that we would normally desire to accomplish.*
> - *Sheriff's office has had to adjust to using lesser-trained personnel.*
> - *When you are short a deputy sheriff it puts a greater burden on the other law enforcement officials by working longer hours, which may cause more accidents.*
> - *Cut back on services due to vacancies.*
>
> *Source*: www.NACo.org

Cities and Towns

Other than the largest cities, most local communities do not have specially designated offices of homeland security, or any other terrorism-specific government office or agency. In general, local communities rely upon the skills and training of their teams of first responders, who include the fire, police, emergency management, emergency medical, and other officials that live within their jurisdictions.

However, these first responders are the heart of the system that the nation depends upon for the protection from and response to terrorist attacks. Local communities are instructed that they may have to manage the aftermath of a terrorist attack for a full 24 to 48 hours on their own before state or federal backup arrives. As should be obvious by the levels of funding that have been described in copious detail throughout this text, the federal government has recognized and responded to such facts.

Local first response still has much catching up to do in order to be able to fulfill the preparedness and response needs of the federal government. Interoperable communication, the condition where all responders and emergency management within and without of each community can talk to each other, is still not possible. Many communities lack the equipment and training necessary to respond to attacks involving weapons of mass destruction. Efforts to vaccinate health care workers from biological weapons such as smallpox have failed, and there are still questions about whether communities could handle an outbreak of one of these diseases even if sufficient vaccines were available to them.

In the larger communities, where the training and equipment are better funded and considered adequate, there are other issues that have presented themselves. Large ports are still not passing minimum security requirements to keep out potential weapons of mass destruction; financial woes are sounded each time the Homeland Security Alert System is raised for specific terrorist threats due to the need for police overtime, and the loss of other essential services to reassigned officials; and contentious battles over the appropriation of both federal and state funding has soured many preexisting relationships.

But, the will to prepare exists, and the growing pains are becoming less severe as more and more funds reach deeper into American communities. Cooperation and intelligence sharing has made the state and local responders a more integral part of the counter-terrorism team that will be necessary to prevent or contain future terrorist attacks, whether they be internationally based or homegrown. The DHS Office of State and Local Coordination was established to serve as a single point of contact for facilitation and coordination of departmental programs that impact state, local, territorial, and tribal governments. Through this office, DHS has brought together many organizations with a long history of interaction with, and support to, state, local, territorial, and tribal government organizations and associations, and the office is working hard to consolidate and coordinate that support. Today, this office facilitates the coordination of DHS-wide programs that impact state, local, territorial, and tribal governments; serves as the primary point-of-contact within DHS for exchanging information with state, local, territorial, and tribal homeland security personnel; identifies homeland security-related activities, best practices, and processes that are most efficiently accomplished at the federal, state, local, or regional levels; and utilizes this information to ensure that opportunities for improvement are provided to our state, territorial, tribal, and local counterparts.

The events of September 11 established the security of community infrastructure as a potential target for terrorist attacks. Community infrastructure has always been vulnerable to natural and other technological disaster events. So much so that FEMA's largest disaster assistance program, Public Assistance, is designed to fund the rebuilding of community infrastructure damaged by a disaster event. Local government officials and local emergency managers must now increase the attention they give to protecting and securing community infrastructure from a terrorist attack. They must also include in these preparedness efforts the local public health system. A checklist designed for the City of Boone (NC) as part of a Technological Annex developed for the town's All-Hazards Planning and Operations Manual in March 2002 is provided next.

GOALS AND QUESTIONS FOR LOCAL GOVERNMENTS PREPARING FOR TERRORISM— BOONE, NORTH CAROLINA

The preparedness and response of terrorist events requires that local governments do the following:

- Identify the types of events that might occur in the community.
- Plan emergency activities in advance to ensure a coordinated response.
- Build capabilities necessary to responsed effectively to the consequences of terrorism.
- Identify the type or nature of an event when it does happen.
- Implement the planned response quickly and efficiently.
- Recover from the incident.

The response to terrorism is similar in many ways to that of other natural or man-made disasters Boone has already prepared for. With additions and modifications, the development of a completely separate system can be avoided. Training and public education are vital, and understanding the federal assistance available will drastically increase local capacity before and during a terrorist attack.

The following are the general types of activities that Boone must undertake to meet the objectives just mentioned:

- Strengthen information and communications technology.
- Establish a well-defined incident command structure that includes the FBI.
- Strengthen local working relationships and communications.
- Educate healthcare and emergency response community about identification of bioterrorist attacks and agents.
- Educate healthcare and emergency response community about medical treatment and prophylaxis for possible biological agents.
- Educate local health department about state and federal requirements and assistance.
- Maintain locally accessible supply of medications, vaccines, and supplies.
- Address healthcare worker safety issues.
- Designate a spokesperson to maintain contact with the public.
- Develop comprehensive evacuation plans.
- Become familiar with state and local laws relating to isolation/quarantine.
- Develop or enhance local capability to prosecute crimes involving weapons of mass destruction or the planning of terrorism events.
- Develop, maintain, and practice an infectious diseases emergency response plan.
- Practice with surrounding jurisdictions/strengthen mutual agreement plans.
- Outline the roles of federal agency assistance in planning and response.
- Educate the public in recognizing events, and how to respond as individuals.
- Stay current.

Source: Town of Boone All-Hazards Planning and Operations Manual, Technological Hazards Annex, March 2002

CONCLUSION

Emergency management in the United States was changed forever by the events of September 11. New focus, new funding, new partners, and new concerns associated with the fight against terrorism are changing the way emergency management functions in this country every day. At the federal government level, the new Department of Homeland Security has been established, which includes FEMA and all the federal government disaster management programs. At the state level, governors and state emergency management directors are calling for better coordination, new communications technologies, and always more and more funding. At the local government level, terrorism is a new threat that greatly expands their facility security requirements and is added to a long list of needs and priorities. But the threat of terrorism is one that can't be ignored. Issues of coordination, communications, and funding concern local governments as well.

The United States has taken its typical response to a new problem. It has reorganized and committed huge amounts of funding to reducing the problem. The ability of the Department of Homeland Security to achieve an enhanced level of coordination is improving, but still has a ways to go. Preventing future terrorism attacks remains mostly outside the purview of DHS, residing with the intelligence community, the military, diplomatic corps, and law enforcement. What DHS can offer is a better prepared and equipped first responder cadre, enhanced transportation and border security, and more money for emergency management programs.

But the question of cost effectiveness will remain to be seen. The likelihood of natural and technological disasters has already proven to be far greater than that of terrorist attacks. In the four years following the September 11 terrorist attacks, the United States has been affected by hurricanes, floods, wildfires, chemical accidents, transportation accidents, volcanoes, ice storms, tornadoes, severe winter weather, avalanches—the list goes on and on. The Department of Homeland Security will need to continually reassess its priorities in terms of terrorism versus other less sinister hazards, and shift funding as appropriate. The terrorist threat will never go away completely, but over time, it should require much less of the attention of the nation's first responders, state responders, and federal government preparedness and response agencies.

CASE STUDY

"REDEFINING READINESS: TERRORISM PLANNING THROUGH THE EYES OF THE PUBLIC," A STUDY BY ROZ D. LASKER OF THE NEW YORK ACADEMY OF MEDICINE (SEPTEMBER 2004)

The Redefining Readiness Study, the first of its kind, measured how Americans might react to protective instructions in two terrorist attacks: a smallpox outbreak and the explosion of a dirty bomb. This information is considered critically important because the plans currently being developed to deal with these situations are based on expert *assumptions* about what people would be concerned about and how they would behave. If planners' assumptions about the public are wrong—

as they have been in the past—the plans being developed will not work as expected, and a large number of people who should be protected will be unnecessarily harmed.

This study included confidential, in-depth conversations with government and private-sector planners and an extensive review of the literature to identify the critical assumptions about the public on which current plans are based. A diverse spectrum of community residents around the country were engaged through 14 group discussions to identify a frame of reference for thinking about terrorism preparedness planning that is meaningful to the general public. Incorporating lessons learned, the research team fielded a telephone survey of 2,545 randomly selected adult residents of households in the continental United States.

The Study

The study uses scenarios that put people in smallpox and dirty bomb situations at a place and time they would be likely to hear about the attack and be told what to do. The smallpox scenario explores how people would react to instructions to go to a public site to be vaccinated if some residents in their community and people in other parts of the country became sick with smallpox after having been exposed to the virus in an attack at a major airport. The dirty bomb scenario explores how the public would react to instructions to stay inside a building other than their own home if a dirty bomb exploded a mile from where they were and a cloud containing radioactive dust were moving in their direction. In addition to these scenarios, the study also explored people's interest in, and perspectives about, their community's terrorism planning activities.

Following are highlighted key findings from the study report, focusing on:

- The public's reactions to the smallpox and dirty bomb situations
- The public's redefinition of readiness in these situations
- The public's role in future planning efforts

The Public's Reactions to the Smallpox and Dirty Bomb Situations

Far fewer people than needed would follow protective instructions in these two terrorist attack situations.

- Only two-fifths of the American people would go to the vaccination site in the smallpox outbreak.
- Only three-fifths of the American people would shelter-in-place for as long as told in the dirty bomb explosion.

One reason for this lack of cooperation is that many people would be seriously worried about something *other* than what planners are trying to protect them from.

- Two-fifths of the American people would be seriously worried about what government officials would say or do. This concern is even more prevalent among members of the public who are Hispanic, African-American, foreign-born, have a low income, lack health insurance coverage, live in New York City, or have not attended college. People's trust in official instructions and actions is important because people who don't have a lot of trust are only half as likely to cooperate in the smallpox and dirty bomb situations as are those who do.

- Three-fifths of the American people would have serious worries about the smallpox vaccine—that's *twice* as many people as would be seriously worried about catching smallpox in the outbreak situation.

  - Worries about vaccine side effects would make one-fifth of the American population afraid to follow instructions to go to the vaccination site. The public's worries appear to be well founded since it is estimated that over 50 million people in this country have conditions that put them at risk of developing serious complications from the vaccine, either from being vaccinated themselves or from accidentally coming in contact with someone who has recently been vaccinated.

  - Half of the American people—and two-thirds of African Americans—would be seriously worried if they were told that the smallpox vaccine is investigational. More people would be seriously worried about this issue than about any other aspect of the smallpox situation. Concern about the investigational status of the vaccine would make one-third of the population decide *not* to get it, even if they were at the vaccination site already.

Many people would face *conflicting* worries and trade-offs in these situations, which would make it very difficult for them to decide what the most protective course of action would be.

- Three-quarters of the people who would be seriously worried about catching smallpox in the outbreak situation *also* would be seriously worried about the vaccine. People who are *only* worried about catching smallpox are three times more likely to cooperate as those who are not. But that increase in cooperation is completely *eliminated* when people are also seriously worried about the vaccine.

- Two-thirds of the American people would try to avoid being in the same place with other people they don't know in the smallpox situation. But going to a public vaccination site *violates* people's inclination toward protective isolation.

- Two-fifths of the population would be afraid of catching smallpox from other people at the site. One-fifth would be afraid of coming in contact with people at the site who shouldn't be exposed to anyone who recently got the vaccine. In the dirty bomb situation, many people face conflicting obligations, and assuring the safety of people who are dependent on them is often more important than assuring their own safety. One-third of the people who would not cooperate fully in this situation would leave the shelter of their building in order to take care of their children; one-quarter would leave to take care of other family members.

A substantial number of people would be able to cooperate with protective instructions if certain conditions were met, but those conditions are *not* met now.

- Three-quarters of the people who said they would not fully cooperate with instructions to stay inside the building in the dirty bomb situation *would* do so if they could communicate with people they care about or if they knew that they and their loved ones were in places that had prepared in advance to take good care of them in this kind of situation. But three-fifths of the American population know only a little or nothing at all about how people would actually be cared for in those places. Overall, the American people are half as likely to cooperate in the dirty bomb situation if they don't know a lot about their building's shelter-in-place plans than if they do. And they are half as likely to cooperate if they lack confidence in their community's preparedness plans than if they don't.

Not surprisingly, considering the serious worries and trade-offs people face, many people would want more information or advice to decide what to do in these situations.

- Members of the public are looking for decision-making support, *not* just facts, and they want to be able to talk with someone beforehand, *not* just during an attack.
- For free telephone support from a trained person in the smallpox situation, considerably *more* people would find it very helpful to talk with someone who they know wants what is best for them (like their health practitioner) than to talk with someone they don't know who works for their local government.

The Public's Redefinition of Readiness in These Two Terrorist Attack Situations

These findings are cause for worry because they suggest that current plans to deal with smallpox and dirty bomb attacks will be far less effective than planners want or the public deserves. Although the study is based on a hypothetical scenario, our findings need to be taken seriously because the way the American people say they would react to instructions in the smallpox outbreak is consistent with the *actual behavior* of health care workers in the CDC Smallpox Vaccination Program. If three-fifths of the American people were reluctant to follow instructions in a smallpox outbreak, the protection of large-scale vaccination might not be achieved, even if planners worked out all of the challenging logistics involved in dispensing the vaccine. If two-fifths of the American population were reluctant to shelter-in-place in a dirty bomb explosion, many people would be unnecessarily exposed to dangerous dust and radiation, and first responders would face excess traffic and congestion in getting to the scene of the explosion.

Planners have been focusing a lot of attention on public education and risk communication, but our study suggests that informing people what they should do in these terrorist attack situations will *not* be sufficient to garner their timely cooperation. On a more optimistic note, the study shows how, by addressing the public's concerns, planners can develop more behaviorally realistic approaches for dealing with smallpox and dirty bomb attacks and, as a result, protect many more people than would otherwise be possible.

The report describes what plans to deal with these two kinds of terrorist attacks would look like if they incorporated the public's perspectives. As readers will see, looking at preparedness planning through the public's eyes redefines the notion of protection.

In the smallpox situation, the public's insights emphasize the importance of developing plans that protect *everyone* at risk: not only the people who are at risk of contracting smallpox, but also the large number of people who are at risk of developing serious complications from the vaccine.

In the dirty bomb situation, the public's insights emphasize that people not only need to be protected from dangerous dust and radiation. They also need to know that they and their loved ones would be safe and cared for in whatever building they happen to be in at the time of an explosion. To make such protection possible, a broad array of places—work sites, shops, malls, schools, day care centers, hospitals, clinics, cultural institutions, recreational and entertainment facilities, government buildings, apartment buildings, and transportation terminals—have to be able to serve as safe havens for people should the need arise. The managers of these places need to recognize that it is as important to prepare to serve as a safe haven as to be able to evacuate people in the event of a fire or an explosion.

The American people's perspectives also redefine how public protection can best be achieved. To a large extent, this involves the development of community and organizational plans that address people's concerns, minimize the conflicts and trade-offs they would face, and support them in choosing the best protective action.

As the plans in the report illustrate, many of these actions need to be taken *now*, well before an attack occurs.

Examples related to the smallpox situation include:

- Strategies to enable everyone in the country to determine their own vaccine risk status and that of the other members of their household
- The training of healthcare practitioners and other community members to provide people with decision-making support
- The involvement of community leaders—particularly among the African-American population—in overseeing the development and testing of vaccines

Examples related to the dirty bomb situation include the development of:

- Confidence-generating safe-haven plans in the broad array of buildings and places where people frequently are
- Back-up systems that can maintain telephone and e-mail service for the general public in the event of a large-scale emergency

The plans also involve changes in actions that would be taken *during* the crisis, as the following strategies from the smallpox plan illustrate.

- Rather than triaging or screening people at public sites, steps would be taken to make sure that anyone who is potentially infected or exposed to smallpox

or who is at risk of developing serious complications from the vaccine stays home and does *not* go to any public vaccination site.
- To provide people with accurate information from people they trust, government-run telephone networks would be supplemented with a more community-embedded telephone support system.

Finally, the plans emphasize the need for communities and the nation to focus on *long-term issues*. Reflecting the public's concerns in the dirty bomb situation, for example, the plans emphasize the need to discuss and address the potential environmental, economic, and health consequences that might ensue.

The Public's Role in Future Planning Efforts

The Redefining Readiness Study documents the value of letting the American people speak for themselves rather than relying on planners' untested assumptions about what the public cares about and how the public will behave. Moreover, the study provides planners in government agencies and private-sector organizations with reliable information from the public that they can use to assess and strengthen their plans to deal with terrorist-initiated smallpox outbreaks and dirty bomb explosions. Because most of the findings in the study are generalizable, they are applicable to planning efforts throughout the country. Some of the strategies in the study's illustrative plans are also applicable to other situations, such as an outbreak of pandemic influenza or SARS, an electrical blackout, or the malfunction of a nuclear reactor.

Planners will need to work with community residents directly, however, to benefit from their insights about responding to many other kinds of terrorist attack and emergency situations. The study documents that involving people in these kinds of planning efforts can accomplish another important objective as well: it can address the trust and confidence issues that currently discourage so many members of the public from following protective instructions.

- The study shows that people are more likely to follow official instructions when they have a lot of trust in what officials tell them to do and are confident that their community is prepared to meet their needs if a terrorist attack occurs. Currently, the American people's trust and confidence levels are disturbingly low. But elected officials, government agencies, and private-sector organizations have a *unique opportunity* to build the public's trust, confidence, and cooperation by involving the public directly in planning. When community members are part of the planning process, they can be more confident that planners are actually aware of their concerns. When community residents play a role in developing protective strategies, they can be more trusting of officials who instruct them to follow those strategies.

Thus far, the public has had little or no direct involvement in community and organizational preparedness planning. The study documents that only a tiny fraction of the American people know very much about the plans that are being developed in their communities, and it paints a mostly discouraging picture about people's perceptions about current planning activities.

- Large proportions of people think their community isn't prepared to deal with these kinds of terrorist attacks, that planners don't know about their concerns and information needs, that people like them can't influence the plans that are being developed, and that neither they nor the people they care about would receive the help they need when they need it if a terrorist attack were to occur.
- People's perceptions about the potential benefits of planning are in stark contrast to the problems they see. Three-fifths of the American population believe that the harm caused by a terrorist attack in their community could be reduced a great deal or a lot by preparing ahead of time to deal with the effects.

Fortunately for planners, the study documents that a large proportion of the American people are interested in community-level planning—not just in learning more about plans, but in being actively involved in developing them.

- In New York City and Washington, D.C., where terrorism is a particularly salient issue, two-fifths of the population are extremely or very interested in *personally* helping a government agency or other community organization develop plans to deal with these kinds of attacks.
- But interest levels are also high in the rest of the country, where people think much less about terrorist attacks and believe the possibility of an attack is much less likely. In those places, one-third of the population has a strong personal interest in participating in planning.

The next challenge is to make it possible for government agencies and private-sector organizations to engage the public in planning efforts. Our study demonstrates that to make participation meaningful and worthwhile to community residents, the process needs to assure them considerable influence in planning and needs to focus their involvement on identifying and addressing the issues they care about a lot. We recognize that this kind of inclusive process would entail a substantial change in the way many planners currently go about their work and that there are a variety of barriers that currently make it difficult for planners to move in this direction. Nonetheless, the stakes are too high to continue the *status quo*. To provide planners with practical models for engaging the public in these kinds of activities, our next step will be to support planning processes in selected sites around the country that demonstrate exactly how community residents can be meaningfully and feasibly engaged.

Source: Lasker, Roz. 2004. "Redefining Readiness: Terrorism Planning Through the Eyes of the Public," Center for the Advancement of Collaborative Strategies in Health. The New York Academy of Medicine, September 14.

10. The Future of Emergency Management

INTRODUCTION

Justice Oliver Wendell Holmes said, "the great thing about this world is not so much where we stand as in what direction we are moving." In the aftermath of September 11, the discipline of emergency management is at a critical crossroads. Emergency managers are faced with new threats, new responsibilities, and new opportunities.

The potential for biochemical terrorist strikes, mass casualty events, and cyberspace attacks still exists. Providing protection to our first responders and to the general public from a myriad of unknown and unpredictable technological hazards is a daunting responsibility. Accepting this responsibility and wisely applying the lessons learned from emergency management practices and policies of the past represents both the challenge and the opportunity for emergency managers.

This chapter explores issues concerning the current political and organizational environment for emergency management. The chapter closes with the authors' opinions on what emergency management must do to survive and grow in this new environment.

ORGANIZATIONAL CHANGES

In the Spring of 2003, the Bush administration and Congress were focused on reorganization and increased appropriations to respond to the threat of terrorism. The new Department of Homeland Security was created, which consolidated various federal agencies and programs with some responsibility for terrorism, including U.S. Border Patrol, Immigration and Naturalization Service, FEMA, Coast Guard, and a few other discrete programs. The reorganization did not include any of the intelligence, diplomatic, or law enforcement programs that are at the center of government efforts at preventing terrorism. By including FEMA, the state and local emergency management structure of the United States has been integrated into the new department.

Significant new appropriations for terrorism and for state and local emergency management dominated the emergency management response to these reorganization initiatives. The U.S. emergency management system, at all levels, has been underfunded for decades.

In the early debate over the Nunn-Lugar antiterrorism legislation, emergency managers and other first responders, particularly the fire community, were lobbying for additional resources to prepare for possible terrorist attacks. It is unfortunate that during these discussions the fire and emergency management communities did not form a partnership to present a collective argument for their needs because it might have worked. Instead, the traditional rivalry between these two groups, both of whom believe they are the most critical first responder, prevailed. The law enforcement community, on the other hand, presented a unified front. As a result, most of the Nunn-Lugar appropriated funds went to support the Department of Justice, FBI, and local law enforcement.

Later, the fire community was successful in establishing a new grant program to upgrade the deteriorating U.S. fire response infrastructure. The fire unions were responsible for getting these funds, and the terrorism threat was only one small part of their rationale. In the post September 11 environment, it was obvious that funding for terrorism-related activities was going to be a high-priority competition.

The National Emergency Management Association (NEMA) and the International Association of Emergency Managers (IAEM) endorsed the inclusion of FEMA in the new Department of Homeland Security. NEMA represents the State Directors of Emergency Management, and IAEM represents the locals. In spite of this endorsement, in most of the states, the governors have designated individuals other that the State Directors of Emergency Management as their agent or czar for Homeland Security. Other than the lure of money, it is hard to understand why the states would take this position. Because FEMA has lost stature and influence since it is no longer an independent agency and the director of FEMA is no longer part of the President's Cabinet, so will the state emergency management organizations. There was no assurance that federal funding for homeland security, when sent to the states, would be controlled by state emergency managers.

WHAT DOES THIS MEAN FOR EMERGENCY MANAGEMENT?

It will take years to sort things out, but let's look at some of the changes that have already taken place. The implications for federal emergency management efforts are numerous. In the new Department of Homeland Security (DHS), FEMA has become a directorate headed by an under secretary who reports up through a deputy secretary to the DHS secretary.

The direct authorities, who in the past were vested in the director of FEMA, are now vested in the DHS secretary. Responsibilities for recommending disaster declarations to the President and for coordination of the federal response to natural and technological disasters or emergencies have been retained by the DHS secretary. This has impacted the timeliness, effectiveness, and operational abilities of the current FEMA operations and staff. The stature and authorities of the leader of federal emergency management activities have been diminished.

The principal mission of DHS is to prevent a terrorist attack from occurring. The focus of the department's major directorates and organizations is to prevent terrorists from entering the United States and attacking its citizens. DHS and the Office of Management and Budget (OMB) have both indicated that FEMA has no role in this mission. FEMA, as configured in the Department's Emergency Preparedness and

Response Directorate (EPR), has no responsibilities in interdicting terrorists. FEMA's mission is to deal with the consequences of future terrorist attacks. This incompatibility of missions puts FEMA at odds with the rest of the department and has led to FEMA being marginalized within DHS.

At the same time, DHS is suffering from the expected growing pains. Most organizations within DHS continue to function independently with little coordination with other DHS organizations. In late 2004, the department suffered its first change in leadership with the departure of the first DHS secretary, Tom Ridge, Deputy Secretary Admiral Jim Loy, and nearly all the undersecretaries responsible for managing DHS's five directorates. It took several months to replace Secretary Ridge and will likely take many more months to fill the remaining senior positions. Such dramatic changes of leadership are difficult for a mature agency, much less one barely two years old. One of the principal causes of FEMA's dysfunction in its early years was near-constant changes in leadership, which led to endless reorganizations and changes in priorities as each new leader attempted to put his or her stamp on the agency. A dysfunctional DHS will continue to have a negative impact on FEMA and emergency management in this country.

Another impact has been the competition for resources among the various organizations within the new department. It is unlikely that the emergency management contingent will be effective in arguing for resources when up against organizations three and four times their size, such as the Immigration and Naturalization Services (INS). The increases in terrorism monies that are potentially flowing to emergency management can evaporate quickly in the absence of terrorist events or rescission of federal spending across the board. These impacts, if they are realized, will certainly extend to the states. They most assuredly will be felt at the local emergency management level, where we already see states using federal support designated for local efforts as offsets to state budget shortfalls.

WHAT IS THE FUTURE OF EMERGENCY MANAGEMENT?

In the first edition of this textbook, published in Spring 2003, we wrote, "We are optimistic that emergency management can survive and thrive in the future if it embraces the lessons learned from the past and moves forward with a progressive agenda that will be valued by the American people."

We believed that the likelihood of a major natural disaster—flood, hurricane, or earthquake—affecting our communities was inevitable. As emergency management systems focus their efforts on preparing for and responding to terrorist events, these efforts should not diminish their capabilities or capacity for dealing with natural hazards if they heeded the following lessons.

Lesson One

Maintain an all-hazards approach to emergency management. Applying this approach takes advantage of the common capabilities necessary to treat any type of disaster or emergency, but allows for incorporating the special needs of terrorism. To abandon the all-hazards approach would be repeating the mistake the emergency management community made in the 1980s. During the era of the cold war, FEMA concentrated more than 75 percent of its financial and human resources on preparing for the next nuclear war. It mandated that states and localities receiving FEMA funding follow suit.

Federal, state, and local capacity to respond to natural disasters was severely diminished. As Hurricanes Hugo, Iniki, and Andrew vividly demonstrated, state and local capacities were quickly overcome. The federal response under FEMA was disorganized and late. In the case of Andrew, the director of FEMA was replaced as the in-charge official and the military provided most of the initial support. This real example of the folly of focusing on any one threat, at the cost of more frequent and widespread threats, provides strong evidence of the wisdom of the all-hazards approach to emergency management.

Lesson Two

The federal response infrastructure, based on the Federal Response Plan, works. Since September 11, many political leaders have called for building a terrorism response structure, forgetting that an effective federal structure already exists. There is no need to build a new infrastructure. This approach was tested in hundreds of natural events and the Oklahoma City bombing—a terrorist event—and it worked. This proven structure is flexible; it needs modification and the addition of new partners to accommodate the unique aspects of terrorism, but the emergency management community should fight any attempts to build a separate structure.

At the state and local levels, state plans and the emergency management compacts that exist between states support this operational approach. Specific lessons learned from September 11, particularly in communications and joint operations, can be readily incorporated into these existing structures.

Lesson Three

Continue to practice the concepts that facilitated the U.S. emergency management system becoming the best system in the world. These concepts are (1) focus on your customers, both internal and external; (2) build partnerships among disciplines and across sectors, including the private sector and the media; (3) support development and application of new technologies to give emergency managers the tools they need to be successful; (4) emphasize communications to partners, the public, and the media; and (5) make mitigation the cornerstone of emergency management.

These simple, commonsense concepts were the key to the respect and success FEMA achieved under Director James Lee Witt and President Clinton. We believe they provide the framework for emergency management to continue to grow and expand its influence and importance to the institutions and people it serves. Emergency management can ensure its place in the future if it focuses on policies, programs, and activities that improve the safety and social and economic security of individuals, institutions, and communities. To do this, emergency management must focus more effort in promoting and implementing mitigation.

Lesson Four

Make mitigation the focus of emergency management in the United States. Mitigation is the positive function that emergency managers can practice everyday, in every community, and not be dependent on an event to prove their value. Mitigation is practiced by all sectors of a community. To be effective, it requires developing partnerships within a community and often brings together disparate parties to solve common problems. Mitigation brings the private sector into the emergency management system because economic sustainability of their businesses depends on risk reduction, so mitigation promotes their support and leadership. Mitigation provides the entry point to involve the private sector in other phases of emergency management and to understand their unique needs in response and recovery.

In the late 1990s, business continuity and mitigation planning was the largest growth area for emergency management. Economic considerations or interest often drive public decisions. Mitigation allows emergency managers to have access and influence to the decision-making process. Mitigation works best at the local level and provides that grassroots constituency that can exert political pressure for continued emergency management support. The Project Impact initiative articulated this concept and made it a reality in more than 225 communities. The Bush administration recognized this by including the words "building disaster-resistant communities" in the objectives for the new Department of Homeland Security.

As this text is being written (in the Spring of 2005), it is apparent that these lessons for the most part have not been heeded by DHS and FEMA. While professing to adopt an all-hazards approach, in reality DHS/FEMA has become focused almost exclusively on the terrorist threat to the near exclusion of traditional natural and technological hazards. Existing funding and staff resources have been reprogrammed at DHS/FEMA to terrorism-based activities, and new resources are being applied almost exclusively to this threat. FEMA's attention has been effectively diverted from any hazard beyond terrorism.

The Federal Response Plan (FRP) has been has been revised into the National Response Plan (NRP). Many of the positive features of the FRP have been retained in the NRP but the focus has been shifted dramatically to responding to a terrorist attack and in doing so, the role of the federal government has been altered dramatically. The NRP places the lead role in responding to major disasters, terrorist attack, or natural disaster, in the hands of federal officials. In fact, the federal government now has the authority to respond to an event in a state without a request by the governor. This alters the traditional role of the federal government, through the FRP, supporting the actions of state and local government. This is a drastic change in the way major disaster events have been handled successfully in the past.

The concepts that made the U.S. emergency management system the best in the world have been compromised severely in recent years. Customer focus, partnerships, and communicating with the public have become secondary to the federal government's response to major disasters, as evidenced by the response to the 2004 hurricanes in Florida. In addition, the resources and programs available to individuals and communities to help them recover have been reduced, and in some cases, eliminated. Technology development has been focused almost exclusively on the terrorist threat.

Finally, mitigation once again has been marginalized. Funding for many of the natural hazard programs has been reduced or taxed to support other DHS functions such as the Transportation Security Administration (TSA). The funding available for the Hazard Mitigation Grant Program (HMGP) has been reduced and the match requirements for the state and local share increased. Project Impact has been eliminated, though it still prospers at the local level in places like Seattle and Tulsa, and the Pre-Disaster Mitigation Program has not proved to be an adequate replacement. FEMA has ceded lead agency responsibility for the National Earthquake Hazard Reduction Program (NEHRP) and it is likely that the National Flood Insurance Program will also migrate from FEMA to another federal agency, possibly the Department of Housing and Urban Development (HUD).

So what happens now? In the coming years, we believe emergency management planners will confront issues in several critical areas, specifically: finding a balance between homeland security and natural disaster management; involving the public in preparedness planning; establishing an effective partnership with the nation's business community; prioritizing resource allocations; and, dealing with the evolving organizational structure of the nation's emergency management system.

BALANCING HOMELAND SECURITY AND NATURAL DISASTER MANAGEMENT

Since the September 11 terrorist attacks, the focus of government emergency management planners, especially at the federal level, has been on terrorism and the new hazards presented by this new threat. This is not the first time that emergency management planners have focused on national security risks.

In the 1950s, the nation's Civil Defense system was developed to address the threat of nuclear attack by the Soviet Union. The government officials who staffed the Civil Defense programs at all levels of government were the nation's first emergency managers. In the 1980s, the newly formed Federal Emergency Management Agency (FEMA), reflecting the priorities of the Reagan Administration, focused its programs and resources almost exclusively on nuclear attack and continuity of government planning.

A series of major natural disasters in the late 1980s and early 1990s (Hurricane Hugo, Loma Prieta Earthquake, Hurricane Iniki, and Hurricane Andrew) exposed the inability of FEMA and the federal government to provide adequate support to state and local emergency managers in responding to large natural disasters.

In the 1990s, FEMA adopted an all-hazards approach to disaster management that resulted in increased resources for natural hazards preparedness and mitigation programs and the development and implementation of the Federal Response Plan that coordinated the efforts of 27 federal agencies and the Red Cross in support of state and local emergency managers.

It appears that history is repeating itself. In spite of a dramatic hurricane season in 2004, evidence of the impacts of global warming, and the forecast for continued severe weather, most of the resources for emergency management planning currently are devoted to terrorism much the way they were to nuclear attack planning in the 1980s. In 2005, emergency management planners at the federal level, at FEMA, and within the Department of Homeland Security must consider how to balance their focus on the terrorist threat with an all-hazards approach to disaster management.

PUBLIC INVOLVEMENT IN PREPAREDNESS PLANNING

Historically, the general public has played a limited, if any, role in the development of emergency management preparedness and mitigation plans. The principal focus of public outreach efforts by the early Civil Defense programs, FEMA's National Flood Insurance Program, and the Red Cross family preparedness programs was to inform and to educate the public. Rarely has the public been included in the actual planning process.

This began to change in the late 1990s, when FEMA launched its national mitigation initiative, Project Impact, which called for the full involvement of all members of the community in developing a community hazard mitigation strategy. Communities such as Tulsa, Oklahoma, and Napa, California, successfully developed and implemented flood mitigation projects with comprehensive public involvement in the planning process.

Recent research conducted by the New York Academy of Medicine indicates that the public is ready to take a more active role in preparedness planning for terrorism events. This research indicates that the current plans will fail because the assumptions about public behavior in the event of a terrorist incident are false. The research found that emergency management planners must engage the public in the planning process in order to fully understand the public's needs and concerns and that the public is vitally interested in getting involved in this process.

PARTNERING WITH THE BUSINESS COMMUNITY

The Department of Homeland Security (DHS) and numerous business groups, such as The Business Roundtable, acknowledge that an effective partnership between government and business must be established as part of the nation's homeland security efforts. This makes complete sense since almost 85 percent of the infrastructure in this country is privately held.

However, in the three and a half years since the September 11 attacks, no such partnership has been established. There has been some progress and cooperation but there is no overall strategy in place to incorporate the business sector into the government's emergency management planning for homeland security.

There are numerous issues that must be resolved before such a strategy can be designed and implemented. A significant issue that must be resolved is how the government protects and uses confidential information it is asking the business community to provide. The business community must be included in the planning process not only for terrorism planning but also for natural disaster management.

PRIORITIZING RESOURCE ALLOCATIONS

The struggle for funding for emergency management programs and activities has only intensified since September 11. At the federal level, funding for traditional natural and technological hazard programs at FEMA have been cut significantly and funding for hazard mitigation programs such as Project Impact have been cut completely. There have been efforts to cut funding to state and local emergency management organizations for personnel and to limit the funds available for post-disaster mitigation projects. On the positive side, there has been increased funding for first responders and the development of community homeland security plans.

At the state and local levels, the struggle to fund emergency management programs and activities continues. Each state has established a homeland security office and in most cases this office is headed by someone other than the State Director for Emergency Management. Numerous large cities also have established homeland security offices that function in parallel to the emergency management office.

The question facing emergency management planners in 2005 is how long resources will be available to sustain two discrete functions—one for homeland security and one for emergency management. How the federal, state, and local governments prioritize and allocate their resources will likely make this decision for them.

ORGANIZATION OF THE NATION'S EMERGENCY MANAGEMENT SYSTEM

Prior to September 11, the nation's emergency management system was composed of a partnership between federal, state, and local government, and a collection of nongovernmental organizations (NGOs) such as the Red Cross and the Salvation Army. The federal government, through FEMA, provided funding, technical assistance, and support to the states and through the states, to the local governments.

With the inclusion of FEMA in the Department of Homeland Security and the focus being placed squarely on terrorism, the structure of the national emergency management system has changed. The director of FEMA no longer reports directly to the President and DHS/FEMA has assumed a more active role in leading the government-wide response to all disasters, terrorism or natural.

At the same time, DHS continues to struggle as an organization and in 2005 will undergo its first change in leadership. It took FEMA nearly 15 years to become a functioning federal agency. How long it takes DHS to become fully functional remains to be seen.

For emergency management planners, this uncertainty in the organizational structure of the system will impact what they do as priorities shift, resources become tighter, and leadership at the top changes. This uncertainty is something they will have to deal with in 2005 and most likely in the years to follow.

A NEW PATH FOR EMERGENCY MANAGEMENT

How emergency managers deal with these new issues will shape the path that emergency management takes as a discipline in the years to come. We still believe that emergency managers must embrace those lessons we noted in Spring 2003, including maintaining an all-hazards approach, maintaining the Federal Response Plan, and practicing the five core principals that made the U.S. emergency management system the best in the world: focusing on customers, building partnerships, developing new technologies, communicating to the public and partners, and making mitigation the cornerstone of emergency management.

With the exception of the new National Response Plan, which is built on the framework and operational structure of the old Federal Response Plan, DHS and FEMA effectively have rejected these lessons. They have once again returned to a single hazard focus as they did in the 1950s and the 1980s with nuclear attack planning. They have made fighting terrorism principally a federal effort that focuses more on the needs of the federal government than on its customers, the American people. It has done little to form new partnerships, especially with the business community, and has adopted its former position of communicating less rather than more to the public and its partners.

For DHS this is consistent with the department's stated mission of preventing future terrorist attacks. This is a direct federal responsibility. However, FEMA's role in DHS and in fighting terrorism is to be prepared to respond to the next event, assist in the recovery, and, most importantly we believe, reduce the impacts of the next terrorist attack through mitigation. DHS has stated that FEMA's mission in not compatible with the DHS mission.

So where does emergency management go from here? At the federal level the trend seems to be the reduction and loss of all natural hazard programs, diminishing investments in preparedness and elimination of hazard mitigation. What will remain are the response and recovery programs that were sorely tested in Florida during the hurricanes in 2004. In 2005, Michael Chertoff, an appellate court judge and former undersecretary at the Department of Justice, was named the new Secretary of Homeland Security. Since his appointment, he has issued statements that indicate he recognizes the important role that FEMA plays in DHS. His first opportunity to back these statements with actions will be during the FY2006 appropriations cycle, in which funding for state and local emergency management has once again been reduced.

If emergency management has a future at all, we believe that it must concentrate and rebuild its constituency at the community level. The time has come for communities to incorporate disaster management and hazard mitigation into its everyday operations, planning, and decision-making. It is also time for communities to establish a local funding source for emergency management.

A new breed of government official will need to be hired to manage this new aspect of community government. This new official should be trained in public policy, public administration, and hazard management. This new official should be responsible for integrating hazard mitigation and disaster management policies and practices in all phases of local government and community life. This new official would report directly to the city or county manager and work closely with and on the same level as other major department heads in the local government. This new official would also be responsible for creating a community partnership for disaster management that includes the business community and all other community stakeholders.

This new official would guide the community through a consensus building process to determine all the risks to the community, to identify what can be done to mitigate these risks, to develop a prioritized plan for mitigating these risks, and to work with government and business leaders, community leaders, and the general public to generate the financial, political, and public resources needed to implement and maintain this plan.

We believe it will be at the community level that local government officials, with support from the business community and other elements of the community, will begin the process of reshaping the emergency management system in this country accounting for all hazards, including terrorism.

CONCLUSION

Whether the emergency management establishment will embrace this path in the future is debatable. Historic trends indicate otherwise; however, throughout the 1990s a new breed of emergency management professionals began to emerge. These individuals were anxious to bring a fresh face to the profession and embraced new strategies for promoting sound emergency management practices, particularly mitigation. The future of emergency management may rest on their ability to balance the new demands of the terrorism threat with the real need to make a difference in the quality of people's lives and their community's sustainability through mitigation.

Appendix A

Acronyms

AAR	After-Action Report
AEC	Agency Emergency Coordinator
AFRO	African Regional Office (WHO)
AOA	Administration on Aging
AOR	Areas of Responsibility (DoD)
ARC	American Red Cross
ARES	Amateur Radio Emergency Services
BHR	Bureau for Humanitarian Response (USAID)
B-NICE	Biological, Nuclear, Incendiary, Chemical, and Explosive (Weapons)
CARE	Cooperative for Assistance and Relief Everywhere
CAT	Crisis Action Team
CBDG	Community Development Block Grant
CBRN	Chemical, Biological, Radiological, and Nuclear (Weapons)
CBRNE	Chemical, Biological, Radiological, Nuclear, and Explosive (Weapons)
CCP	Crisis Counseling Assistance and Training Program
CCP	Casualty Collection Point
CCP	Citizens Corps Program
CDC	Centers for Disease Control and Prevention, U.S. Public Health Service
CDRG	Catastrophic Disaster Response Group
CENTCOM	Central Command (DoD)
CEPPO	Chemical Emergency Preparedness and Prevention Office
CERCLA	Comprehensive Environmental Response, Compensation, and Liability Act
CFA	Catalog of Federal Domestic Assistance
CHE	Complex Humanitarian Emergency
CJTF	Commander for the Joint Task Force (DoD)
CMHS	Center for Mental Health Services
CMOC	Civil/Military Operations Center (DoD)
CMT	Crisis Management Team
CNN	Cable News Network
CRC	Convention on the Rights of the Child
CRC	Crisis Response Cell
CRM	Crisis Resource Manager
CRS	Catholic Relief Services

DAE	Disaster Assistance Employee
DART	Disaster Assistance Response Team (USAID)
DEA	Drug Enforcement Agency
DCE	Defense Coordinating Element
DCO	Defense Coordinating Officer
DCSA	Defense Support of Civil Authorities
DEST	Domestic Emergency Support Team
DFO	Disaster Field Office
DHHS	Department of Health and Human Services
DHS	Department of Homeland Security
DMAT	Disaster Medical Assistance Team
DMORT	Disaster Mortuary Response Team, National Disaster Medical System
DMTP	Disaster Management Training Programme
DoD	United States Department of Defense
DOJ	Department of Justice
DOL	Department of Labor
DOT	Department of Transportation
DRC	Disaster Recovery Center
DRD	Disaster Response Division
DRRP	Disaster Reduction and Recovery Programme
DUA	Disaster Unemployment Assistance
EAS	Emergency Alert System
EC	Emergency Coordinator
ECHO	European Community Humanitarian Organization
ECS	Emergency Communications Staff
EDA	Economic Development Administration
EGOM	Empowered Group of Ministers (India)
EICC	Emergency Information and Coordination Center
EMPG	Emergency Management Performance Grants
EMRO	Eastern-Mediterranean Regional Office (WHO)
EMS	Emergency Medical Services
EOC	Emergency Operations Center
ERC	Emergency Response Coordinator (UN)
ERCG	Emergency Response Coordination Group, Public Health Service/Centers for Disease Control and Agency for Toxic Substances and Disease Registry
ERD	Emergency Response Division (UNDP)
ERL	Emergency Recovery Loan (WBG)
ERT	Emergency Response Team
ERT-A	Emergency Response Team Advance Element
ERT-N	National Emergency Response Team
ERU	Emergency Response Unit (IFRC)
ESF	Emergency Support Function
EST	Emergency Support Team
EUCOM	European Command (DoD)

EURO	Regional Office for Europe (WHO)
FAA	Federal Aviation Administration
FACT	Field Assessment and Coordination Team (IFRC)
FAO	Food and Agriculture Organization
FBI	Federal Bureau of Investigation
FCO	Federal Coordinating Officer
FECC	Federal Emergency Communications Coordinator
FEMA	Federal Emergency Management Agency
FERC	FEMA Emergency Response Capability
FESC	Federal Emergency Support Coordinator
FFP	Office of Food for Peace (BHR)
FHA	Foreign Humanitarian Assistance (DOD)
FHWA	Federal Highway Administration
FIRST	Federal Incident Response Support Team
FOC	FEMA Operations Center
FRC	Federal Resource Coordinator
FRERP	Federal Radiological Emergency Response Plan
FRN	FEMA Radio Network
FRP	Federal Response Plan
FSA	Farm Service Agency
GSN	Global Seismographic Network
HAO	Humanitarian Assistance Operations (DoD)
HAST	Humanitarian Assistance Survey Team (DoD)
HAZUS	Hazards—US (FEMA Consequence Modeling System)
HET-ESF	Headquarters Emergency Transportation Emergency Support Function
HHS	Department of Health and Human Services
HSAS	Homeland Security Advisory System
HSEEP	Homeland Security Exercise and Evaluation Program (ODP)
HSOC	Homeland Security Operations Center
HSPD	Homeland Security Presidential Directive
HUD	Department of Housing and Urban Development
IAEM	International Association of Emergency Managers
IASC	Inter-Agency Standing Committee
IBRD	International Bank for Reconstruction and Development (WBG)
ICPAE	Interagency Committee on Public Affairs in Emergencies
ICRC	International Committee of the Red Cross
ICP	Incident Command Post
ICS	Incident Command System
ICVA	International Council for Voluntary Agencies
IDA	International Development Association (WBG)
IDNDR	International Decade for Natural Disaster Reduction (UN)
IDP	Internally Displaced Persons
IFC	International Finance Corporation (WBG)
IFG	Individual and Family Grant
IFRC	International Federation of Red Cross/Red Crescent Societies

IHP	Individuals and Households Program
IIMG	Interagency Incident Management Group
IMD	Indian Meteorological Department
IMF	International Monetary Fund
IMT	Incident Management Team
INS	Immigration and Naturalization Service
IO	International Organization
ISCID	International Centre for Settlement of Investment Disputes (WBG)
ISDR	International Strategy for Disaster Reduction (UN)
JCS	Joint Chiefs of Staff (DoD)
JFO	Joint Field Office
JIC	Joint Information Center
JOC	Joint Operations Center
JTF	Joint Task Force (DoD)
JTTF	Joint Terrorism Task Force
MACC	Multi-agency Command Center
MIGA	Multilateral Investment Guarantee Agency (WBG)
MMRS	Metropolitan Medical Response System
MOA	Memorandum of Agreement
MOU	Memorandum of Understanding
MSF	Medecin sans Frontiers
NACo	National Association of Counties
NASA	National Aeronautics and Space Agency
NCA	National Command Authority (DoD)
NDMOC	National Disaster Medical Operations Center
NDMS	National Disaster Medical System
NDMSOSC	National Disaster Medical System Operations Support Center
NEHRP	National Earthquake Hazard Reduction Program
NEIC	National Earthquake Information Center
NEMA	National Emergency Management Association
NEP	National Exercise Program (ODP)
NEPEC	National Earthquake Prediction Evaluation Council
NGO	Nongovernmental Organization
NIMS	National Incident Management System
NIRT	Nuclear Incident Response Team
NIST	National Institute of Standards and Technology
NMRT	National Medical Response Team
NOAA	National Oceanic and Atmospheric Administration
NPSC	National Processing Service Center
NRC	Nuclear Regulatory Commission
NRCC	National Response Coordination Center
NRT	National Response Team
NRP	National Response Plan
NSF	National Science Foundation
NSSE	National Security Special Event
NSEP	National Security Emergency Preparedness

NVOAD	National Voluntary Organizations Active in Disaster
OCHA	Office for the Coordination of Humanitarian Affairs
ODP	Office for Domestic Preparedness
OEP	Office of Emergency Preparedness, U.S. Public Health Service
OET	Office of Emergency Transportation
OFDA	Office of U.S. Foreign Disaster Assistance
OPA	Office of Public Affairs
OS	Operation Support (OFDA)
OSC	On-Scene Coordinator
OSTP	White House Office of Science and Technology Policy
OTI	Office of Transition Initiatives (BHR)
PACOM	Pacific Command (DoD)
PAHO	Pan-American Health Organization (WHO)
PAO	Public Affairs Officer
PFO	Principal Federal Official
PK/HA	Office of Peacekeeping and Humanitarian Affairs (DoD)
PM	Office of Political/Military Affairs (DoD)
PMPP	Prevention, Mitigation, Preparedness, and Planning (OFDA)
PNP	Private Nonprofit
PRM	Bureau of Population, Refugees, and Migration (USAID)
PS	Program Support (OFDA)
PSA	Public Service Announcement
PSYOPS	Psychological Operations (DoD)
PVO	Private Voluntary Organization
QIP	Quick Impact Project (UNHCR)
RACES	Radio Amateur Civil Emergency Services
RDD	Radiological Dispersion Device
REACT	Radio Emergency Associated Communication Team
REC	Regional Emergency Coordinator
RECC	Regional Emergency Communications Coordinator
RECP	Regional Emergency Communications Plan
RET	Regional Emergency Transportation
RETCO	Regional Emergency Transportation Coordinator
RMT	Response Management Team (OFDA)
ROC	Regional Operations Center
ROE	Rules of Engagement (DoD)
ROST	Regional Operations Support Team
RRT	Regional Response Team
SAMHSA	Substance Abuse and Medical Health Services Administration
SAR	Search and Rescue
SAC	FBI Senior Agent-in-Charge
SBA	U.S. Small Business Administration
SCO	State Coordinating Officer
SEARO	South-East Asia Regional Office (WHO)
SFHA	Special Flood Hazard Areas
SFLEO	Senior Federal Law Enforcement Official

SHSP	State Homeland Security Program (ODP)
SIOC	Strategic Information and Operations Center
SITREP	Situation Report
SOCOM	Special Operations Command (DoD)
SOUTHCOM	Southern Command (DoD)
START	Scientific and Technical Advisory and Response Team
TAG	Technical Assistance Group (OFDA)
TOPOFF	Top Officials Terrorism Exercise (biennial)
TRADE	ODP Training and Data Exchange Group
TRANSCOM	Transportation Command (DoD)
UASI	Urban Areas Security Initiative
UN	United Nations
UNDAC	UN Disaster Assessment and Coordination
UNDP	United Nations Development Programme
UNFPA	United Nations Populations Fund
UNHCR	United Nations High Commissioner for Refugees
UNHRD	UN Humanitarian Response Depot
UNICEF	United Nations Children's Fund
US&R/USAR	Urban Search and Rescue
USACE	United States Army Corps of Engineers
USACOM	United States Atlantic Command (DoD)
USAID	United States Agency for International Development
USDA	United States Department of Agriculture
USGS	United States Geological Survey
VMAT	Veterinarian Medical Assistance Team
WB	World Bank
WBG	World Bank Group
WFP	World Food Programme
WHO	World Health Organization
WMD	Weapons of Mass Destruction
WTC	World Trade Center
ZECP	Zone Emergency Communications Planner

Appendix B

Emergency Management Web Sites

Emergency Type	Contact Agency	Web Site
All hazards	The Federal Emergency Management Agency (FEMA)	www.fema.gov
Hazardous materials, chemical accidents, oil spills	Environmental Protection Agency (EPA) National Response Center	www.epa.gov/ceppo www.nrc.uscg.mil
Land-based natural hazards, earthquakes, floods, volcanoes	U.S. Geological Survey (USGS)	www.usgs.gov
Earthquakes	National Earthquake Information Center (NEIC) Earthquake Engineering Research Institute	www.neic.cr.usgs.gov www.eeri.org
Floods	Association of State Floodplain Managers	www.floods.org
Hurricanes and meteorological hazards	National Oceanographic and Atmospheric Administration (NOAA)	www.nhc.noaa.gov/ www.noaa.gov
Wildland fires National security hazard information	National Interagency Fire Center American Red Cross Department of Health and Human Services (DHHS): Anthrax and Biological Incidents Department of Transportation (DOT)	www.nifc.gov www.redcross.org www.hhs.gov/hottopics/ healing/biological.html http://ntl.bts.gov/faz/ sept11.html
	Environmental Protection Agency (EPA) Federal Bureau of Investigation (FBI)	www.epa.gov/safewater www.fbi.gov/pressrel/ attack/attacks.htm
	Federal Consumer Information Center	www.pueblo.gsa.gov/ crisis.htm
	Office of Homeland Security (OHS)	www.whitehouse.gov/ homeland
	Small Business Administration Economic Injury Disaster Loans	www.sba.gov/news/ current01/ economicinjuryfactsheet. html
	U.S. Postal Service Updates	www.usps.com/news/ 2001/press/ serviceupdates.htm
	White House Federal Recovery Action	www.whitehouse.gov/ response/fedresponse. html

continued

Emergency Type	Contact Agency	Web Site
Disaster services	American National Red Cross Disaster Relief	www.redcross.org www.disasterrelief.org
	Pacific Disaster Center (Information Technology for Disaster Response)	www.pdc.gov
	National Voluntary Organizations Active in Disaster (NVOAD)	www.nvoad.org
State and local organizations	National Emergency Management Association (NEMA)—State Emergency Managers Association	www.nemaweb.org
	International Association of Emergency Managers (IAEM)—Local Emergency Managers Association	www.iaem.com
	Extension Disaster Education Network	www.agctr.lsu.edu/eden
	University of Colorado National Hazards Center	www.colorado.edu/ hazards
	University of Delaware Disaster Research Center	www.udel.edu/DRC
Federal government Web sites	Health and Human Services Administration on Aging	www.aoa.dhhs.gov/ default.htm
	Health and Human Services (DHHS)— Center for Mental Health Services	www.mentalhealth.org/ cmhs/ EmergencyServices/ default.asp
	Department of Commerce—Economic Development Administration (EDA)	www.doc.gov/eda
	Department of Labor (DOL)—Disaster Unemployment Assistance	http://workforcesecurity. doleta.gov/unemploy/ disaster.asp
	Department of Transportation (DOT)— Federal Highway Administration (FHA)	http://workforcesecurity. doleta.gov/unemploy/ disaster.asp
	Department of Housing and Urban Development (HUD)	www.hud.gov/disassit. cfm
	Small Business Administration (SBA)	www.sba.gov/disaster
	U.S. Army Corps of Engineers	www.usace.army.mil/ index.html
	U.S. Department of Agriculture (USDA)— Farm Service Agency (FSA)	www.fsa.usda.gov/pas/ default.asp
International Web sites	International Federation of Red Cross/Red Crescent Societies	www.ifrc.org
	International Committee of the Red Cross	www.icrc.org
	United Nations Children's Fund	www.unicef.org
	International Monetary Fund (IMF)	www.imf.org
	The World Bank	www.worldbank.org
	United Nations Development Programme (UNDP)	www.undp.org
	International Strategy for Disaster Reduction (UN)	www.unisdr.org
	World Food Programme (WFP)	www.wfp.org
	World Health Organization (WHO)	www.who.org

Emergency Type	Contact Agency	Web Site
	United Nations High Commissioner for Refugees	www.unhcr.ch
	U.S. Agency for International Development (USAID)	www.usaid.gov
	U.S. Department of Defense (DoD)	www.dod.mil
	Interaction	www.interaction.org
Terror Related Web sites	American Red Cross	www.redcross.org/ services/disaster/ 0,1082,0_589_,00.html
	CDC Emergency Preparedness	www.bt.cdc.gov/
	Central Intelligence Agency Terrorism Handbook	www.cia.gov/cia/ reports/cbr_handbook/ cbrbook.htm
	Department of Homeland Security	www.dhs.gov
	Department of State Terrorism Page	www.state.gov/m/ds/ terrorism/
	Federal Bureau of Investigation	www.fbi.gov
	FEMA Terrorism Page	www.fema.gov/hazards/ terrorism/
	FirstGov Terrorism Page	www.firstgov.gov/ Topics/Usgresponse. shtml
	Food and Drug Administration Bioterror Info	www.fda.gov/oc/ opacom/hottopics/ bioterrorism.html
	Johns Hopkins School for Advanced International Studies 9/11 page	//www.sais-jhu.edu/ centers/cse/links/ september11links.html
	National Commission on Terrorist Attacks Upon the U.S.	www.9-11commission. gov/
	National Terrorism Preparedness Institute	terrorism.spcollege.edu /NEW/index.aspx
	Nuclear Control Institute	www.nci.org/nci-nt.htm
	Ready.gov	www.ready.gov
	Terrorism Research Center	www.terrorism.com/

Appendix C

Emergency Management Agency Addresses

FEMA HEADQUARTERS

500 C Street SW
Washington, DC 20472
(Note: Please use this address to reach the following offices by inserting the Office Name as the second line of the address:)

- Office of the Director
- Office of National Security Affairs
- Office of Intergovernmental Affairs
- Office of Policy and Regional Operations
- Office of the General Counsel
- Office of Public Affairs
- Office of Human Resources
- Office of Equal Rights
- Office of Financial Management
- Office of the Inspector General
- Mitigation Directorate
- Preparedness, Training, and Exercises Directorate
- Response and Recovery Directorate
- Federal Insurance Administration
- Operations Support Directorate
- Information Technology Directorate

National Emergency Training Center
16825 South Seton Avenue
Emmitsburg, MD 21727

U.S. Fire Administration
16825 South Seton Avenue
Emmitsburg, MD 21727

Emergency Management Institute
16825 South Seton Avenue
Emmitsburg, MD 21727

Mount Weather Emergency Assistance Center
19844 Blue Ridge Mountain Road
State Route 601
Bluemont, VA 20135

REGIONAL OFFICES

FEMA Region I
442 J.W. McCormack POCH
Boston, MA 02109-4595
This office serves the states of Maine, New Hampshire, Vermont, Rhode Island, Connecticut, and the Commonwealth of Massachusetts.

FEMA Region II
26 Federal Plaza, Room 1337
New York, NY 10278-0002
This office serves the states of New York, New Jersey, the Commonwealth of Puerto Rico, and the Territory of the U.S. Virgin Islands.

FEMA Region III
615 Chestnut Street
One Independence Mall, Sixth Floor
Philadelphia, PA 19106-4404
This office serves the states of Delaware, Maryland, Pennsylvania, Virginia, West Virginia, and the District of Columbia.

FEMA Region IV
3003 Chamblee Tucker Road
Atlanta, GA 30341
This office serves the states of Alabama, Florida, Georgia, Kentucky, Mississippi, North Carolina, South Carolina, and Tennessee.

FEMA Region V
536 South Clark St., 6th Floor
Chicago, IL 60605
This office serves the states of Illinois, Indiana, Michigan, Minnesota, Ohio, and Wisconsin.

FEMA Region VI
FRC 800 North Loop 288
Denton, TX 76209
This office serves the states of Arkansas, Louisiana, New Mexico, Oklahoma, and Texas.

FEMA Region VII
2323 Grand Boulevard, Suite 900
Kansas City, MO 64108-2670
This office serves the states of Iowa, Kansas, Missouri, and Nebraska.

FEMA Region VIII
Denver Federal Center
Building 710, Box 25267
Denver, CO 80255-0267
This office serves the states of Colorado, Montana, North Dakota, South Dakota, Utah, and Wyoming.

FEMA Region IX
1111 Broadway, Suite 1200
Oakland, CA 94607
This office serves the states of Arizona, California, Hawaii, and Nevada; and the Territory of American Samoa, the Territory of Guam, the Commonwealth of the Northern Mariana Islands, the Republic of the Marshall Islands, the Federated States of Micronesia, and the Republic of Palau.

FEMA Region X
Federal Regional Center
130 228th Street, SW
Bothell, WA 98021-9796
This office serves the states of Alaska, Idaho, Oregon, and Washington.

STATE OFFICES AND AGENCIES OF EMERGENCY MANAGEMENT

Alabama Emergency Management Agency
5898 County Road 41
P.O. Drawer 2160
Clanton, AL 35046-2160
(205) 280-2200
(205) 280-2495 fax
www.aema.state.al.us

Alaska Division of Emergency Services
P.O. Box 5750
Fort Richardson, AK 99505-5750
(907) 428-7000
(907) 428-7009 fax
www.ak-prepared.com

American Samoa Territorial Emergency Management Coordination (TEMCO)
American Samoa Government
P.O. Box 1086
Pago Pago, American Samoa 96799
(011)(684) 699-6415
(011)(684) 699-6414 fax

Arizona Division of Emergency Services
5636 East McDowell Road
Phoenix, AZ 85008
(602) 231-6245
(602) 231-6356 fax
www.state.az.us/es

Arkansas Department of Emergency Management
P.O. Box 758
Conway, AR 72033

(501) 730-9750
(501) 730-9754 fax
www.adem.state.ar.us

California Governor's Office of Emergency Services
P.O. Box 419047
Rancho Cordova, CA 95741-9047
(916) 845-8510
(916) 845-8511 fax
www.oes.ca.gov

Colorado Office of Emergency Management
Division of Local Government
Department of Local Affairs
15075 South Golden Road
Golden, CO 80401-3979
(303) 273-1622
(303) 273-1795 fax
www.dola.state.co.us/oem/oemindex.htm

Connecticut Office of Emergency Management
Military Department
360 Broad Street
Hartford, CT 06105
(860) 566-3180
(860) 247-0664 fax
www.mil.state.ct.us/OEM.htm

Delaware Emergency Management Agency
165 Brick Store Landing Road
Smyrna, DE 19977
(302) 659-3362
(302) 659-6855 fax
www.state.de.us/dema/index.htm

District of Columbia Emergency Management Agency
2000 14th Street NW, 8th Floor
Washington, DC 20009
(202) 727-6161
(202) 673-2290 fax
www.dcema.dc.gov

Florida Division of Emergency Management
2555 Shumard Oak Blvd.
Tallahassee, FL 32399-2100
(850) 413-9969
(850) 488-1016 fax
www.floridadisaster.org

Georgia Emergency Management Agency
P.O. Box 18055
Atlanta, GA 30316-0055
(404) 635-7000
(404) 635-7205 fax
www.State.Ga.US/GEMA

Office of Civil Defense, Government of Guam
P.O. Box 2877
Hagatna, Guam 96932
(011)(671) 475-9600
(011)(671) 477-3727 fax
http://ns.gov.gu

Hawaii State Civil Defense
3949 Diamond Head Road
Honolulu, HI 96816-4495
(808) 734-4246
(808) 733-4287 fax
http://scd.state.hi.us

Idaho Bureau of Disaster Services
4040 Guard Street, Bldg. 600
Boise, ID 83705-5004
(208) 334-3460
(208) 334-2322 fax
www.state.id.us/bds/bds.html

Illinois Emergency Management Agency
110 East Adams Street
Springfield, IL 62701
(217) 782-2700
(217) 524-7967 fax
www.state.il.us/iema

Indiana State Emergency Management Agency
302 West Washington Street
Room E-208 A
Indianapolis, IN 46204-2767
(317) 232-3986
(317) 232-3895 fax
www.ai.org/sema/index.html

Iowa Division of Emergency Management
Department of Public Defense
Hoover Office Building
Des Moines, IA 50319
(641) 281-3231

(641) 281-7539 fax
www.state.ia.us/government/dpd/emd/index.htm

Kansas Division of Emergency Management
2800 S.W. Topeka Boulevard
Topeka, KS 66611-1287
(785) 274-1401
(785) 274-1426 fax
www.ink.org/public/kdem

Kentucky Emergency Management
EOC Building
100 Minuteman Parkway, Bldg. 100
Frankfort, KY 40601-6168
(502) 607-1682
(502) 607-1614 fax
http://kyem.dma.state.ky.us

Louisiana Office of Emergency Preparedness
P.O. Box 44217
Baton Rouge, LA 70804
(225) 342-5470
(225) 342-5471 fax
www.loep.state.la.us

Maine Emergency Management Agency
State Office Building, Station 72
Augusta, ME 04333
(207) 626-4503
(207) 626-4499 fax
www.state.me.us/mema/memahome.htm

CNMI Emergency Management Office
Office of the Governor
Commonwealth of the Northern Mariana Islands
P.O. Box 10007
Saipan, Mariana Islands 96950
(670) 322-9529
(670) 322-7743 fax
www.cnmiemo.org

National Disaster Management Office
Office of the Chief Secretary
P.O. Box 15
Majuro, Republic of the Marshall Islands 96960-0015
(011)(692) 625-5181
(011)(692) 625-6896 fax

Maryland Emergency Management Agency
Camp Fretterd Military Reservation
5401 Rue Saint Lo Drive
Reistertown, MD 21136
(410) 517-3600
(877) 636-2872 toll free
(410) 517-3610 fax
www.mema.state.md.us

Massachusetts Emergency Management Agency
400 Worcester Road
Framingham, MA 01702-5399
(508) 820-2000
(508) 820-2030 fax
www.state.ma.us/mema

Michigan Division of Emergency Management
4000 Collins Road
P.O. Box 30636
Lansing, MI 48909-8136
(517) 333-5042
(517) 333-4987 fax
www.msp.state.mi.us/division/emd/emdweb1.htm

National Disaster Control Officer

Federated States of Micronesia
P.O. Box PS-53
Kolonia, Pohnpei Micronesia 96941
(011)(691) 320-8815
(001)(691) 320-2785 fax

Minnesota Division of Emergency Management
Department of Public Safety
444 Cedar Street, Suite 223
St. Paul, MN 55101-6223
(615) 296-0450
(615) 296-0459 fax
www.dps.state.mn.us/emermgt

Mississippi Emergency Management Agency
P.O. Box 4501, Fondren Station
Jackson, MS 39296-4501
(601) 352-9100
(800) 442-6362 toll free
(601) 352-8314 fax
www.mema.state.ms.us

Missouri Emergency Management Agency
P.O. Box 16
2302 Militia Drive
Jefferson City, MS 65102
(573) 526-9100
(573) 634-7966 fax
www.sema.state.mo.us/semapage.htm

Montana Division of Disaster & Emergency Services
1100 North Main
P.O. Box 4789
Helena, MT 59604-4789
(406) 841-3911
(406) 444-3965 fax
www.state.mt.us/dma/des/index.shtml

Nebraska Emergency Management Agency
1300 Military Road
Lincoln, NE 68508-1090
(402) 471-7410
(402) 471-7433 fax
www.nebema.org

Nevada Division of Emergency Management
2525 South Carson Street
Carson City, NV 89711
(775) 687-4240
(775) 687-6788 fax
http://dem.state.nv.us

Governor's Office of Emergency Management
State Office Park South
107 Pleasant Street
Concord, NH 03301
(603) 271-2231
(603) 225-7341 fax

New Jersey Office of Emergency Management
P.O. Box 7068
West Trenton, NJ 08628-0068
(609) 538-6050
(609) 538-0345 fax
www.state.nj.us/oem/county

Emergency Management Bureau
Department of Public Safety
P.O. Box 1628
13 Bataan Boulevard

Santa Fe, NM 87505
(505) 476-9606
(505) 476-9650
www.dps.nm.org/emc.htm

New York State Emergency Management Office
1220 Washington Avenue
Building 22, Suite 101
Albany, NY 12226-2251
(518) 457-2222
(518) 457-9995 fax
www.nysemo.state.ny.us

North Carolina Division of Emergency Management
116 West Jones Street
Raleigh, NC 27603
(919) 733-3867
(919) 733-5406 fax
www.dem.dcc.state.nc.us

North Dakota Division of Emergency Management
P.O. Box 5511
Bismarck, ND 58506-5511
(701) 328-8100
(701) 328-8181 fax
www.state.nd.us/dem

Ohio Emergency Management Agency
2855 W. Dublin Granville Road
Columbus, OH 43235-2206
(614) 889-7150
(614) 889-7183 fax
www.state.oh.us/odps/division/ema

Office of Civil Emergency Management
Will Rogers Sequoia Tunnel
2401 N. Lincoln
Oklahoma City, OK 73152
(405) 521-2481
(405) 521-4053 fax
www.odcem.state.ok.us

Oregon Emergency Management
Department of State Police
595 Cottage Street NE
Salem, OR 97310
(503) 378-2911 ext. 225
(503) 588-1378
www.osp.state.or.us/oem/oem.htm

Palau NEMO Coordinator
Office of the President
P.O. Box 100
Koror, Republic of Palau 96940
(011)(680) 488-2422
(011)(680) 488-3312

Pennsylvania Emergency Management Agency
P.O. Box 3321
Harrisburg, PE 17105-3321
(717) 651-2001
(717) 651-2040 fax
www.pema.state.pa.us

Puerto Rico Emergency Management Agency
P.O. Box 966597
San Juan, PR 00906-6597
(787) 724-0124
(787) 725-4244 fax

Rhode Island Emergency Management Agency
645 New London Ave
Cranston, RI 02920-3003
(401) 946-9996
(401) 944-1891 fax
www.state.ri.us/riema/riemaaa.html

South Carolina Emergency Management Division
1100 Fish Hatchery Road
West Columbia, SC 29172
(803) 737-8500
(803) 737-8570 fax
www.state.sc.us/epd

South Dakota Division of Emergency Management
500 East Capitol
Pierre, SD 57501-5070
(605) 773-6426
(605) 773-3580 fax
www.state.sd.us/state/executive/military/sddem.htm

Tennessee Emergency Management Agency
3041 Sidco Drive
Nashville, TN 37204-1502
(615) 741-4332
(615) 242-9635 fax
www.tnema.org

Texas Division of Emergency Management
5805 N. Lamar
Austin, TX 78752
(512) 424-2138
(512) 424-2444 or 7160 fax
www.txdps.state.tx.us/dem

Utah Division of Comprehensive Emergency Management
1110 State Office Building
P.O. Box 141710
Salt Lake City, UT 84114-1710
(801) 538-3400
(801) 538-3770 fax
www.cem.ps.state.ut.us

Vermont Emergency Management Agency
Department of Public Safety
Waterbury State Complex
103 South Main Street
Waterbury, VT 05671-2101
(802) 244-8721
(802) 244-8655 fax
www.dps.state.vt.us

Virgin Islands Territorial Emergency Management (VITEMA)
2-C Contant, A-Q Building
Virgin Islands 00820
(304) 774-2244
(304) 774-1491

Virginia Department of Emergency Management
10501 Trade Court
Richmond, VA 23236-3713
(804) 897-6502
(804) 897-6506
www.vdem.state.va.us

State of Washington Emergency Management Division
Building 20, M/S: TA-20
Camp Murray, WA 98430-5122
(253) 512-7000
(253) 512-7200 fax
www.wa.gov/wsem

West Virginia Office of Emergency Services
Building 1, Room EB-80
1900 Kanawha Boulevard
East Charleston, WV 25305-0360
(304) 558-5380

(304) 344-4538 fax
www.state.wv.us/wvoes

Wisconsin Emergency Management
2400 Wright Street
P.O. Box 7865
Madison, WI 53707-7865
(608) 242-3232
(608) 242-3247 fax
http://badger.state.wi.us/agencies/dma/wem/index.htm

Wyoming Emergency Management Agency
5500 Bishop Blvd.
Cheyenne, WY 82009-3320
(307) 777-4920
(307) 635-6017 fax
http://wema.state.wy.us

DEPARTMENT OF HOMELAND SECURITY (DHS)

U.S. Department of Homeland Security
Washington, D.C. 20528
Operator Number: 202-282-8000
Comment Line: 202-282-8495

State Offices of Homeland Security

Alabama
Homeland Security Director
James Walker
Alabama Office of Homeland Security
401 Adams Ave.—Suite 560
Montgomery, AL 36103-5690
334-271-7200
http://homelandsecurity.alabama.gov/

Alaska
Adjutant General
BG Craig Campbell
PO Box 5800
Ft. Richardson, AK 99505-0800
907-428-6003
http://www.ak-prepared.com/homelandsecurity/default.htm

American Samoa
Special Assistant to the Governor
Leiataua Birdsall V. Ala'ilima
Office of Territory Emergency Mgmt.

American Somoa Government
Pago Pago, AS 96799
011-684-633-4116

Arizona
Director of Homeland Security
Frank Navarette
1700 West Washington Street, 3rd Floor
Phoenix, AZ 85007
602-542-4331
http://www.homelandsecurity.az.gov/

Arkansas
Director, Dept. of Emergency Management
Wayne Ruthven
PO Box 758
Conway, AR 72033
501-730-9750
http://www.adem.state.ar.us

California
Chief Deputy Director, Office of Homeland Security
Rick Martinez
State Capitol, 1st Floor
Sacramento, CA 95814
916-324-8908
http://www.oes.ca.gov/Operational/OESHome.nsf/1?OpenForm

Colorado
Acting Deputy Director
Pamela Sillars
700 Kipling Street
Denver, CO 80215
303-273-1770
http://www.dlg.oem2.state.co.us/oem/oemindex.htm

Connecticut
Deputy Commissioner, Division of Homeland Security
Wayne Sandford
55 West Main St., Suite 500
Waterbury, CT 06702
203-805-6600
http://www.state.ct.us/dps/HS/Index.htm
HLS@po.state.ct.us

Delaware
Homeland Security Director
Phil Cabaud
Office of the Governor

Tatnall Building—2nd Floor
William Penn Street
Dover, DE 19901
302-744-4242
http://www.state.de.us/dema/pages/terror_prep.htm

District of Columbia
Deputy Mayor for Public Safety and Justice
Margret Nedelkoff Kellems
202-727-4036
http://washingtondc.gov

Florida
Commissioner, Florida Dept. of Law Enforcement
Guy Tunnell
PO Box 1489
Tallahassee, FL 32302-1489
850-410-7233
http://www.fdle.state.fl.us

Georgia
Director of Homeland Security
Bill Hitchens
PO Box 1456
Atlanta, GA 30371
404-624-7030
http://www.gahomelandsecurity.com/

Guam
Guam Homeland Security/Office of Civil Defense
Homeland Security Advisor
Frank Blas
221B Chalan Palasyo
Agana Heights, Guam 96910
671-475-9600
Fax: 671-477-3727
http://www.guamhs.org

Hawaii
Adjutant General, State Civil Defense
BG Robert Lee
3949 Diamond Head Rd.
Honolulu, HI 96816-4495
808-733-4246
http://www.scd.state.hi.us

Idaho
Adjutant General
MG Jack Kane

4040 West Guard Street
Boise, ID 83705-5004
208-422-5242
http://www.accessidaho.org/health_safety/homeland_security.html

Illinois
Deputy Chief of Staff for Public Safety
Carl Hawkinson
214 State House
Springfield, IL 62706
217-782-2700
http://www.illinois.gov/security/

Indiana
Director, Indiana Counter-Terrorism and Security Council
Earl S. Morgan, Sr.
100 North Senate Avenue
Rm. N240
Indianapolis, IN 46204
317-232-8303
http://www.in.gov/c-tasc

Iowa
Administrator, Emergency Management
Ellen Gordon
Hoover State Office Building
1305 E. Walnut, Level A
Des Moines, IA 50319
515-281-3231
http://www.iowahomelandsecurity.org/

Kansas
Administrator for Emergency Management
MG Gene Krase
2800 SW Topeka Blvd.
Topeka, KS 66611
785-274-1121
http://www.kansashomelandsecurity.org/

Kentucky
Adjutant General
MG D. Allen Youngman
Kentucky Office of Homeland Security
200 Mero Street
Frankfort, KY 40601
502-564-2081
http://homeland.state.ky.us

Louisiana
Adjutant General and Director
Louisiana Office of Emergency Preparedness
MG Bennett C. Landreneau
7667 Independence Blvd.
Baton Rouge, LA 70806
225-925-7333
http://www.loep.state.la.us/homeland/default.htm

Maine
Adjutant General
Homeland Security
MG Joseph Tinkham, II
1 State House Station
Augusta, ME 04333-0001
207-626-4440
http://www.maine.gov/portal/homelandsec.html

Maryland
Homeland Security Director
Thomas J. Lockwood
State House, 100 State Circle
Annapolis, MD 21401
410-974-3901
http://www.mema.state.md.us

Massachusetts
Executive Office of Public Safety
Ed Flynn
1 Ashburton Place, Rm. 2133
Boston, MA 02108
617-727-3600 ×556
http://www.state.ma.us/eops/

Michigan
COL Tadarial Sturdivant
Director of State Police
Contact: Capt. John Ort
713 South Harrison Rd
E. Lansing, MI 48823
517-336-6198
www.msp.state.mi.us

Minnesota
Commissioner of Public Safety and Homeland Security Director
Rich Stanek
DPS, North Central Life Tower
445 Minnesota St., Ste. 1000

St. Paul, MN 55101
http://www.dps.state.mn.us/homsec/mohshome.asp

Mississippi
Executive Director, Mississippi Emergency Management Agency
Robert Latham
PO Box 4501
Jackson, MS 39296-4501
601-960-9999
http://www.homelandsecurity.ms.gov/

Missouri
Special Adviser for Homeland Security
Col. Tim Daniel
PO Box 809
Jefferson City, MO 65102
573-522-3007
http://www.homelandsecurity.state.mo.us

Montana
Administrator, Disaster and Emergency Services
Department of Military Affairs—HAFRC
Montana Disaster and Emergency Services
Jim Greene
1900 Williams Street
PO Box 4789
Helena, MT 59604-4789
406-841-3911
http://www.discoveringmontana.com/homelandsecurity/css/default.asp

Nebraska
Lieutenant Governor Dave Heineman
PO Box 94848
Lincoln, NE 68509-4848
402-471-2256
http://www.nol.org/homeland/

Nevada
Homeland Security Director
Jerry Bussell
2525 S. Carson St
Carson City, NV 89710
775-687-7320
http://homelandsecurity.nv.gov/

New Hampshire
Director, Emergency Management and State Fire Marshal
Donald Bliss

10 Hazen Drive
Concord, NH 03305
603-271-3294
http://www.nhoem.state.nh.us/Terrorism/terrorism.asp

New Jersey
N.J. Office of Counter—Terrorism
Sidney Caspersen, Director
PO Box 091
Trenton, NJ 08625
609-341-3434
http://www.state.nj.us/njhomelandsecurity/index.html

New Mexico
Homeland Security Director
R.L. Stockard
New Mexico Office of Public Safety
P.O. Box 1628
Santa Fe, NM 87507-1628
505-827-3370
http://www.dps.nm.org/emergency/index.htm

New York
Director, Office of Public Security
James McMahon
Executive Chamber
633 3rd Ave, 38th Floor
NYC, NY 10017
212-867-7060
http://www.state.ny.us/security/
info@security.state.ny.us

North Carolina
Secretary, Dept of Crime Control and Public Safety
Bryan Beatty
4701 Mail Service Center
Raleigh, NC 27699
919-733-2126
http://www.ncgov.com/asp/subpages/safety_security.asp

North Dakota
Homeland Security Coordinator/Emergency Management Director
Doug Friez
Fraine Barracks Ln, Bldg 35
Fraine Barracks
Bismarck, ND 58504
701-328-8100
http://www.state.nd.us/dem/homeland/

Northern Mariana Islands
Special Advisor for Homeland Security
Jerry Crisostomo
Caller Box 10007
Saipan, MP 96950
670-664-2280

Ohio
Director of Public Safety
Kenneth L. Morckel
1970 W. Broad St.
Columbus, OH 43223-1102
614-466-4344
http://www.state.oh.us/odps/default.htm

Oklahoma
Secretary of Safety and Security
Bob Ricks
State Capitol Bldg, Room 212
Oklahoma City, OK 73105
405-425-2001
http://www.youroklahoma.com/homelandsecurity/

Oregon
Superintendent of Oregon State Police
Ronald C. Ruecker
400 Public Service Bldg
Salem, OR 97310
503-378-3725
http://www.osp.state.or.us/oem/

Pennsylvania
Director, Pennsylvania Office of Homeland Security
Keith Martin
2605 Interstate Drive
Harrisburg, PA 17110
717-651-2715
http://www.homelandsecurity.state.pa.us

Puerto Rico
Attorney General
Annabelle Rodriguez
La Fortaleza
PO Box 9020082
San Juan, PR 00902-0082
787-723-7924

Rhode Island
Adjutant General
MG Reginald Centracchio
222 State House
Providence, RI 02903
401-275-4102
http://www.state.ri.us/riema/

South Carolina
Chief, S.C. Law Enforcement Division (SLED)
Robert M. Stewart
PO Box 21398
Columbia, SC 29221-1398
803-737-9000
http://www.myscgov.com/SCSGPortal/static/law_enforcement_tem1.html

South Dakota
Director of Homeland Security
John Berheim
500 East Capitol Avenue
Pierre, SD 57501
1-866-homland
http://www.state.sd.us/homeland/

Tennessee
MG (Ret.) Jerry Humble
215 Eighth Avenue, North
Nashville, TN 37203
615-532-7825
http://www.state.tn.us/homelandsecurity/

Texas
Deputy Attorney General for Criminal Justice
Jay Kimbrough
PO Box 12428
Austin, TX 78711
512-936-1882
http://www.texashomelandsecurity.com/

Utah
Division Director, Comprehensive Emergency Management
Scott Behunin
210 State Capitol
Salt Lake City, UT 84114
801-538-3400
http://cem.utah.gov/HLS/

Vermont
Commissioner, VT State Police
Kerry Sleeper
103 South Main Street
Waterbury, VT 05671-2101
802-244-8775
http://www.dps.state.vt.us/homeland/

Virgin Islands
Adjutant General
MG Cleave A. McBean
21-22 Kongens Gade
St. Thomas, VI 00802
340-712-7711

Virginia
Assistant to the Governor for Commonwealth Preparedness
John Hager
202 N. 9th Street, 5th Floor
Richmond, VA 23219
804-225-3826
http://www.commonwealthpreparedness.state.va.us

Washington
Adjutant General and Director
State Military Department
MG Timothy J. Lowenberg
Washington Military Dept., Bldg 1
Camp Murray, WA 98430-5000
253-512-8201
http://www.wa.gov/wsem/

West Virginia
Secretary, Dept. of Military Affairs and Public Safety
Joe Martin
State Capitol Complex, Bldg 6, Rm B-122
Charleston, WV 25305
304-558-2930
http://www.state.wv.us/wvhomeland/

Wisconsin
Administrator, Emergency Management
Johnnie Smith
2400 Wright Street
PO Box 7865
Madison, WI 53707-7865
608-242-3210
http://emergencymanagement.wi.gov/

Wyoming
Wyoming Office of Homeland Security
Director
Joe More
122 West 25th Street
Herschler Bldg. 1st Floor East
Cheyenne, WY 82002
307-777-4663
http://wyohomelandsecurity.state.wy.us/

READY.GOV CITIZEN PREPAREDNESS RECOMMENDATIONS

Step 1—Get a Kit of Emergency Supplies

Be prepared to improvise and use what you have on hand to make it on your own for *at least* three days, maybe longer. Though there are many things that might make you more comfortable, think first about fresh water, food, and clean air. Consider putting together two kits. In one, put everything needed to stay where you are and make it on your own. The other should be a lightweight, smaller version you can take with you if you have to get away.

You'll need a gallon of water per person per day. Include in the kits canned and dried foods that are easy to store and prepare. If you live in a cold weather climate, include warm clothes and a sleeping bag for each member of the family.

Start now by gathering basic emergency supplies—a flashlight, a battery-powered radio, extra batteries, a first aid kit, toilet articles, prescription medicines, and other special things your family may need. Many potential terrorist attacks could send tiny microscopic "junk" into the air. Many of these materials can hurt you only if they get into your body, so think about creating a barrier between yourself and any contamination. It's smart to have something for each member of the family that covers their mouth and nose.

Plan to use two to three layers of a cotton t-shirt, handkerchief, or towel. Or, consider filter masks, readily available in hardware stores, which are rated based on how small a particle they filter. It is very important that the mask or other material fit your face snugly so that most of the air you breathe comes through the mask, not around it. Do whatever you can to make the best fit possible for children.

Also, include duct tape and heavyweight garbage bags or plastic sheeting that can be used to seal windows and doors if you need to create a barrier between yourself and any potential contamination outside.

Step 2—Make a Plan for What You Will Do in an Emergency

Be prepared to assess the situation, use common sense and whatever you have on hand to take care of yourself and your loved ones. Depending on your circumstances and the nature of the attack, the first important decision is deciding whether to stay or go. You should understand and plan for both possibilities.

Develop a Family Communications Plan. Your family may not be together when disaster strikes, so plan how you will contact one another and review what you will do in different situations. Consider a plan where each family member calls, or e-mails, the same friend or relative in the event of an emergency. It may be easier to make a long-distance phone call than to call across town, so an out-of-state contact may be in a better position to communicate among separated family members. You may have trouble getting through, or the phone system may be down altogether, but be patient.

Staying Put. There are circumstances when staying put and creating a barrier between yourself and potentially contaminated air outside, a process known as *shelter-in-place*, can be a matter of survival. Choose an interior room or one with as few windows and doors as possible. Consider precutting plastic sheeting to seal windows, doors, and air vents. Each piece should be several inches larger than the space you want to cover so that you can duct tape it flat against the wall. Label each piece with the location of where it fits.

If you see large amounts of debris in the air, or if local authorities say the air is badly contaminated, you may want to shelter-in-place. Quickly bring your family and pets inside, lock doors, and close windows, air vents, and fireplace dampers. Immediately turn off air conditioning, forced air heating systems, exhaust fans, and clothes dryers. Take your emergency supplies and go into the room you have designated. Seal all windows, doors, and vents. Watch TV, listen to the radio, or check the Internet for instructions.

Getting Away. Plan in advance how you will assemble your family and anticipate where you will go. Choose several destinations in different directions so you have options in an emergency. If you have a car, keep at least a half tank of gas in it at all times. Become familiar with alternate routes as well as other means of transportation out of your area. If you do not have a car, plan how you will leave if you have to. Take your emergency supply kit and lock the door behind you. If you believe the air may be contaminated, drive with your windows and vents closed and keep the air conditioning and heater turned off. Listen to the radio for instructions.

At Work and School. Think about the places where your family spends time: school, work, and other places you frequent. Talk to your children's schools and your employer about emergency plans. Find out how they will communicate with families during an emergency. If you are an employer, be sure you have an emergency preparedness plan. Review and practice it with your employees. A community working together during an emergency also makes sense. Talk to your neighbors about how you can work together.

continues

Step 3—Be Informed about What Might Happen

Some of the things you can do to prepare for the unexpected, such as assembling a supply kit and developing a family communications plan, are the same for both a natural or man-made emergency. However, there are important differences among potential terrorist threats that will impact the decisions you make and the actions you take.

Specific Terrorist Threats

A biological attack is the deliberate release of germs or other substances that can make you sick. Many agents must be inhaled, enter through a cut in the skin, or be eaten to make you sick.

A chemical attack is the deliberate release of a toxic gas, liquid, or solid that can poison people and the environment.

A nuclear blast is an explosion with intense light and heat, a damaging pressure wave, and widespread radioactive material that can contaminate the air, water, and ground surfaces for miles around.

A radiation threat or *dirty bomb* is the use of common explosives to spread radioactive materials over a targeted area.

Be prepared to adapt this information to your personal circumstances and make every effort to follow instructions received from authorities on the scene. Above all, stay calm, be patient, and think before you act. With these simple preparations, you can be ready for the unexpected.

Source: www.ready.gov

READY.GOV RECOMMENDATIONS FOR TERRORISM PREPAREDNESS

Biological Threat

A biological attack is the deliberate release of germs or other biological substances that can make you sick. Many agents must be inhaled, enter through a cut in the skin, or be eaten to make you sick. Some biological agents, such as anthrax, do not cause contagious diseases. Others, like the smallpox virus, can result in diseases you can catch from other people.

If There Is a Biological Threat

Unlike an explosion, a biological attack may or may not be immediately obvious. Although it is possible that you will see signs of a biological attack, as was sometimes the case with the anthrax mailings, it is perhaps more likely that local health care workers will report a pattern of unusual illness or there will be a wave of sick people seeking emergency medical attention. You will probably learn of the danger through an emergency radio or TV broadcast, or some other signal used in your community. You might get a telephone call or emergency response workers may come to your door.

In the event of a biological attack, public health officials may not immediately be able to provide information on what you should do. It will take time to determine exactly what the illness is, how it should be treated, and who is in danger. However, you should watch TV, listen to the radio, or check the Internet for official news including the following:

- Are you in the group or area authorities consider in danger?
- What are the signs and symptoms of the disease?
- Are medications or vaccines being distributed?
- Where? Who should get them?
- Where should you seek emergency medical care if you become sick?

During a Declared Biological Emergency

1. If a family member becomes sick, it is important to be suspicious.
2. Do not assume, however, that you should go to a hospital emergency room or that any illness is the result of the biological attack. Symptoms of many common illnesses may overlap.
3. Use common sense, practice good hygiene and cleanliness to avoid spreading germs, and seek medical advice.
4. Consider if you are in the group or area authorities believe to be in danger.
5. If your symptoms match those described and you are in the group considered at risk, immediately seek emergency medical attention.

If You Are Potentially Exposed

1. Follow instructions of doctors and other public health officials.
2. If the disease is **contagious** expect to receive **medical evaluation and treatment**. You may be advised to stay away from others or even deliberately **quarantined**.
3. For **noncontagious** diseases, expect to receive **medical evaluation and treatment**.

If You Become Aware of an Unusual and Suspicious Substance Nearby

1. Quickly get away.
2. Protect yourself. Cover your mouth and nose with layers of fabric that can filter the air but still allow breathing. Examples include two to three layers of cotton such as a t-shirt, handkerchief, or towel. Otherwise, several layers of tissue or paper towels may help.

continues

3. Wash with soap and water.
4. Contact authorities.
5. Watch TV, listen to the radio, or check the Internet for official news and information including what the signs and symptoms of the disease are, if medications or vaccinations are being distributed, and where you should seek medical attention if you become sick.
6. If you become sick, seek emergency medical attention.

Chemical Threat

A chemical attack is the deliberate release of a toxic gas, liquid, or solid that can poison people and the environment.

Possible Signs of Chemical Threat

- Many people suffering from watery eyes, twitching, choking, having trouble breathing, or losing coordination
- Many sick or dead birds, fish, or small animals are also cause for suspicion

If you see signs of chemical attack, find clean air quickly.

- Quickly try to **define the impacted area** or where the chemical is coming from, if possible.
- Take immediate action to **get away**.
- If the chemical is inside a building where you are, get out of the building without passing through the contaminated area, if possible.
- If you can't get out of the building or find clean air without passing through the area where you see signs of a chemical attack, it may be better to move as far away as possible and shelter-in-place.
- If you are outside, quickly decide what is the fastest way to find clean air. Consider if you can get out of the area or if you should go inside the closest building and shelter-in-place.

If You Think You Have Been Exposed to a Chemical

If your eyes are watering, your skin is stinging, and you are having trouble breathing, you may have been exposed to a chemical.

- If you think you may have been **exposed to a chemical, strip immediately** and **wash**.
- Look for a hose, fountain, or any source of **water**, and wash with **soap** if possible, being sure not to scrub the chemical into your skin.
- Seek emergency **medical attention**.

Explosions

If There Is an Explosion

- Take shelter against your desk or a sturdy table.
- Exit the building ASAP.
- Do not use elevators.
- Check for fire and other hazards.
- Take your emergency supply kit if time allows.

If There Is a Fire

- Exit the building ASAP.
- Crawl low if there is smoke.
- Use a wet cloth, if possible, to **cover** your nose and mouth.
- Use the back of your hand to feel the upper, lower, and middle parts of closed doors.
- If the door is not hot, bracc yourself against it and open slowly.
- If the door is hot, do not open it. Look for another way out.
- Do not use elevators.
- If you catch fire, do not run. Stop-drop-and-roll to put out the fire.
- If you are at home, go to a previously designated meeting place.
- Account for your family members and carefully supervise small children.
- Never go back into a burning building.

If You Are Trapped in Debris

- If possible, use a flashlight to signal your location to rescuers.
- Avoid unnecessary movement so that you don't kick up dust.
- Cover your nose and mouth with anything you have on hand. (Dense-weave cotton material can act as a good filter. Try to breathe through the material.)
- Tap on a pipe or wall so that rescuers can hear where you are.
- If possible, use a whistle to signal rescuers.
- Shout only as a last resort. Shouting can cause a person to inhale dangerous amounts of dust.

Nuclear Blast

A nuclear blast is an explosion with intense light and heat, a damaging pressure wave, and widespread radioactive material that can contaminate the air, water, and ground surfaces for miles around. During a nuclear incident, it is important to avoid radioactive material, if possible. Although experts may predict at this

continues

time that a nuclear attack is less likely than other types, terrorism by its nature is unpredictable.

If There Is a Nuclear Blast

If There Is Advanced Warning of an Attack

Take cover immediately, as far below ground as possible, though any shield or shelter will help protect you from the immediate effects of the blast and the pressure wave.

If There Is No Warning

1. Quickly assess the situation.
2. Consider if you can get out of the area or if it would be better to go inside a building to limit the amount of radioactive material you are exposed to.
3. If you take shelter go as far below ground as possible, close windows and doors, turn off air conditioners, heaters, or other ventilation systems. Stay where you are, watch TV, listen to the radio, or check the Internet for official news as it becomes available.
4. To limit the amount of radiation you are exposed to, think about shielding, distance, and time:
   - **Shielding:** If you have a thick shield between yourself and the radioactive materials more of the radiation will be absorbed, and you will be exposed to less.
   - **Distance:** The farther away you are away from the blast and the fallout the lower your exposure.
   - **Time:** Minimizing time spent exposed will also reduce your risk.

Use available information to assess the situation. If there is a significant radiation threat, health care authorities may or may not advise you to take potassium iodide. Potassium iodide is the same stuff added to your table salt to make it iodized. It may or may not protect your thyroid gland, which is particularly vulnerable, from radioactive iodine exposure. Plan to speak with your health care provider in advance about what makes sense for your family.

Radiation Threat

A radiation threat, commonly referred to as a *dirty bomb* or *radiological dispersion device (RDD)*, is the use of common explosives to spread radioactive materials over a targeted area. It is not a nuclear blast. The force of the explosion and radioactive contamination will be more localized. Although the blast will be immediately obvious, the presence of radiation will not be clearly defined until trained personnel with specialized equipment are on the scene. As with any radiation, you want to try to limit exposure. It is important to avoid breathing radiological dust that may be released in the air.

If There is a Radiation Threat or Dirty Bomb

1. If you are outside and there is an explosion or authorities warn of a radiation release nearby, cover your nose and mouth and quickly go inside a building that has not been damaged. If you are already inside, check to see if your building has been damaged. If your building is stable, stay where you are. Close windows and doors; turn off air conditioners, heaters, or other ventilation systems.
2. If you are inside and there is an explosion near where you are or you are warned of a radiation release inside, cover your nose and mouth and go outside immediately. Look for a building or other shelter that has not been damaged and quickly get inside.

    Once you are inside, close windows and doors; turn off air conditioners, heaters, or other ventilation systems.
3. If you think you have been exposed to radiation, take off your clothes and wash as soon as possible.
4. Stay where you are, watch TV, listen to the radio, or check the Internet for official news as it becomes available.
5. Remember: To limit the amount of radiation you are exposed to, think about shielding, distance, and time:
    * **Shielding**: If you have a thick shield between yourself and the radioactive materials more of the radiation will be absorbed, and you will be exposed to less.
    * **Distance**: The farther away you are away from the blast and the fallout the lower your exposure.
    * **Time**: Minimizing time spent exposed will also reduce your risk.

As with any emergency, local authorities may not be able to immediately provide information on what is happening and what you should do. However, you should watch TV, listen to the radio, or check the Internet often for official news and information as it becomes available.

Natural Disasters

Some of the things you can do to prepare for the unexpected, such as making an emergency supply kit and developing a family communications plan, are the same for both a natural or man-made emergency. However, there are important differences among natural disasters that will impact the decisions you make and the actions you take. Some natural disasters are easily predicted, others happen

continues

without warning. Planning what to do in advance is an important part of being prepared.

Find out what natural disasters are most common in your area. You may be aware of some of your community's risks: others may surprise you. Historically, flooding is the nation's single most common natural disaster. Flooding can happen in every U.S. state and territory. Earthquakes are often thought of as a West Coast phenomenon, yet 45 states and territories in the United States are at moderate to high risk from earthquakes and are located in every region of the country. Other disasters may be more common in certain areas. Tornados are nature's most violent storms and can happen anywhere. However, states located in "Tornado Alley," as well as areas in Pennsylvania, New York, Connecticut, and Florida are at the highest risk for tornado damage. Hurricanes are severe tropical storms that form in the southern Atlantic Ocean, Caribbean Sea, Gulf of Mexico, and in the eastern Pacific Ocean. Scientists can now predict hurricanes, but people who live in coastal communities should plan what they will do if they are told to evacuate.

Source: www.ready.gov

A DAY IN THE LIFE OF HOMELAND SECURITY

Today, U.S. Customs and Border Protection agents will:

- Process over 1.1 million passengers arriving into our nation's airports and seaports
- Inspect over 57,006 trucks and containers, 580 vessels, 2,459 aircraft, and 323,622 vehicles coming into this country
- Execute over 64 arrests
- Seize 4,639 pounds of narcotics in 118 narcotics seizures
- Seize an average of $715,652 in currency in 11 seizures
- Seize an average of $23,083 in arms and ammunition and $467,118 in merchandise
- Deploy 1200 dog teams to aid inspections
- Make 5,479 predeparture seizures of prohibited agricultural items
- Apprehend 2,617 people crossing illegally into the United States
- Rescue three people illegally crossing the border in dangerous conditions
- Deploy 35,000 vehicles, 108 aircraft, 118 horses on equestrian patrol, and 480 all-terrain vehicles
- Utilize 238 Remote Video Surveillance Systems, each system using 1 to 4 cameras to transmit images to a central location; and maintain the integrity of 5,525 miles of border with Canada and 1,989 miles of border with Mexico

Today, Transportation Security Administration employees will:

- Screen approximately 1.5 million passengers before they board commercial airlines

Today, the Federal Law Enforcement Training Center will:

- Provide law enforcement training for more than 3.500 federal officers and agents from 75 different federal agencies

Today, the Office for Domestic Preparedness will:

- Disburse millions of dollars to states and cities across the country

continues

Today, U.S. Coast Guard units will:

- Save 10 lives and assist 192 people in distress
- Protect $2.8 million in property
- Interdict 14 illegal migrants at sea
- Conduct 109 search and rescue cases
- Seize $9.6 million of illegal drugs
- Respond to 20 oil and hazardous chemical spills
- Conduct 50 Port Security Patrols
- Conduct 20 Homeland Security Air Patrols
- Board two high interest vessels
- Escort eight vessels (i.e., cruise ships or high interest ships) in and out of port
- Maintain over 90 security zones around key infrastructure in major ports or coastal areas
- Educate 502 people in Boating Safety Courses

Today the U.S. Citizenship and Immigration Services will:

- Provide information and services to approximately 225,000 customers in one of its 250 field locations
- Respond to 75,000 calls to its 1-800 customer service number that helps to assist our customers navigate the immigration process
- Naturalize approximately 1,900 new citizens
- Process approximately 19,000 applications for a variety of immigration related benefits

Today, U.S. Immigration and Customs Enforcement agents will:

- Make 217 arrests on immigration-related violations
- Make 41 arrests on customs violations
- Remove 407 criminal aliens and other illegal aliens
- Investigate 12 cases involving unauthorized employment threatening critical infrastructure
- Participate in 24 drug seizures resulting in the seizure of 5,311 pounds of marijuana, 774 pounds of cocaine, and 16 pounds of heroin
- Make seven currency seizures, totaling $478,927
- Make gland jury appearances resulting in the indictment of a combination of 32 people and companies
- Launch 20 vessels in support of marine operations protecting the territorial seas of Puerto Rico, South Florida, the Gulf of Mexico, and Southern California
- Fly 25 surveillance flights supporting criminal investigations in Puerto Rico and the continental United States
- Disseminate 80 criminal investigative leads to field offices
- Review 1,200 classified intelligence cables; protect over 8,000 federal facilities

- Screen over one million federal employees and visitors entering federal facilities
- Make six arrests for criminal offenses on federal property
- Intercept 18 weapons from entering federal facilities to include firearms, knives, and box cutters; and deploy federal air marshals to protect the skies

Today, Department of Homeland Security Information Analysis and Infrastructure Protection employees will:

- Distribute four information bulletins or warning products to critical infrastructure about vulnerability assessments, risk reduction, and protective measures
- Receive and review 500 cyber security reports from Internet security firms, government organizations, private companies, and foreign governments
- Review more than a 1,000 pieces of intelligence from the intelligence community and law enforcement agencies

Today, the U.S. Secret Service will:

- Protect high profile government officials including the President, the Vice President, visiting heads of state, and former Presidents
- Provide protection to traveling protectees in 17 different cities
- Screen over 4,000 people entering protective sites
- Examine 1,500 protective intelligence reports to assess potential threats to protectees
- Complete 11 protective intelligence investigations to assess potential risk to protectees from individuals or groups
- Open over 90 new cases involving financial and electronic crime, identity theft, counterfeiting, and personnel security investigations
- Prevent over $6 million in financial crime losses to the American public; and seize (on average) $172,000 in counterfeit currency

Today, DHS Science and Technology employees will:

- Engage the best and brightest minds—along with the most advanced technologies—through three distinct Centers of Excellence, which enlist academics, businesses, and scientists as partners with government to boost our efforts to develop an enduring national research capability in homeland protection
- Develop and implement technical standards for chemical, biological, radiological, and nuclear countermeasures
- Deploy radiation sensors to detect the illicit transport of radioactive materials, and experiment with capabilities to similarly protect our cities
- Receive approximately 27 new homeland security technology proposals from large and small businesses
- Receive an average of six Homeland Security technology proposals submitted via the science.technology@dhs.gov e-mail address.

continues

Today, Federal Emergency Management Agency (FEMA) employees will:

- Improve the effectiveness of 220 fire service personnel through courses offered by FEMA's National Fire Academy
- Help protect 1,000 students at risk for tornadoes by providing their school administrators with information about how to properly construct tornado shelters
- Provide 4,000 people volunteer opportunities to help better prepare their communities through Citizen Corps at its Web site, www.citizencorps.gov (the site receives 36,000 hits per day)
- Help save $2.7 million in damages from flooding across the country through the department's flood plain management
- Spend $10.6 million to help communities respond and recover from disasters
- Help protect an additional 104 homes from the devastating effects of flooding through flood insurance policies issued by the National Flood Insurance Program
- Help 224 Americans recover from disasters by providing direct federal disaster relief assistance in the forms of low-interest loans, unemployment insurance, crisis counseling and temporary housing
- Distribute $45,243 to state and local governments through FEMA's Emergency Management Performance Grants to help develop, maintain, and improve their emergency management capabilities
- Distribute $51,506 through FEMA's Community Emergency Response Team grants to help state emergency managers initiative, organize, train, and maintain teams of citizens who are qualified to assist in responding to disasters
- Provide an average of $917,808 in grants to America's fire departments through the Assistance to Firefighter Grant program; distribute (on average) $221,917 through FEMA's Emergency Operations Center grants to state governments to help them develop and improve emergency management facilities
- Distribute (on average) $218,493 through FEMA's Interoperable Communications Equipment grants to help develop and support communications interoperability among first responders and public safety emergency officials.

Source: Department of Security Results Agenda—August 2004

References

"After-Action Report on the Response to the September 11 Terrorist Attack of the Pentagon." Washington, D.C.: Titan Systems Inc., July 2002.

American Red Cross Disaster Operations Center Visitor's Guide.

Bakhet, Omar. *Linking Relief to Development*. UNDP Rwanda, June 1998.

Bagli, Charles V. "Seeking Safety, Downtown Firms Are Scattering," *The New York Times*, January 29, 2002, p. A-1.

Burke, Robert. *Counter Terrorism for Emergency Responders*. CRC/Lewis Publishers, 2000.

Center for Disaster Management and Humanitarian Assistance. *NGOs and Disaster Response Who Are These Guys and What Do They Do Anyways?*, (n.d.), www.cdmha.org/ppt/%20presentation.ppt.

Coyle, David. *The United Nations and How it Works*. New York: Columbia University Press, 1969.

Department of Defense, CCRP. *The Complex Process of Responding to Crisis*, (n.d.), www.dodccrp.org/ngoCh2.html.

Emergency Management Institute (EMI). 2001–2002 Catalog of Activities, Emmitsburg, MD, www.fema.gov.

Erickson, Paul A. *Emergency Response Planning for Corporate and Municipal Managers*.

Federal Emergency Management Agency. Report on Costs and Benefits of Natural Hazard Mitigation. Washington, D.C.: FEMA, March 1997.

Federal Emergency Management Agency. *Partnerships in Preparedness, A Compendium of Exemplary Practices in Emergency Management, Volume IV*. Washington, D.C.: FEMA, January 2000.

Federal Emergency Management Agency. *Partnerships in Preparedness, A Compendium of Exemplary Practices in Emergency Management, Volume II*. Washington, D.C.: FEMA, May 1997.

Federal Emergency Management Agency. *International Technical Assistance Activities of the United States Federal Emergency Management Agency*. Washington, D.C.: FEMA, 2001.

Federal Emergency Management Agency. *Federal Response Plan*. Washington, D.C.: FEMA, April 1999.

Federal Emergency Management Agency, Office of the Inspector General. *FEMA's Disaster Management Program: A Performance Audit After Hurricane Andrew*. H-01-93. Washington, D.C.: FEMA, January 1993.

FEMA Emergency Information Field Guide (condensed). October 1998.

FEMA Web site, www.fema.gov.

Flynn, Kevin. "After the Attack: The Firefighters; Department's Cruel Toll: 350 Comrades," *New York Times*, September 13, 2001, Section: National Desk.

Gandel, Stephen. "Consultants push Wall Street to leave; Downtown's losses are huge, but some companies shrug off fears, concentrate workers in midtown," *Crain's New York Business*, March 4, 2002, p. 1.

Gilbert, Alorie. "Out of the Ashes," *Information Week*, March 2002. [Online], 3 pages, www.informationweek.com/story/IWK20020104S0008.

Gilbert, Roy, and Alcira Kreimer. *Learning from the World Bank's Experience of Natural Disaster Related Assistance*. Washington, D.C.: The World Bank, 1999.

GlobalCorps. *OFDA's Evolving Role*, (n.d.), www.globalcorps.com/ofda/ofdarole.html.

Hawley, Chris. "Globalization and Sept. 11 are pushing Wall Street off Wall Street, analysts say." The Associated Press State & Local Wire, February 1 2002, State and Regional

Section. *Lexis-Nexis Universe: World Trade Center, Firm and Tenant.* Online, March 20, 2002.

Hedges, Chris. "Monday Counting Losses, Department Rethinks Fighting Every Fire," *New York Post*, October 1, 2001.

Houston, Allen. "Crisis Communications: Confused Messages Spur Switch to Sole Spokesman," *PR Week*, November 5, 2001, p. 11.

Houston, Allen. "Crisis Communications: The Readiness is All: How the Red Cross Responded," *PR Week*, October 22, 2001, p. 13.

Intergovernmental Panel on Climate Change. *Special Issues in Developing Countries*, 2001, www.ipcc.ch/pub/tar/wg2/340.htm.

International Monetary Fund. *IMF Emergency Assistance Related to Natural Disasters and Post-Conflict Situations: A Factsheet*, 2001, www.imf.org/external/np/eexr/facts/conflict.htm.

Madison County, North Carolina. "Multi-Hazard Plan for Madison County."

Mancino, Kimberly. *Development Relief: NGO Efforts to Promote Sustainable Peace and Development in Complex Humanitarian Emergencies*. Washington, D.C.: Interaction, 2001.

Maynard, Kimberly. *Healing Communities in Conflict: International Assistance in Complex Emergencies*, (n.d.), www.ciaonet.org/book/maynard/maynard07.html.

McKinsey & Company. "Improving NYPD Emergency Preparedness and Response," August 19, 2002.

Mileti, Dennis S. *Disasters by Design: A Reassessment of Natural Hazards in the United States*. Washington, D.C.: John Henry Press, 1999.

Mitchell, James K., *et al. Crucibles of Hazard: Mega-Cities and Disasters in Transition*. New York: United Nations University Press, 1999.

National Association of Counties. "Counties and Homeland Security: Policy Agenda to Secure the People of America's Counties." August 2002, online, www.naco.org/programs/homesecurity?policyplan.cfm.

National Emergency Management Association. "NEMA Reports on State Homeland Security Structures," June 2002, online, www.nemaweb.org/ShowExtendedNewscfm?ID=171.

National Emergency Management Association. "White Paper on Domestic Preparedness." October 1, 2001.

Natsios, Andrew S. *U.S. Foreign Policy and the Four Horsemen of the Apocalypse*. Westport, CT: Praeger Publishers, 1997.

Office of Homeland Security. "State and local Actions for Homeland Security, July 2002, online, www.whitehouse.gov/homeland/stateandlocal.

Office for the Coordination of Humanitarian Affairs. *Information Summary on Military and Civil Defence Assets (MCDA) and the Military and Civil Defence Unit (MCDU)*, (n.d.), www.reliefweb.int/ocha_ol/programs/response/mcdunet/0mcduinf.html.

Office for the Coordination of Humanitarian Affairs. *Coordination of Humanitarian Response*, (n.d.), www.reliefweb.int/ocha_ol/programs/response/service.html.

Otero, Juan. "Congress, Administration Examines Emergency Communications Systems," *Nation's Cities Weekly*, October 22, 2001, p. 5.

Pan American Health Organization. *Natural Disasters: Protecting the Public's Health*, (n.d.), PAHO Scientific Publication No. 575.

Powell, Michael, and Christine Haughney. "A Towering Task Lags in New York, City Debates Competing Visions for Rebuilding Devastated Downtown," *The Washington Post.com*, online, Page A03. Available: www.washingtonpost.com/ac2/. . ./A22040-2002Feb16?language=printe [2002, February 25].

Ranganath, Priya. *Mitigation and the Consequences of International Aid in Postdisaster Reconstruction*. Centre d'Etude et de Cooperation Internationale, 2000.

Rendleman, John. "Back Online, Despite its losses, Verizon went right back to work restoring communications services," October 29, 2001, *InformationWeek.com*, online, 2 pages. Available: www.informationwee.com/story/IWK2001105S0017.

Salomons, Dirk. *Building Regional and National Capacities for Leadership in Humanitarian Assistance*. New York: The Praxis Group, 1998.

Site One. *History of the Red Cross*, (n.d.), www.siteone.com/redcross/anniv.htm.

Smith, Lloyd R. "Lessons Learned from Oklahoma City: Your Employees . . . Their Needs, Their Role in Response and Recovery," www.recovery.com.

The Oklahoma Department of Civil Emergency Management. "After Action Report, Alfred P. Murrah Building Bombing. Lessons Learned." 1997.

"The Look Back at Mitch's Rampage," *USA Today*, June 8, 1999, www.usatoday.com/weather/news/1998/wmrmpage.htm.

The United States Conference of Mayors. "A National Action Plan for Safety and Security in America's Cities," December 2001. www.usmayors.org/uscm/home.asp.

The World Bank. "Assistance to Post-Conflict Countries and the HIPC Framework," April 2000, www.imf.org/external/np/hipc/2001/pc/042001.htm.

Town of Boone, North Carolina. "All Hazards Planning and Operations Manual," July 1999.

"Tsunami Education a Priority in Hawaii and West Coast States," *Bulletin of the American Meteorological Society*, June 2001, p. 1207.

U.S. Agency for International Development. *Field Operations Guide for Disaster Assessment and Response, Version 3.0*. Washington, D.C.: USAID, 1998.

U.S. Agency for International Development. *OFDA Annual Report 2000*, (n.d.), Washington, D.C.: USAID, 2000.

U.S. Agency for International Development. "Rebuilding Postwar Rwanda: The Role of the International Community," 1996, www.usaid.gov/pubs/usaid_eval/ascii/pnaby212.txt.

U.S. Department of State. "Bureau of Population, Refugees, and Migration," (n.d.), www.state.gov/g/prm.

UN Rwanda. "The United Nations Development Programme in Rwanda," (n.d.),www.un.rw/UNRwa/UNDP.shtml.

UNDP. "About ERD," (n.d.), www.undp.org/erd/about.htm.

UNDP. "Building Bridges Between Relief and Development: A Compendium of the UNDP in Crisis Countries," (n.d.), www.undp.org/erd/archives/bridges.htm.

UNDP. *Disaster Profiles of the Least Developed Countries*, United Nations, May 2001.

UNDP. "Further Elaboration on Follow-Up to Economic and Social Council Resolution 1995/96: Strengthening of the Coordination of Emergency Humanitarian Assistance," Second Regular Session, March 10–14 1997, www.undp.org/erd/archives/executiv.htm.

UNDP. "Reintegration & Rehabilitation," (n.d.), www.undp.org/erd/randr.htm.

UNDP. "Strengthening National Disaster Management," (n.d.), www.undp.org/erd/archives/brochures/drrp/countries/Botswana.htm.

UNDP. "The Disaster Reduction and Recovery Program: An Introduction," (n.d.), www.undp.org/erd/archives/brochures/drrp/main.htm.

UNDP. "The United Nations Development Programme Mission Statement," (n.d.), www.undp.org/info/discover/mission.html.

UNDP. "Working for Solutions to Crises: The Development Response," (n.d.), www.undp.org/erd/archives/bridges2.htm.

United Nations. General Assembly Economic and Social Council, United Nations, May 8, 2001.

U.S. Army. "About NGOs," (n.d.), http://call.army.mil/fmso/ngos/indtroduction.html.

"VIPs, Disaster Service Calls May Get Priority." *The Daily Yomiuri* (Tokyo), The Yomiuri Shimbun, November 26, 2001, p. 1.

Walsh, Edward. "National Response to Terror; FEMA leads Effort; Borders Tightened," *The Washington Post*, September 12, 2001, p. A-1.

Waugh, William Jr. *Living with Hazards—Dealing with Disasters: An Introduction to Emergency Management*. New York: M.E. Sharpe, 2000.

Wax, Alan J., and Julie Claire Diop. "Return to Downtown; Office leases are being signed again, but revival will take a while," *Newsday*, March 11, 2002, p. D13.

"Wireless System Improves Communications," *American City & County*, August 2001.

Zevin, Rona. "Tapping Web Power in Emergencies," *American City & County*, August 2001.

About the Authors

George D. Haddow currently serves as an Adjunct Professor at the Institute for Crisis, Disaster, and Risk Management at The George Washington University, Washington, D.C. He is also a Principal in Bullock & Haddow LLC, a Washington-based disaster management and homeland security consulting firm. Bullock & Haddow's client list includes the Federal Emergency Management Agency (FEMA), the World Bank, the Corporation for National and Community Service, DRI International and several rural counties in North Carolina. From 1993–2001, Mr. Haddow served in the Office of the Director of the Federal Emergency Management Agency (FEMA) as the White House Liaison and the deputy Chief of Staff. In these positions, Mr. Haddow was involved in the day-to-day management of FEMA, responsible for the Director's communications; policy formulation in the areas of disaster response, public/private partnerships, public information, environmental protection and disaster mitigation including the design and implementation of FEMA's national disaster mitigation initiative entitled Project Impact: Building Disaster Resistant Communities. Mr. Haddow managed FEMA's disaster management and mitigation projects in Argentina, Honduras, El Salvador, Nicaragua, Guatemala, Dominican Republic, Haiti, Ecuador and the Bahamas.

Jane A. Bullock currently serves as an Adjunct Professor at the Institute for Crisis, Disaster, and Risk Management at The George Washington University, Washington, D.C. Ms. Bullock has worked in emergency management for over twenty years, culminating in serving as Chief of Staff to James Lee Witt the Director of the Federal Emergency Management Agency (FEMA). In this position Ms. Bullock served as principal advisor to the Director, reorganized and streamlined the Agency operations, managed the day-to-day operations of the Agency; directed, monitored, and evaluated Agency strategic and communication processes; and oversaw administration of the Agency's resources, including the disaster relief fund, supervising the response and recovery to over 300 disasters. She represented the Administration with Congress, State and municipal governments, foreign officials, constituent groups and the media. She was Chief architect of FEMA's Project Impact: Building Disaster Resistant Communities, a nationwide effort by communities and businesses to implement prevention and risk reduction programs. She was principal on a project to create National Disaster Response and Mitigation system for Argentina and in six Central American and Caribbean countries. In response to the emerging threat of terrorism, she served as part of the Clinton Administration's Interagency Committee on National Security and Critical Infrastructure Protection. Since leaving FEMA, she is a partner in a disaster management/homeland security consulting firm with clients in the United States and worldwide. She is the coauthor of a series of books on planning and building communities in hazardous areas.

Index